Partial Differential Equations

for
Scientists and
Engineers

Partial Differential Equations

for
Scientists and
Engineers

Stanley J. Farlow
Professor of Mathematics
The University of Maine

DOVER PUBLICATIONS
Garden City, New York

Bibliographical Note

This Dover edition, first published in 1993, is an unabridged, corrected and enlarged republication of the work first published by John Wiley & Sons, New York, 1982. For this edition the author has corrected a number of errors and added a new section, "Answers to Selected Problems." The tables originally on the front endpapers have been transferred to two new appendixes (3 and 4).

Library of Congress Cataloging-in-Publication Data

Farlow, Stanley J., 1937–
    Partial differential equations for scientists and engineers / Stanley J. Farlow.
        p.   cm.
    Originally published: New York : Wiley, c1982.
    Includes bibliographical references and index.
    ISBN-13: 978-0-486-67620-3 (pbk.)
    ISBN-10: 0-486-67620-X (pbk.)
    1. Differential equations, Partial. I. Title.
QA377.F37 1993
515′.353—dc20
                                           93-4556
                                             CIP

Manufactured in the United States by LSC Communications Book LLC
67620X23   2021
www.doverpublications.com

Preface

Over the last few years, there has been a significant increase in the number of students studying partial differential equations at the undergraduate level, and many of these students have come from areas other than mathematics, where intuition rather than mathematical rigor is emphasized. In writing *Partial Differential Equations for Scientists and Engineers*, I have tried to stimulate intuitive thinking, while, at the same time, not losing too much mathematical accuracy. At one extreme, it is possible to approach the subject on a high mathematical epsilon-delta level, which generally results in many undergraduate students not knowing what's going on. At the other extreme, it is possible to wave away all the subtleties until neither the student *nor the teacher knows what's going on.* I have tried to steer the mathematical thinking somewhere between these two extremes.

Partial Differential Equations for Scientists and Engineers evolved from a set of lecture notes I have been preparing for the last five years. It is an unconventional text in one regard: It is organized in 47 semi-independent lessons in contrast to the more usual chapter-by-chapter approach.

Separation of variables and integral transforms are the two most important analytic tools discussed. Several nonstandard topics, such as Monte Carlo methods, calculus of variations, control theory, potential theory, and integral equations, are also discussed because most students will eventually come across these subjects at some time in their studies. Unless they study these topics here, they will probably never study them formally.

This book can be used for a one- or two-semester course at the junior or senior level. It assumes only a knowledge of differential and integral calculus and ordinary differential equations. Most lessons take either one or two days, so that a typical one-semester syllabus would be: Lessons 1–13, 15–17, 19–20, 22–23, 25–27, 30–32, 37–39. All 47 lessons can easily be covered in two semesters, with plenty of time to work problems.

The author wishes to thank the editors at Wiley for their invitation to write this book as well as the reviewers, Professor Chris Rorres and Professor M. Kursheed Ali, who helped me greatly with their suggestions. Any further suggestions for improvement of this book, either from students or teachers, would be greatly appreciated. Thanks also to Dorothy, Susan, Alexander, and Daisy Farlow.

Stanley J. Farlow

Contents

4 ELLIPTIC-TYPE PROBLEMS / 243

5 NUMERICAL AND APPROXIMATE METHODS / 299

Partial Differential Equations

for
Scientists and
Engineers

PART 1

Introduction

Introduction to Partial Differential Equations

PURPOSE OF LESSON: To show what partial differential equations are, why they are useful, and how they are solved; also included is a brief discussion on how they are classified as various kinds and types. An overview is given of many of the ideas that will be studied in detail later.

Most physical phenomena, whether in the domain of fluid dynamics, electricity, magnetism, mechanics, optics, or heat flow, can be described in general by partial differential equations (PDEs); in fact, most of mathematical physics are PDEs. It's true that simplifications can be made that reduce the equations in question to ordinary differential equations, but, nevertheless, the complete description of these systems resides in the general area of PDEs.

What Are PDEs?

A partial differential equation is an equation that contains partial derivatives. In contrast to ordinary differential equations (ODEs), where the unknown function depends only on *one variable,* in PDEs, the unknown function depends on several variables (like temperature $u(x,t)$ depends both on location x and time t).

Let's list some well-known PDEs; note that for notational simplicity we have called

$$u_t = \frac{\partial u}{\partial t} \qquad u_x = \frac{\partial u}{\partial x} \qquad u_{xx} = \frac{\partial^2 u}{\partial x^2} \qquad \cdots$$

A Few Well-Known PDEs

$u_t = u_{xx}$ (heat equation in one dimension)

$u_t = u_{xx} + u_{yy}$ (heat equation in two dimensions)

$u_{rr} + \dfrac{1}{r}u_r + \dfrac{1}{r^2}u_{\theta\theta} = 0$ (Laplace's equation in polar coordinates)

Introduction to Partial Differential Equations 3

$$u_{tt} = u_{xx} + u_{yy} + u_{zz} \quad \text{(wave equation in three dimensions)}$$
$$u_{tt} = u_{xx} + \alpha u_t + \beta u \quad \text{(telegraph equation)}$$

Note on the Examples

The unknown function u always depends on *more* than one variable. The variable u (which we differentiate) is called the **dependent** variable, whereas the ones we differentiate *with respect to* are called the **independent** variables. For example, it is clear from the equation

$$u_t = u_{xx}$$

that the dependent variable $u(x,t)$ is a function of two independent variables x and t, whereas in the equation

$$u_t = u_{rr} + \frac{1}{r}u_r + \frac{1}{r^2}u_{\theta\theta}$$

$u(r,\theta,t)$ depends on r, θ, and t.

Why Are PDEs Useful?

Most of the **natural laws of physics,** such as Maxwell's equations, Newton's law of cooling, the Navier-Stokes equations, Newton's equations of motion, and Schrodinger's equation of quantum mechanics, are stated (or can be) in terms of PDEs, that is, these laws describe physical phenomena by relating **space and time derivatives.** Derivatives occur in these equations because the derivatives represent *natural things* (like velocity, acceleration, force, friction, flux, current). Hence, we have equations relating partial derivatives of some unknown quantity that we would like to find.

The purpose of this book is to show the reader two things

1. How to *formulate* the PDE from the physical problem (constructing the mathematical model).
2. How to *solve* the PDE (along with initial and boundary conditions).

We wait a few lessons before we start the modeling problem; now, a brief overview on how PDEs are solved.

How Do You Solve a Partial Differential Equation?

This is a good question. It turns out that there is an entire arsenal of methods available to the practitioner; the most important methods are those that change PDEs into ODEs. Ten useful techniques are

1. *Separation of Variables.* This technique reduces a PDE in n variables to n ODEs.
2. *Integral Transforms.* This procedure reduces a PDE in n independent variables to one in $n - 1$ variables; hence, a PDE in two variables could be changed to an ODE.
3. *Change of Coordinates.* This method changes the original PDE to an ODE or else another PDE (an easier one) by changing the coordinates of the problem (rotating the axis and things like that).
4. *Transformation of the Dependent Variable.* This method transforms the unknown of a PDE into a new unknown that is easier to find.
5. *Numerical Methods.* These methods change a PDE to a system of *difference equations* that can be solved by means of iterative techniques on a computer; in many cases, this is the only technique that will work. In addition to methods that replace PDEs by difference equations, there are other methods that attempt to approximate solutions by polynomial surfaces (spline approximations).
6. *Pertubation Methods.* This method changes a nonlinear problem into a sequence of *linear ones* that approximates the nonlinear one.
7. *Impulse-response Technique.* This procedure decomposes initial and boundary conditions of the problem into *simple impulses* and finds the response to each impulse. The overall response is then found by adding these simple responses.
8. *Integral Equations.* This technique changes a PDE to an **integral equation** (an equation where the unknown is inside the integral). The integral equation is then solved by various techniques.
9. *Calculus of Variations Methods.* These methods find the solution to PDEs by reformulating the equation as a *minimization problem*. It turns out that the minimum of a certain expression (very likely the expression will stand for total energy) is also the solution to the PDE.
10. *Eigenfunction Expansion.* This method attempts to find the solution of a PDE as an infinite sum of *eigenfunctions*. These eigenfunctions are found by solving what is known as an eigenvalue problem corresponding to the original problem.

Kinds of PDEs

Partial differential equations are classified according to many things. Classification is an important concept because the general theory and methods of solution usually apply only to a given class of equations. Six basic classifications are
1. *Order of the PDE.* The order of a PDE is the order of the *highest partial derivative* in the equation, for example,

$$u_t = u_{xx} \quad \text{(second order)}$$

$$u_t = u_x \quad \text{(first order)}$$
$$u_t = uu_{xxx} + \sin x \quad \text{(third order)}$$

2. *Number of Variables.* The number of variables is the *number of independent* variables, for example,

$$u_t = u_{xx} \quad \text{(two variables: } x \text{ and } t)$$
$$u_t = u_{rr} + \frac{1}{r}u_r + \frac{1}{r^2}u_{\theta\theta} \quad \text{(three variables: } r, \theta, \text{ and } t)$$

3. *Linearity.* Partial differential equations are either *linear* or *nonlinear*. In the linear ones, the dependent variable u and all its derivatives appear in a linear fashion (they are not multiplied together or squared, for example). More precisely, a **second-order linear equation in two variables** is an equation of the form

(1.1)
$$\boxed{Au_{xx} + Bu_{xy} + Cu_{yy} + Du_x + Eu_y + Fu = G}$$

where A, B, C, D, E, F, and G can be *constants* or given *functions* of x and y; for example,

$$u_{tt} = e^{-t}u_{xx} + \sin t \quad \text{(linear)}$$
$$uu_{xx} + u_t = 0 \quad \text{(nonlinear)}$$
$$u_{xx} + yu_{yy} = 0 \quad \text{(linear)}$$
$$xu_x + yu_y + u^2 = 0 \quad \text{(nonlinear)}$$

4. *Homogeneity.* The equation (1.1) is called **homogeneous** if the right-hand side $G(x,y)$ is identically zero for all x and y. If $G(x,y)$ is not identically zero, then the equation is called **nonhomogeneous**.
5. *Kinds of Coefficients.* If the coefficients A, B, C, D, E, and F in equation (1.1) are constants, then (1.1) is said to have **constant coefficients** (otherwise, variable coefficients).
6. *Three Basic Types of Linear Equations.* All linear PDEs like equation (1.1) are either
    (a) parabolic
    (b) hyperbolic
    (c) elliptic

Parabolic. **Parabolic equations** describe heat flow and diffusion processes and satisfy the property $B^2 - 4AC = 0$.

Hyperbolic. **Hyperbolic equations** describe vibrating systems and wave motion and satisfy the property $B^2 - 4AC > 0$.

Elliptic. **Elliptic equations** describe *steady-state* phenomena and satisfy the property $B^2 - 4AC < 0$.

Examples.

(a) $u_t = u_{xx}$ $B^2 - 4AC = 0$ (parabolic)

(b) $u_{tt} = u_{xx}$ $B^2 - 4AC = 4$ (hyperbolic)

(c) $u_{\xi\eta} = 0$ $B^2 - 4AC = 1$ (hyperbolic)

(d) $u_{xx} + u_{yy} = 0$ $B^2 - 4AC = -4$ (elliptic)

(e) $yu_{xx} + u_{yy} = 0$ $B^2 - 4AC = -4y$ $\begin{cases} \text{elliptic for } y > 0 \\ \text{parabolic for } y = 0 \\ \text{hyperbolic for } y < 0 \end{cases}$

(In the case of variable coefficients, the situation can change from point to point.)

NOTES

1. In general, $B^2 - 4AC$ is a *function* of the independent variables; hence, an equation can change from one basic type to another throughout the domain of the equation (although it's not common).
2. The general linear equation (1.1) was written with independent variables x and y. In many problems, one of the two variables stands for time and hence would be written in terms of x and t.
3. A *general classification* diagram is given in Figure 1.1.

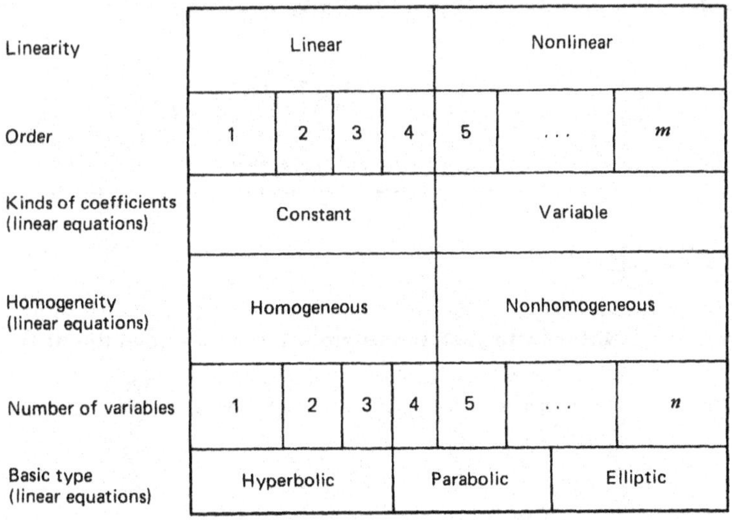

Linearity	Linear				Nonlinear		
Order	1	2	3	4	5	...	m
Kinds of coefficients (linear equations)	Constant				Variable		
Homogeneity (linear equations)	Homogeneous				Nonhomogeneous		
Number of variables	1	2	3	4	5	...	n
Basic type (linear equations)	Hyperbolic		Parabolic		Elliptic		

FIGURE 1.1 Classification diagram for partial differential equations.

PROBLEMS

1. Classify the following equations according to all the properties we've discussed in Figure 1.1:

   (a) $u_t = u_{xx} + 2u_x + u$
   (b) $u_t = u_{xx} + e^{-t}$
   (c) $u_{xx} + 3u_{xy} + u_{yy} = \sin x$
   (d) $u_{tt} = uu_{xxxx} + e^{-t}$

2. How many solutions to the PDE $u_t = u_{xx}$ can you find? Try solutions of the form $u(x,t) = e^{ax + bt}$.
3. If $u_1(x,y)$ and $u_2(x,y)$ satisfy equation (1.1), then is it true that the sum satisfies it?; if yes, prove it.
4. Probably the easiest of all PDEs to solve is the equation

$$\frac{\partial u(x,y)}{\partial x} = 0$$

Can you solve this equation? (Find all functions $u(x,y)$ that satisfy it.)
5. What about the PDE

$$\frac{\partial^2 u(x,y)}{\partial x \partial y} = 0$$

Can you find all solutions $u(x,y)$ to this equation? (How many are there?) How does this compare with an ODE like

$$\frac{d^2 y}{dx^2} = 0$$

insofar as the number of solutions is concerned?

OTHER READING

1. *Elementary Partial Differential Equations* by P. W. Berg and J. L. McGregor. Holden-Day, 1966. Clearly written with several nice problems; a nice book to own.

2. *Analysis and Solution of Partial Differential Equations* by R. L. Street. Brooks-Cole, 1973. A well-written text covering many of the topics we will cover in this book.

PART 2

Diffusion-Type Problems

Diffusion-Type Problems (Parabolic Equations)

PURPOSE OF LESSON: To show how parabolic PDEs are used to model heat-flow and diffusion-type problems. The physical meaning of different terms (such as u_t, u_x, u_{xx}, and u) are explained and a few examples of parabolic equations presented.

The idea of an *initial-boundary-value problem* is introduced along with an example. One of the major goals of this lesson is to give the reader an intuitive feeling for parabolic-type problems.

We begin this lesson by introducing a simple physical problem and showing how it can be described by means of a mathematical model (which will involve a PDE). We then complicate the problem and show how new partial differential equations can describe the new situations. The partial differential equations in this lesson are not derived or solved now, but will be in later lessons.

A Simple Heat-Flow Experiment

Suppose we have the following simple experiment that we break into steps:

STEP 1   We start with a reasonably long (say $L = 2$ m) rod (say copper) 2 cm in diameter whose lateral sides (but not the ends) we wrap with insulation. We could even use copper tubing provided we pour some sort of insulation down the inside. In other words, heat can flow in and out of the rod *at the ends*, but not across the lateral boundary.

STEP 2   Next, we place this rod in an environment whose temperature is fixed at some temperature T_0 (degrees °C) for a sufficiently long time, so that the temperature of the entire rod comes to a steady-state temperature similar to the environment. For simplicity, we let the temperature of the environment $T_0 = 10$°C.

STEP 3   We take the rod out of the environment at a time that we call $t = 0$ and attach two *temperature elements* to the ends of the rod. The purpose of these elements is to keep the ends at specific temperatures T_1 and T_2 (say $T_1 = 0°C$ and $T_2 = 50°C$). In other words, two thermostats constantly monitor the temperature at the ends of the rod, and if the temperatures differ from their prescribed values T_1 and T_2, strong heating (or cooling) elements come into operation to adjust the temperature accordingly. Our experiment is illustrated in Figure 2.1.

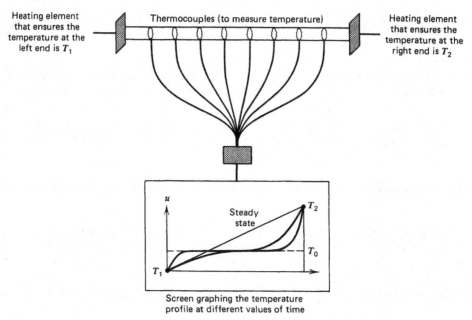

FIGURE 2.1   Schematic diagram of the experiment.

STEP 4   We now monitor the temperature profile of the rod on some type of display. (Why we want to perform an experiment of this kind is another question; we will talk about that later.) This completes our discussion of the experiment. The main purpose of this lesson is to show how this physical problem (and variations of it) can be explained (modeled) by parabolic PDEs.

The Mathematical Model of the Heat-Flow Experiment

The description of our physical problem requires three types of equations
1. The *PDE* describing the physical phenomenon of heat flow.
2. The *boundary conditions* describing the physical nature of our problem on the boundaries.
3. The *initial conditions* describing the physical phenomenon at the start of the experiment.

The Heat Equation

The basic equation of *one-dimensional* heat flow is the relationship

(2.1) | PDE $\quad u_t = \alpha^2 u_{xx} \quad 0 < x < L \quad 0 < t < \infty$

which relates the quantities

u_t = the *rate of change* in temperature with respect to time (measured in deg/sec)

and

u_{xx} = the *concavity* of the temperature profile $u(x,t)$ (which essentially compares the temperature at one point to the temperature at neighboring points).

This equation will be derived from the basic *conservation of heat equation* in later lessons, but for the time being, we examine it by itself. This equation simply says that the temperature $u(x,t)$ (at some point along the rod x and at some point in time t) is increasing ($u_t > 0$) or decreasing ($u_t < 0$) according to whether u_{xx} is positive or negative. Figure 2.2 illustrates the change in temperature at different points along the rod.

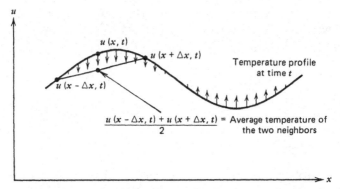

FIGURE 2.2   Arrows indicating change in temperature according to $u_t = \alpha^2 u_{xx}$.

To see how u_{xx} can be interpreted to measure heat flow, suppose we approximate u_{xx} by the difference quotient

$$u_{xx}(x,t) \cong \frac{1}{\Delta x^2} [u(x + \Delta x, t) - 2u(x,t) + u(x - \Delta x, t)]$$

Since this can be rewritten

$$u_{xx}\,(x,t) \cong \frac{2}{\Delta x^2} \left[\frac{u(x + \Delta x, t) + u(x - \Delta x, t)}{2} - u(x,t) \right]$$

we have the following interpretation of u_{xx}:

1. If the temperature $u(x,t) <$ average of the two neighboring temperatures, then $u_{xx} > 0$ (here, the net flow of heat into x is positive).
2. If the temperature $u(x,t) =$ average of the two neighboring temperatures, then $u_{xx} = 0$ (here the net flow of heat into x is zero).
3. If the temperature $u(x,t) >$ average of the two neighboring temperatures, then $u_{xx} < 0$ (here the net flow of heat into x is negative).

This is illustrated in Figure 2.2. In other words, if the temperature at a point x is greater than the average of the temperature at two nearby points $x - \Delta x$ and $x + \Delta x$, then the temperature at x will be decreasing. Furthermore, the exact rate of decrease u_t is proportional to this difference. The proportionality constant α^2 is a property of the material, and we will discuss this constant more in the next few lessons.

Boundary Conditions

All physical problems have boundaries of some kind, so we must describe mathematically what goes on there in order to adequately describe the problem. In our experiment, the boundary conditions (BCs) are quite easy. Since the temperature u was fixed for all time $t > 0$ at T_1 and T_2 at the two ends $x = 0$ and $x = L$, we would simply say

(2.2) $\qquad$ BCs $\begin{cases} u(0,t) = T_1 \\ u(L,t) = T_2 \end{cases}$ $0 < t < \infty$

Initial Conditions

All physical problems must start from some value of time (generally called $t = 0$), so we must specify the physical apparatus at this time. Since we started monitoring the rod temperature in our example from the time the rod had achieved a constant temperature of T_0, we have

(2.3) $\boxed{\text{IC} \quad u(x,0) = T_0 \quad 0 \leqslant x \leqslant L}$

We have now mathematically described the experiment. By writing equations (2.1), (2.2), and (2.3) together, we have what is called an *initial-boundary-value problem* (IBVP)

$$
\begin{array}{lll}
\text{PDE} & u_t = \alpha^2 u_{xx} & 0 < x < L \qquad 0 < t < \infty
\end{array}
$$

(2.4)   BCs $\begin{cases} u(0,t) = T_1 \\ u(L,t) = T_2 \end{cases}$ $0 < t < \infty$

$$
\text{IC} \qquad u(x,0) = T_0 \qquad 0 \leqslant x \leqslant L
$$

The interesting thing here, which is not at all obvious, is that there is *only one function* $u(x,t)$ that satisfies the problem (2.4), and that function will describe the temperature of the rod. Hence, our goal in the near future will be to find that unique solution $u(x,t)$ to (2.4).

Before finishing this lesson, we will discuss some variations of this basic problem. We start with a few modifications of the heat equation $u_t = \alpha^2 u_{xx}$.

More Diffusion-Type Equations

Lateral Heat Loss Proportional to the Temperature Difference

The equation

$$
u_t = \alpha^2 u_{xx} - \beta (u - u_0) \qquad \beta > 0
$$

describes heat flow in the rod with both diffusion $\alpha^2 u_{xx}$ along the rod and heat loss (or gain) across the *lateral* sides of the rod. Heat loss ($u > u_0$) or gain ($u < u_0$) is proportional to the difference between the temperature $u(x,t)$ of the rod and the surrounding medium u_0 (with β the proportionality constant). If β is very large in contrast to α^2, then the flow of heat *back and forth* along the rod will be small in contrast to the flow *in and out the sides,* and, hence, the heat will drain out the sides (at each point) according to the approximate equation $u_t = -\beta (u - u_0)$.

In chemistry where u may stand for concentration, the equation

$$
u_t = \alpha^2 u_{xx} - \beta(u - u_0)
$$

says that the rate of change (u_t) of the substance is due both to the diffusion $\alpha^2 u_{xx}$ (in the x-direction) and to the fact that the substance is being created ($u < u_0$) or destroyed ($u > u_0$) by a chemical reaction proportional to the difference between two concentrations u and u_0.

Internal Heat Source

The *nonhomogeneous* equation

$$
u_t = \alpha^2 u_{xx} + f(x, t)
$$

corresponds to the situation where the rod is being supplied with an internal heat source (everywhere along the rod and for all time t). It may be that a wire carrying electrical current passes through the rod and the resistance generates a constant heat source $f(x, t) = K$.

Diffusion-convection Equation

Suppose a pollutant is being carried along in a stream moving with velocity v. It is obvious that the concentration $u(x,t)$ of the substance changes as a function of both x (positive x measures the distance downstream) and time t. The rate of change u_t is measured by the *diffusion-convection equation*

$$u_t = \alpha^2 u_{xx} - vu_x$$

The term $\alpha^2 u_{xx}$ is the diffusion contribution and $-vu_x$ is the convection component. Whether the pollutant primarily diffuses or convects depends on the relative size of the two coefficients α^2 and v. You have probably seen smoke rising from a smoke stack. Here, the smoke particles are *convected* upward with the hot air and, at the same time, *diffuse* within the air currents.

In addition to these modifications in the heat equation, the *boundary conditions* of the rod can also be changed to correspond to other physical situations. We will discuss some of these modifications in Lesson 3.

NOTES

The heat equation $u_t = \alpha^2(x)u_{xx}$ with a variable coefficient $\alpha(x)$ would correspond to a problem where the diffusion within the rod depends on x (the material is *nonhomogeneous*). For example, if copper and steel slabs were placed next to each other (see Figure 2.3) and if the left side of the copper slabs were fixed at

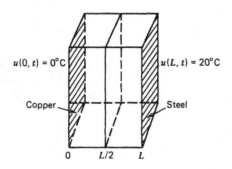

$u(0, t) = 0°C$ $u(L, t) = 20°C$

Copper   Steel

0    $L/2$ L

$\longrightarrow x$ (one-dimensional heat flow)

FIGURE 2.3

$u(0,t) = 0°C$ and the right side of the steel sheet were fixed at $u(L,t) = 20°C$, then the PDE that describes the heat flow would be

$$u_t = \alpha^2(x)u_{xx} \qquad 0 < x < L$$

where $\alpha(x) = \begin{cases} \alpha_1 \text{ (diffusion coefficient of copper)} & 0 < x < L/2 \\ \alpha_2 \text{ (diffusion coefficient of steel)} & L/2 < x < L \end{cases}$

PROBLEMS

1. If the initial temperature of the rod were

$$u(x,0) = \sin \pi x \qquad 0 \leqslant x \leqslant 1$$

and if the BCs were

$$u(0,t) = 0$$
$$u(1,t) = 0$$

what would be the behavior of the rod temperature $u(x,t)$ for later values of time?
HINT Use the physical interpretation of the heat equation $u_t = \alpha^2 u_{xx}$.

2. Suppose the rod has a constant internal heat source, so that the basic equation describing the heat flow within the rod is

$$u_t = \alpha^2 u_{xx} + 1 \qquad 0 < x < 1$$

Suppose we fix the boundaries' temperatures by $u(0,t) = 0$ and $u(1,t) = 1$. What is the steady-state temperature of the rod? In other words, does the temperature $u(x,t)$ converge to a constant temperature $U(x)$ independent of time?
HINT Set $u_t = 0$. It would be useful to graph this temperature. Also start with an initial temperature of zero and draw some temperature profiles.

3. Suppose a metal rod loses heat across the lateral boundary according to the equation

$$u_t = \alpha^2 u_{xx} - \beta u \qquad 0 < x < 1$$

and suppose we keep the ends of the rod at $u(0,t) = 1$ and $u(1,t) = 1$. Find the steady-state temperature of the rod (graph it). Where is heat flowing in this problem?

4. Suppose a laterally insulated metal rod of length $L = 1$ has an initial temperature of $\sin (3\pi x)$ and has its left and right ends fixed at temperatures zero and 10°C. What would be the IBVP that describes this problem?
*Note that the boundary and initial data do not match up in this problem.

OTHER READING

1. *Equations of Mathematical Physics* by A. N. Tikhonov and A. A. Samarskii. Macmillan, 1963; Dover, 1990. An encyclopedia of information; contains many good examples and problems.

Boundary Conditions for Diffusion-Type Problems

PURPOSE OF LESSON: To show how heat-flow and diffusion-type problems can give rise to a variety of boundary conditions and to introduce the important concept of *flux*.

Three important types of BCs discussed are

1. $u = g(t)$ (temperature specified on the boundary).

2. $\dfrac{\partial u}{\partial n} + \lambda u = g(t)$ (temperature of the *surrounding medium* is specified; n is the *outward normal* direction to the boundary).

3. $\dfrac{\partial u}{\partial n} = g(t)$ (heat flow across the boundary specified).

When describing the various types of boundary conditions that can occur for heat-flow problems, three basic types generally come to mind. Lesson 3 discusses these three kinds of BCs and gives an example of how they occur in experiments.

Type 1 BC (Temperature specified on the boundary)

Consider the heat flow in the one-dimensional rod illustrated in Figure 3.1 and suppose we make the ends of the rod follow the temperature curves $g_1(t)$ and $g_2(t)$.

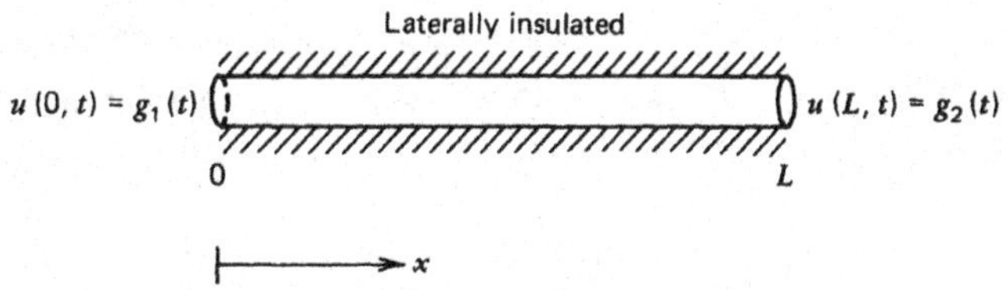

Laterally insulated

$u(0, t) = g_1(t)$ $u(L, t) = g_2(t)$

0   L

x

FIGURE 3.1   Temperature specified on the boundary.

As we mentioned in the previous lesson, an apparatus that keeps the ends at specified temperatures requires a thermostat at each end and heating elements to adjust the temperature accordingly. Problems with BCs of this kind are fairly common. It may even be that the goal of the problem is to find the *boundary temperatures* (boundary control) $g_1(t)$ and $g_2(t)$ that will force the temperature to behave in a suitable manner. In the steel industry, it is often necessary to determine the *boundary controls* so that the temperature of the metal inside the furnace changes over time but the temperature *gradient* from one point to another is small.

Similar types of BCs also apply to higher dimensional domains, for example, in two dimensions, we could imagine the interesting problem of finding the temperature inside the circular disc (of radius R) when the boundary temperature is specified in polar coordinates to be

$$u(R,\theta,t) = \cos t \sin \theta$$

See Figure 3.2.

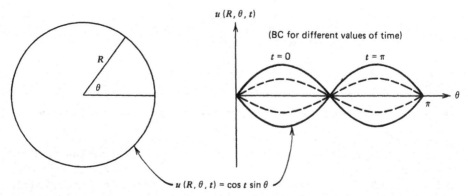

FIGURE 3.2 Oscillating boundary temperature.

Of course, we'd have to have an initial temperature to get this experiment started, but in this case, the effects of our IC would vanish after a short period of time, and the resulting temperature inside the circle would depend on the boundary temperature.

Type 2 BC (Temperature of the surrounding medium specified)

Suppose we consider again our laterally insulated copper rod, but now instead of requiring the two boundaries to be specified at temperatures $g_1(t)$ and $g_2(t)$, we only bring them in contact with surrounding mediums that have those temperatures. In other words, suppose the left side of the rod is enclosed in a container of liquid that has a changing temperature $g_1(t)$, while the right end is enclosed in another liquid with temperature $g_2(t)$ (Figure 3.3).

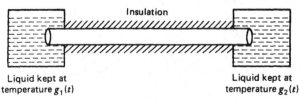

FIGURE 3.3 Convection cooling at the boundaries.

By specifying these types of BCs, we cannot say the boundary temperatures of the rod will be the same as the liquid temperatures $g_1(t)$ and $g_2(t)$, but we do know (Newton's law of cooling) that whenever the rod temperature at one of the boundaries is *less* than the respective liquid temperatures, then heat will flow into the rod at a rate proportional to this difference. In other words, for the one-dimensional rod with boundaries at $x = 0$ and L, Newton's law of cooling states

(3.1) $\begin{cases} \text{Outward flux of heat (at } x = 0) = h[u(0,t) - g_1(t)] \\ \text{Outward flux of heat (at } x = L) = h[u(L,t) - g_2(t)] \end{cases}$

where h is a **heat-exchange coefficient,** which is a measure of how many calories flow across the boundary per unit of temperature difference per second per cm and the **outward flux of heat** is the number of calories crossing the ends of the rod per second. Note that the outward flux of heat will be positive at either end provided the temperature of the rod is greater than the surrounding medium. Equations (3.1) can now be used in conjunction with what is known as Fourier's Law of Cooling to arrive at our BCs. Fourier's law gives us another representation (the first one is 3.1) for the outward flux of heat and by setting these two representations equal to each other, we get our BCs. First, we state Fourier's law (proven experimentally):

(3.2) | Outward flux of heat across a boundary is proportional to the inward normal derivative across the boundary.

This law says that if the temperature is increasing rapidly in the direction *outward* from the boundary of D (Figure 3.4), then heat will flow *from* the surrounding medium *into* the domain D.

In our one-dimensional problem, Fourier's law takes the form:

(3.3) $\begin{cases} \text{Outward flux of heat (at } x = 0) = k\,\dfrac{\partial u(0,t)}{\partial x} \\ \text{Outward flux of heat (at } x = L) = -k\,\dfrac{\partial u(L,t)}{\partial x} \end{cases}$

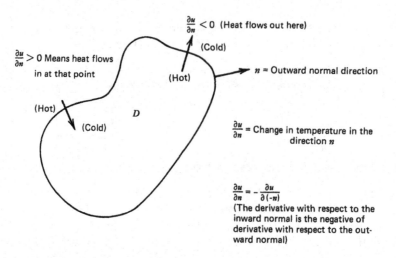

FIGURE 3.4   Illustration of Fourier's law.

where k is the **thermal conductivity** of the metal, which is a measure of how well the material conducts heat. (Poorly conducting materials have values near zero in cgs. units, while copper and aluminum have values close to one.)

Fourier's law (3.3) actually holds anywhere inside the rod and not just at the boundary; for example,

(3.4)
$$\boxed{\text{Flux of heat crossing } x_0 \text{ (from left to right)} \; = \; -kA\frac{\partial u}{\partial x}(x_0,t)}$$

See Figure 3.5.

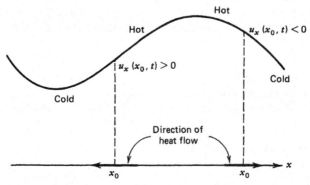

FIGURE 3.5   Another illustration of Fourier's law.

Fourier's law (3.4) says that if $u_x(x_0,t) < 0$, then heat will flow from *left to right*; if $u_x(x_0,t) > 0$, then the flow of heat through point x_0 will be from *right to left* (heat always flows from high to low temperatures).

Finally, if we use the two expressions (3.1) and (3.3) for heat flux, we have our desired BCs for Figure 3.3 in purely mathematical terms; namely,

$$\text{BCs} \quad \begin{cases} \dfrac{\partial u(0,t)}{\partial x} = \dfrac{h}{k}\,[u(0,t) - g_1(t)] \\[3mm] \dfrac{\partial u(L,t)}{\partial x} = -\dfrac{h}{k}\,[u(L,t) - g_2(t)] \end{cases} \qquad 0 < t < \infty$$

Quite often, the constant h/k is simply written as λ, and so we have the BCs for heat flow across the boundary

$$
\begin{aligned}
(3.5) \qquad\qquad u_x(0,t) &= \lambda\,[u(0,t) - g_1(t)] \\
u_x(L,t) &= -\lambda\,[u(L,t) - g_2(t)]
\end{aligned}
$$

In higher dimensions, we have similar BCs; for example, if the boundary of a circular disc is interfaced with a moving liquid that has a temperature $g(\theta,t)$, our BC would be

$$\frac{\partial u}{\partial r}(R,\theta,t) = -\frac{h}{k}\,[u(R,\theta,t) - g(\theta,t)]$$

Here, $\dfrac{\partial u}{\partial r}(R,\theta,t)$ represents the outward normal derivative (in the positive r-direction) of u evaluated at a point (R,θ) on the boundary. This type of BC would be called a *linear* BC (since it is linear in u and u_r) but nonhomogeneous due to the right-hand side $g(\theta,t)$.

Type 3 BC (Flux specified—including the special case of insulated boundaries)

Insulated boundaries are those that do not allow any flow of heat to pass, and, hence, the normal derivative (inward or outward) must be zero on the boundary (since the normal derivative is proportional to the flux). In the case of the one-dimensional rod with insulated ends at $x = 0$ and $x = L$, the BCs are

$$
\begin{aligned}
u_x(0,t) &= 0 \\
u_x(L,t) &= 0
\end{aligned} \qquad 0 < t < \infty
$$

In two-dimensional domains, an insulated boundary would mean that the *normal derivative* of the temperature across the boundary is zero. For example, if the circular disc were insulated on the boundary, then the BC would be $u_r(R,\theta,t) = 0$ for all $0 \leq \theta < 2\pi$ and all $0 < t < \infty$.

On the other hand, if we specify the amount of heat entering across the boundary of our disc, the BC is

$$u_r(R,\theta,t) = f(\theta,t)$$

where $f(\theta,t)$ would represent the amount of heat crossing *into* the circular disc from an outside heating source.

We now illustrate different types of BCs.

Typical BCs for One-Dimensional Heat Flow

Suppose we have a copper rod 200 cm long that is laterally insulated and has an initial temperature of 0°C. Suppose the top of the rod ($x = 0$) is insulated, while the bottom ($x = 200$) is immersed in moving water that has a constant temperature of $g_2(t) = 20$°C (Figure 3.6).

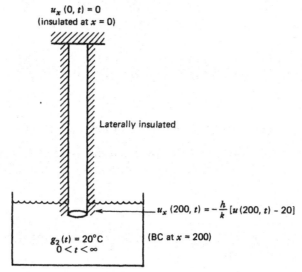

$u_x(0, t) = 0$
(insulated at $x = 0$)

Laterally insulated

$u_x(200, t) = -\frac{h}{k}[u(200, t) - 20]$
(BC at $x = 200$)

$g_2(t) = 20$°C
$0 < t < \infty$

FIGURE 3.6   Initial-boundary-value problem.

The mathematical model for this problem would be the following four equations:

$$\text{PDE} \qquad u_t = \alpha^2 u_{xx} \qquad 0 < x < 200 \qquad 0 < t < \infty$$

(3.6)   BCs $\begin{cases} u_x(0,t) = 0 \\ u_x(200,t) = -\dfrac{h}{k}[u(200,t) - 20] \end{cases} \qquad 0 < t < \infty$

$$\text{IC} \qquad u(x,0) = 0\text{°C} \qquad 0 \leq x \leq 200$$

where

α^2 = 1.16 cm²/sec   (diffusivity constant for copper)

k = 0.93 cal/cm-sec°C   (thermal conductivity of copper)

h = heat exchange coefficient. To find h is a hard problem in itself. It measures the rate that heat is being exchanged between the bottom of the rod and the surrounding water. It is a function of how fast the water is being circulated, the nature of the interface, and so forth. The reader would have to carry out an experiment to determine its value.

NOTES

1. A typical heat-flow problem inside a square is shown in Figure 3.7.

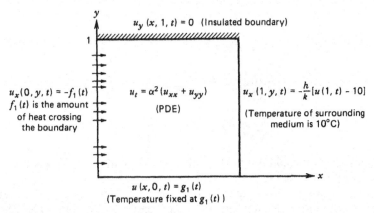

$u_y(x, 1, t) = 0$ (Insulated boundary)

$u_x(0, y, t) = -f_1(t)$
$f_1(t)$ is the amount
of heat crossing
the boundary

$u_t = \alpha^2 (u_{xx} + u_{yy})$
(PDE)

$u_x(1, y, t) = -\frac{h}{k}[u(1, t) - 10]$

(Temperature of surrounding
medium is 10°C)

$u(x, 0, t) = g_1(t)$
(Temperature fixed at $g_1(t)$)

FIGURE 3.7 Typical BCs for diffusion problems inside a square.

In this problem, after we specify the initial temperature $u(x,y,0)$ at $t = 0$ inside the square, the PDE and BCs in the diagram will take over for $0 < t < \infty$ and determine the subsequent temperature values $u(x,y,t)$. Whatever the temperature is, however, it must satisfy the BCs in Figure 3.7.

2. Note that the BC

$$u_r(R,\theta,t) = -\frac{h}{k}[u(R,\theta,t) - g(\theta,t)]$$

on the circle will not require the boundary temperature to be $g(\theta,t)$, but when the heat-exchange coefficient h is large, then the BC essentially says that the boundary temperature $u(R,\theta,t)$ is almost equal to $g(\theta,t)$.

PROBLEMS

1. Draw rough sketches of the solution to the IBVP (3.6) for different values of time. Do your sketches satisfy the BCs? What is the steady-state temperature of the rod? Is this obvious based on your intuition?
2. What is your interpretation of the initial-boundary-value problem?

$$\text{PDE} \qquad u_t = \alpha^2 u_{xx} \qquad 0 < x < 1 \qquad 0 < t < \infty$$

$$\text{BCs} \qquad \begin{cases} u(0,t) = 0 \\ u_x(1,t) = 1 \end{cases} \qquad 0 < t < \infty$$

$$\text{IC} \qquad u(x,0) = \sin(\pi x) \qquad 0 \leqslant x \leqslant 1$$

Can you draw rough sketches of the solution for different values of time? Will the solution come to a steady state; is this obvious?

3. What is your physical interpretation of the problem?

$$\text{PDE} \qquad u_t = \alpha^2 u_{xx} \qquad 0 < x < 1 \qquad 0 < t < \infty$$

$$\text{BCs} \qquad \begin{cases} u_x(0,t) = 0 \\ u_x(1,t) = 0 \end{cases} \qquad 0 < t < \infty$$

$$\text{IC} \qquad u(x,0) = \sin(\pi x) \qquad 0 \leqslant x \leqslant 1$$

Can you draw rough sketches of this solution for various values of time? What about the steady-state temperature?

4. Suppose a metal rod laterally insulated has an initial temperature of 20°C but immediately thereafter has one end fixed at 50°C. The rest of the rod is immersed in a liquid solution of temperature 30°C. What would be the IBVP that describes this problem?

OTHER READING

1. *Conduction of Heat in Solids* by H. S. Carslaw and J. C. Jaeger. Oxford University Press, 1959. An excellent reference that discusses BCs of many physical problems.

2. *Partial Differential Equations in Biology* by C. S. Peskin. Courant Institute of Mathematical Sciences, 1976. Several biological phenomena such as nerve cells, the inner ear, and the cardiovascular system are modeled by PDEs.

Derivation of the Heat Equation

PURPOSE OF LESSON: To show how the one-dimensional heat equation

$$u_t = \alpha^2 u_{xx} + f(x,t)$$

is derived from the basic principle of *conservation of heat*. Physical concepts such as *thermal conductivity*, *thermal capacity*, and *density* are discussed, and it is shown how the rate of heat transfer depends on these three basic physical parameters. A few variations of the basic heat equation are also discussed.

In all areas of science, we begin with a given set of assumptions that are taken to be self-evident and from which all other ideas are derived. Of course, what is self-evident to one person may hold doubts for others. The history of science consists of pushing back the basic axioms further and further so that there is a universally agreed upon starting point.

For example, one person may think that all relevant facts will spring from a basic assumption, say assumption *B*. From assumption *B*, he or she may prove theorem *C*, which in turn proves theorem *D*, which in turn proves others (Figure 4.1).

? ⟶ Assumption *A* ⟶ Assumption *B* ⟶ Theorem *C* ⟶ Theorem *D* ⟶ ?

FIGURE 4.1 The axiomatic method.

This of course is progress—the more new results a person can prove, the better. Physicists, chemists, and biologists all proceed in this basic manner.

On the other hand, instead of proving new theorems we may ask if it is possible to find a new assumption, say assumption *A*, more basic than assumption *B*, so that assumption *B* can be proven from *A*. In this way, we are pushing back the frontiers of knowledge. In the general area of heat flow, the concept of *conservation of energy* (heat energy) is the basis from which other principles are derived (Figure 4.2).

| Conservation of energy (Assumption) | ⟶ | $u_t = \alpha^2 u_{xx} + f(x, t)$ | ⟶ | Other properties of heat flow |

FIGURE 4.2   Conservation of energy: the cornerstone of heat-flow problems.

We could, of course, forget this lesson and use the heat equation as the starting point (some people may think it is self-evident in itself), but this would be shortchanging serious students, since conservation of energy assumptions are basic to science. Scientists often begin modeling specific problems by writing conservation of energy relationships and then rewriting them as partial differential equations.

We now turn to the goal of the lesson—to derive the heat equation from the conservation of heat equation.

Derivation of the Heat Equation

Suppose we have a one-dimensional rod of length L for which we make the following assumptions:
1.   The rod is made of a single homogeneous conducting material.
2.   The rod is laterally insulated (heat flows only in the x-direction).
3.   The rod is thin (the temperature at all points of a cross section is constant).

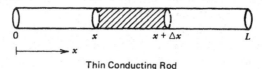

Thin Conducting Rod

FIGURE 4.3   Thin conducting rod.

If we apply the principle of conservation of heat to the segment $[x, x + \Delta x]$, we can claim

(4.1)
> Net change of heat inside $[x, x + \Delta x]$
> = Net flux of heat across the boundaries
> + Total heat generated inside $[x, x + \Delta x]$

Now, inasmuch as the total amount of heat (in calories) inside $[x, x + \Delta x]$ at any time t is measured by (see reference 1 in this lesson)

$$\text{Total heat inside } [x, x + \Delta x] = \int_x^{x + \Delta x} c\rho A u(s, t)\, ds$$

where

c = thermal capacity of the rod (measures the ability of the rod to store heat).

ρ = density of the rod.

A = cross-section area of the rod

we can write the conservation of energy equation (4.1) via calculus as

(4.2)
$$\frac{d}{dt} \int_x^{x+\Delta x} c\rho A u(s,t) \, ds = c\rho A \int_x^{x+\Delta x} u_t(s,t) \, ds$$

$$= kA \left[u_x(x + \Delta x, t) - u_x(x,t) \right] + A \int_x^{x+\Delta x} f(s,t) \, ds$$

where

k = thermal conductivity of the rod (measures the ability to conduct heat).

$f(x,t)$ = external heat source (calories per cm per sec).

The problem now is to replace equation (4.2) by one that does not contain integrals. The reader may recall the Mean Value Theorem from calculus.

Mean Value Theorem

If $f(x)$ is a continuous function on $[a,b]$, then there exists at least one number ξ, $a < \xi < b$ that satisfies

$$\int_a^b f(x) \, dx = f(\xi) \, (b - a)$$

Applying this result to equation (4.2), we arrive at the following equation:

$$c\rho A u_t(\xi_1, t)\Delta x = kA[u_x(x + \Delta x, t) - u_x(x,t)] + Af(\xi_2, t) \, \Delta x \quad x < \xi < x + \Delta x$$

or

$$u_t(\xi, t) = \frac{k}{c\rho} \left\{ \frac{u_x(x + \Delta x, t) - u_x(x,t)}{\Delta x} \right\} + \frac{1}{c\rho} f(\xi, t)$$

Finally letting $\Delta x \to 0$, we have the desired result

(4.3)
$$\boxed{u_t(x,t) = \alpha^2 u_{xx}(x,t) + F(x,t)}$$

where

$$\alpha^2 = \frac{k}{c\rho} \qquad \text{(called the diffusivity of the rod)}$$

$$F(x,t) = \frac{1}{c\rho}f(x,t) \qquad \text{(heat source density)}$$

This completes our discussion. Before we close, however, suppose the rod were not laterally insulated and that heat can flow in and out across the lateral boundary at a rate proportional to the difference between the temperature $u(x,t)$ and the surrounding medium that we keep at zero. In this case, the conservation of heat principle will give.

(4.4) $$u_t = \alpha^2 u_{xx} - \beta u + F(x;t)$$

where β = rate constant for the lateral heat flow ($\beta > 0$).

NOTES

1. The constant k is the **thermal conductivity** of the rod and a measure of the heat flow (in calories) that is transmitted per second through a plate 1 cm thick across an area of 1 cm² when the temperature difference is 1°C; values for k can be found in *The Handbook of Chemistry and Physics*. Typical values of k are close to 1 for copper and near zero for insulating-type materials.

    If the material of the rod is uniform, then k will not depend on x. For some materials, the value of k depends on the temperature u and hence the heat equation

$$u_t = \frac{1}{c\rho} \frac{\partial}{\partial x} \{k(u)u_x\}$$

    is *nonlinear*. Most of the time, however, k changes very slowly with u and this nonlinearity is neglected.

2. The constant c is known as the **thermal capacity** (or specific heat) of the substance and measures the amount of energy the substance can store. For example, a baked potato would have a large thermal capacity, since it can store a large amount of heat per unit mass of potato (that's why it takes a long time to heat). Technically, the thermal capacity is the amount of heat (in calories) necessary to produce a 1°C change in temperature of 1 g of the substance. For most of our problems, c is taken as a constant independent of x and u; typical values can be found in *The Handbook of Chemistry and Physics*.

3. The units of some of the basic quantities of heat flow (in the cgs. measurement system) are

    u = temperature (degrees centigrade).

    u_t = rate of change in temperature (°C/sec).

u_x = slope of temperature curve (°C/cm).
u_{xx} = concavity of temperature curve (°C/cm²).
c = thermal capacity (cal/g-°C).
k = thermal conductivity (cal/cm-sec-°C).
ρ = density (g/cm³).
α^2 = diffusivity (cm²/sec).

4. Note that the diffusivity $\alpha^2 = \dfrac{k}{c\rho}$ of a material is proportional to the conductivity k of the material and inversely proportional to the density ρ and
thermal capacity c; this should have some intuitive appeal to the reader.

PROBLEMS

1. Substitute the units of each quantity $u, u_t, \ldots$ into the equation

$$u_t = \alpha^2 u_{xx} - \beta u$$

to see that every term has the same units of °C/sec.
2. Substitute the units of each quantity into the equation

$$u_t = \alpha^2 u_{xx} - v u_x$$

where v has units of velocity to see that every term has the same units.
3. Derive the heat equation

$$u_t = \frac{1}{c\rho} \frac{\partial}{\partial x} [k(x) u_x] + f(x,t)$$

for the situation where the thermal conductivity $k(x)$ depends on x.
4. Suppose $u(x,t)$ measures the concentration of a substance in a moving stream
(moving with velocity v). Suppose the concentration $u(x,t)$ changes both by
diffusion and convection; derive the equation

$$u_t = \alpha^2 u_{xx} - v u_x$$

from the fact that at any instant of time, the total mass of the material is
not created or destroyed in the region $[x, x + \Delta x]$.
HINT   Write the conservation equation

Change of mass inside $[x, x + \Delta x]$
= Change due to *diffusion* across the boundaries
+ Change due to the material being *carried* across the boundaries

OTHER READING

1. *Applied Mathematics for the Engineer and Physicist* by L. A. Pipes. McGraw-Hill, 1958. An older reference, but still a good one for the practicing scientist.

2. *Equations of Mathematical Physics* by A. N. Tikhonov and A. A. Samarskii. Macmillan, 1963; Dover, 1990. A good text for derivations of equations. A companion volume, *A Collection of Problems in* [*on*] *Mathematical Physics*, by B. M. Budak, A. A. Samarskii, and A. N. Tikhonov (Pergamon, 1964; Dover, 1988) is available, which will give the student additional experience with solving problems in partial differential equations.

Separation of Variables

PURPOSE OF LESSON: To introduce the powerful method of separation of variables and to show how this method can be used to solve a well-known diffusion problem. Inasmuch as the method is not well understood by some students due to its complicated algebraic nature, several intuitive explanations are given along the way.

The basic idea is to break down the *initial conditions* of the problem into simple components, find the response to each component, and then add up these individual responses. This gives the response to the *arbitrary initial condition*.

The actual step-by-step methodology of separation of variables somewhat hides this basic interpretation, but that's what's going on nevertheless.

Separation of variables is one of the oldest techniques for solving initial-boundary-value problems (IBVPs) and applies to problems where

1. The PDE is linear and homogeneous (not necessarily constant coefficients).
2. The boundary conditions are of the form

$$\alpha u_x(0,t) + \beta u(0,t) = 0$$
$$\gamma u_x(1,t) + \delta u(1,t) = 0$$

where α, β, γ, and δ are constants (boundary conditions of this form are called **linear homogeneous BCs**).

It dates back to the time of Joseph Fourier (in fact, it's occasionally called *Fourier's method*) and is probably the most widely used method of solution (when applicable).

Instead of showing how the method works in general, let's apply it to a specific problem (later we will discuss it in more generality). Consider the IBVP (diffusion problem)

PDE $\qquad u_t = \alpha^2 u_{xx} \qquad 0 < x < 1 \qquad 0 < t < \infty$

BCs $\qquad \begin{cases} u(0,t) = 0 \\ u(1,t) = 0 \end{cases} \quad 0 < t < \infty$

$$\text{IC} \qquad u(x,0) \; = \; \phi(x) \qquad 0 \leqslant x \leqslant 1$$

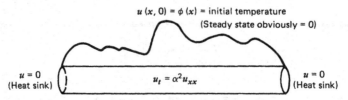

FIGURE 5.1   Diagram of the diffusion problem.

Before getting to separation of variables, let's first think about our problem. Here we have a finite rod where temperature at the ends is fixed at zero (suppose it's a temperature problem where zero means so many degrees). We are also given data for the problem in the form of an initial condition; our goal is to find the temperature $u(x,t)$ at later points in time.

Now for the method itself—but first an overview.

Overview of Separation of Variables

Separation of variables looks for simple-type solutions to the PDE of the form

$$u(x,t) \; = \; X(x)T(t)$$

where $X(x)$ is some function of x and $T(t)$ is some function of t. The solutions are simple because any temperature $u(x,t)$ of this form will retain its basic "shape" for different values of time t (Figure 5.2).

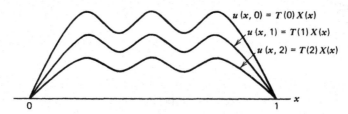

FIGURE 5.2   Graph of X(x)T(t) for different values of t.

The general idea is that it is possible to find an infinite number of these solutions to the PDE (which, at the same time, also satisfy the BCs). These simple functions $u_n(x,t) = X_n(x)T_n(t)$ (called **fundamental solutions**) are the building blocks of our problem, and the solution $u(x,t)$ we are looking for is found by adding the simple fundamental solutions $X_n(x)T_n(t)$ in such a way that the resulting sum

$$\sum_{n=1}^{\infty} A_n X_n(x) T_n(t)$$

satisfies the initial conditions. Inasmuch as this sum still satisfies the PDE and the BCs, we now have the solution to our problem. Let's now carry this out in detail.

Separation of Variables

STEP 1 (Finding elementary solutions to the PDE)

We wish to find the function $u(x,t)$ that satisfies the following four conditions:

$$\text{PDE} \qquad u_t = \alpha^2 u_{xx} \qquad 0 < x < 1 \qquad 0 < t < \infty$$

$$\text{BCs} \qquad \begin{cases} u(0,t) = 0 \\ u(1,t) = 0 \end{cases} \qquad 0 < t < \infty$$

$$\text{IC} \qquad u(x,0) = \phi(x) \qquad 0 \leq x \leq 1$$

To begin, we look for solutions of the form $u(x,t) = X(x)T(t)$ by substituting $X(x)T(t)$ into the PDE and solving for $X(x)T(t)$. Making this substitution gives

$$X(x)T'(t) = \alpha^2 X''(x)T(t)$$

Now, here is the part that makes all this work: If we *divide* each side of this equation by $\alpha^2 X(x)T(t)$, we have

$$\frac{T'(t)}{\alpha^2 T(t)} = \frac{X''(x)}{X(x)}$$

and obtain what is called **separated variables**, that is, the left side of the equation depends only on t and the right side, only on x. Inasmuch as x and t are *independent of each other*, each side must be a fixed constant (say k); hence, we can write

$$\frac{T'}{\alpha^2 T} = \frac{X''}{X} = k$$

or

$$T' - k\alpha^2 T = 0$$
$$X'' - kX = 0$$

So now we can solve each of these two ODEs, multiply them together to get a solution to the PDE (note that we have essentially changed a second-order PDE to two ODEs). However, we now make an important observation, namely, that we want the separation constant k to be *negative* (or else the $T(t)$ factor doesn't go to zero as $t \to \infty$). With this in mind, it is general practice to rename $k = -\lambda^2$, where λ is nonzero ($-\lambda^2$ is guaranteed to be negative). Calling our separation constant by its new name, we can now write the two ODEs as

$$T' + \lambda^2\alpha^2 T = 0$$
$$X'' + \lambda^2 X = 0$$

We will now solve these equations. Both equations are standard-type ODEs and have solutions

$$T(t) = Ae^{-\lambda^2\alpha^2 t} \quad (A \text{ an arbitrary constant})$$
$$X(x) = A \sin(\lambda x) + B \cos(\lambda x) \quad (A, B \text{ arbitrary})$$

and hence all functions

$$u(x,t) = e^{-\lambda^2\alpha^2 t} [A \sin(\lambda x) + B \cos(\lambda x)]$$

(with A, B, and λ arbitrary) will satisfy the PDE $u_t = \alpha^2 u_{xx}$; this verification is problem 1 in the problem set. At this point, we have an infinite number of functions that satisfy the PDE.

STEP 2   (Finding solutions to the PDE and the BCs)

We are now to the point where we have many solutions to the PDE but not all of them satisfy the BCs or the IC. The next step is to choose a certain *subset* of our current crop of solutions

(5.1) $e^{-\lambda^2\alpha^2 t}[A \sin(\lambda x) + B \cos(\lambda x)]$

that satisfy the boundary conditions

$$u(0,t) = 0$$
$$u(1,t) = 0$$

To do this, we substitute our solutions (5.1) into these BCs, getting

$$u(0,t) = Be^{-\lambda^2\alpha^2 t} = 0 \Rightarrow B = 0$$
$$u(1,t) = Ae^{-\lambda^2\alpha^2 t} \sin \lambda = 0 \Rightarrow \sin \lambda = 0$$

This last BC restricts the separation constant λ from being any nonzero number, it must be a root of the equation $\sin \lambda = 0$. In other words, in order that $u(1,t) = 0$, it is necessary to *pick*

$$\lambda = \pm\pi, \ \pm 2\pi, \ \pm 3\pi, \ \ldots$$

36   Diffusion-Type Problems

or

$$\lambda_n = \pm n\pi \qquad n = 1, 2, 3, \ldots$$

Note that the last BC could also imply $A = 0$, but if we choose this, we would get the zero solution in (5.1).

We have now finished the second step; we have found an infinite number of functions

(5.2) $\boxed{u_n(x,t) = A_n e^{-(n\pi\alpha)^2 t} \sin(n\pi x) \qquad n = 1, 2, \ldots}$

each one satisfying the PDE and the BCs.* These are the building blocks of the problem, and our desired solution will be a certain sum of these simple functions; the specific sum will depend on the initial conditions. See Figure 5.3 for the graphs of these fundamental solutions $u_n(x,t)$:

STEP 3 (Finding the solution to the PDE, BCs, and the IC)

The last step (and probably the most interesting from a mathematical point of view) is to add the fundamental solutions

$$u(x,t) = \sum_{n=1}^{\infty} A_n e^{-(n\pi\alpha)^2 t} \sin(n\pi x)$$

in such a way (pick the coefficients A_n) that the initial condition

$$u(x,0) = \phi(x)$$

is satisfied. Substituting the sum into the IC gives

(5.3) $$\phi(x) = \sum_{n=1}^{\infty} A_n \sin(n\pi x)$$

This equation leads us to the interesting question asked by the French mathematician Joseph Fourier, is it possible to expand the initial temperature $\phi(x)$ as the sum of the elementary functions as follows:

$$A_1 \sin(\pi x) + A_2 \sin(2\pi x) + A_3 \sin(3\pi x) + \ldots$$

The answer to this question is yes provided $\phi(x)$ is a reasonably nice function— continuous. Hence, the question now becomes how to find the coefficients A_n.

* Notice that the functions u_n and u_{-n} are essentially the same except for a minus sign.

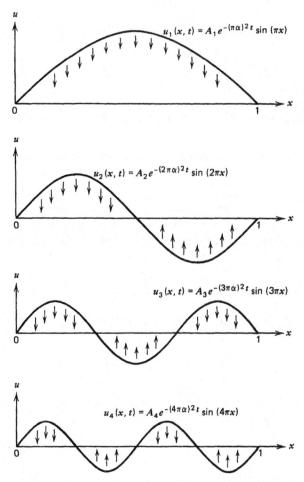

FIGURE 5.3 Fundamental solutions $u_n(x,t) = A_n e^{-(n\pi\alpha)^2 t} \sin(n\pi x)$.

This is actually very easy: One uses a property of the functions

$$\{\sin(n\pi x); \quad n = 1, 2, \ldots\}$$

known as **orthogonality**. It turns out (see problem 2) that these functions are orthogonal to each other in the sense

$$\int_0^1 \sin(m\pi x) \sin(n\pi x)\, dx = \begin{cases} 0 & m \neq n \\ 1/2 & m = n \end{cases}$$

This property can be illustrated by looking at the graphs of these functions (Figure 5.4).

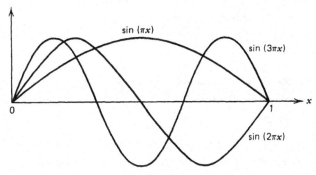

FIGURE 5.4 Orthogonal sequence of functions.

So, we are now in position to solve for the coefficients in the expression

$$\phi(x) = A_1 \sin(\pi x) + A_2 \sin(2\pi x) + A_3 \sin(3\pi x) + A_4 \sin(4\pi x) + \ldots$$

We *multiply* each side of this equation by $\sin(m\pi x)$ (m, an arbitrary integer) and *integrate* from zero to one; doing this, we get

$$\int_0^1 \phi(x) \sin(m\pi x)\, dx = A_m \int_0^1 \sin^2(m\pi x)\, dx = \frac{1}{2} A_m$$

(all other terms drop out due to orthogonality). Solving for A_m gives

$$A_m = 2 \int_0^1 \phi(x) \sin(m\pi x)\, dx$$

We're done; the solution is

(5.4)
$$u(x,t) = \sum_{n=1}^{\infty} A_n e^{-(n\pi\alpha)^2 t} \sin(n\pi x)$$

where the coefficients A_n are given by

(5.5)
$$A_n = 2 \int_0^1 \phi(x) \sin(n\pi x)\, dx$$

We can check this answer to see that it satisfies all four of our original conditions in the problem. This ends step 3.

Many students are disappointed when they finally discover that the solution is this complicated, and many hardly give the solution a second look (that's too bad). The solution is not all that difficult if one takes the time to analyze it; in fact, the more complicated it is, the more information it contains. Here are a few notes that will help you interpret this solution.

NOTES

1. Observe that the only difference between the *Fourier sine expansion* of $\phi(x)$ in (5.3) and the solution (5.4) is the insertion of the time factor

$$e^{-(n\pi\alpha)^2 t}$$

in each term. Hence, if our IC were a very simple expression like

$$\phi(x) = \sin(\pi x) + \frac{1}{2}\sin(3\pi x)$$

then the solution would simply be

$$u(x,t) = e^{-(\pi\alpha)^2 t}\sin(\pi x) + \frac{1}{2}e^{-(3\pi\alpha)^2 t}\sin(3\pi x)$$

In this case, it's obvious that if we expanded $\phi(x)$ as a Fourier sine series, we would get

$$A_1 = 1$$
$$A_2 = 0$$
$$A_3 = \frac{1}{2}$$
$$A_4 = A_5 = \ldots = 0$$

2. We can interpret the solution (5.4) in the following manner: We expand the initial temperature $\phi(x)$ as a sum of simple functions, $A_n \sin(n\pi x)$ and then find the response to each of these (which is $A_n e^{-(n\pi\alpha)^2 t}\sin(n\pi\alpha)$; and then add these individual responses to get the solution corresponding to the IC $u(x,0) = \phi(x)$.

3. The terms in the solution

$$u(x,t) = A_1 e^{-(\pi\alpha)^2 t}\sin(\pi x) + A_2 e^{-(2\pi\alpha)^2 t}\sin(2\pi x) + \ldots$$

are functions of x and t. Note that the terms further out in the series get small very fast due to the factor

$$e^{-(n\pi\alpha)^2 t}$$

Hence, for long time periods, the solution is approximately equal to the first term

$$u(x,t) \cong A_1 e^{-(\pi\alpha)^2 t} \sin(\pi x)$$

which is the shape of a damped sine curve (Figure 5.5).

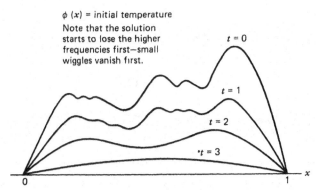

$\phi(x)$ = initial temperature
Note that the solution
starts to lose the higher
frequencies first—small
wiggles vanish first.

$t = 0$

$t = 1$

$t = 2$

$t = 3$

0   1

x

FIGURE 5.5 Higher-order terms damp faster in diffusion problems.

PROBLEMS

1. Show that $u(x,t) = e^{-\lambda^2\alpha^2 t} [A \sin(\lambda x) + B \cos(\lambda x)]$ satisfies the PDE $u_t = \alpha^2 u_{xx}$ for arbitrary A, B, and λ.

2. Show $\displaystyle\int_0^1 \sin(\pi m x) \sin(\pi n x)\, dx = \begin{cases} 0 & m \neq n \\ 1/2 & m = n \end{cases}$

    HINT Use the identity

$$\sin(mx)\sin(nx) = \frac{1}{2}[\cos(m-n)x - \cos(m+n)x]$$

3. Find the Fourier sine expansion of $\phi(x) = 1 \qquad 0 \leqslant x \leqslant 1$. Draw the first three or four terms.

4. Using the results of problem 3, what is the solution to the IBVP

$$\begin{aligned}
\text{PDE} \qquad & u_t = u_{xx} \qquad 0 < x < 1 \\[2mm]
\text{BCs} \qquad & \begin{cases} u(0,t) = 0 \\ u(1,t) = 0 \end{cases} \qquad 0 < t < \infty \\[2mm]
\text{IC} \qquad & u(x,0) = 1 \qquad 0 \leqslant x \leqslant 1
\end{aligned}$$

(Note that this problem is physically impossible, since we are pulling the temperature from one to zero instantaneously. In most problems, if the BCs are zero, then the initial temperature $\phi(x)$ should also be zero at $x = 0$ and $x = 1$.)

5. What is the solution to problem 4 if the IC is changed to

$$u(x,0) = \sin (2\pi x) + \frac{1}{3} \sin (4\pi x) + \frac{1}{5} \sin (6\pi x)$$

6. What would be the solution to problem 4 if the IC were

$$u(x,0) = x - x^2 \qquad 0 < x < 1$$

OTHER READING

Partial Different Equations of Mathematical Physics by Tyn Myint-U. Elsevier, 1973. A well-written text slightly more advanced than the current one; Chapter 6. A large chapter on separation of variables with several good problems.

Transforming Nonhomogeneous BCs into Homogeneous Ones

PURPOSE OF LESSON: To show how the initial-boundary-value problem

$$\text{PDE} \qquad u_t - \alpha^2 u_{xx} = f(x,t)$$

$$\text{BCs} \qquad \begin{cases} \alpha_1 u_x(0,t) + \beta_1 u(0,t) = g_1(t) \\ \alpha_2 u_x(L,t) + \beta_2 u(L,t) = g_2(t) \end{cases}$$

$$\text{IC} \qquad u(x,0) = \phi(x)$$

can be transformed into a new one (with *zero* BCs) like

$$U_t - \alpha^2 U_{xx} = F(x,t)$$
$$\alpha_1 U_x(0,t) + \beta_1 U(0,t) = 0$$
$$\alpha_2 U_x(L,t) + \beta_2 U(L,t) = 0$$
$$U(x,0) = \phi(x)$$

This new problem can then be solved by
1. Separation of variables if the new PDE just happens to be homogeneous [$F(x,t) = 0$].
2. Integral transforms and eigenfunction expansions if $F(x,t) \neq 0$.

Although the method of separation of variables that we discussed in the last lesson is very powerful and gives us a nice series solution, the reader should realize it doesn't apply to all problems. In order for separation of variables to apply, the BCs must be of the following form (*linear homogeneous* BCs):

(6.1) $$\alpha_1 u_x(0,t) + \beta_1 u(0,t) = 0$$
$$\alpha_2 u_x(L,t) + \beta_2 u(L,t) = 0$$

The purpose of this lesson is to show how problems with *nonhomogeneous* BC like

PDE $\qquad u_t = \alpha^2 u_{xx}$

(6.2) BCs $\qquad \begin{cases} \alpha_1 u_x(0,t) + \beta_1 u(0,t) = g_1(t) \\ \alpha_2 u_x(L,t) + \beta_2 u(L,t) = g_2(t) \end{cases}$ (nonhomogeneous BCs)

IC $\qquad u(x,0) = \phi(x)$

can be solved by transforming them into others with zero BCs. The new problem can then be solved by other methods (like eigenfunction expansions). We start our discussion by transforming an extremely simple problem with nonhomogeneous BCs into one with zero BCs.

Transforming Nonhomogeneous BCs to Homogeneous Ones

Consider heat flow in an insulated rod where the two ends are kept at constant temperatures k_1 and k_2; that is,

PDE $\qquad u_t = \alpha^2 u_{xx} \qquad 0 < x < L \qquad 0 < t < \infty$

(6.3) BCs $\qquad \begin{cases} u(0,t) = k_1 \\ u(L,t) = k_2 \end{cases} \qquad 0 < t < \infty$

IC $\qquad u(x,0) = \phi(x) \qquad 0 \leqslant x \leqslant L$

The difficulty here is that since the BCs are not homogeneous, we cannot solve this problem by separation of variables. However, it is obvious that the solution will have a steady-state solution (solution when $t = \infty$) that varies *linearly* (in x) between the boundary temperatures k_1 and k_2 (Figure 6.1).

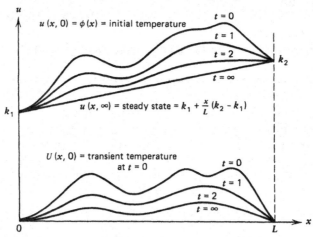

FIGURE 6.1 Solution of (6.3) for various values of time.

In other words, it seems reasonable to think of our temperature $u(x,t)$ as the sum of two parts

$$u(x,t) = \text{steady state} + \text{transient}$$

Eventual solution for large time

Part of the solution that depends on the IC (and will go to zero)

$$= [k_1 + \frac{x}{L}(k_2 - k_1)] + U(x,t)$$

This being the case, our goal is to find the *transient* $U(x,t)$. By substituting

$$u(x,t) = [k_1 + \frac{x}{L}(k_2 - k_1)] + U(x,t)$$

in the original problem (6.3), we will arrive at a new problem in $U(x,t)$. We can then solve this new one for $U(x,t)$ and add it to the steady state to get $u(x,t)$. Carrying out this simple substitution in (6.3) gives us

(6.4)

$$\text{PDE} \qquad U_t = \alpha^2 U_{xx} \qquad 0 < x < L$$

$$\text{BCs} \qquad \begin{cases} U(0,t) = 0 \\ U(L,t) = 0 \end{cases} \qquad 0 < t < \infty$$

$$\text{IC} \qquad U(x,0) = \phi(x) - [k_1 + \frac{x}{L}(k_2 - k_1)]$$

$$\overline{\phi}(x) = \text{new IC—but known}$$

This problem (fortunately) has a homogeneous PDE as well as homogeneous BCs, and so we can solve it by separation of variables; in fact, the reader probably remembers the solution:

(6.5)

$$U(x,t) = \sum_{n=1}^{\infty} a_n e^{-(n\pi\alpha)^2 t} \sin(n\pi x/L)$$

where

$$a_n = \frac{2}{L} \int_0^L \overline{\phi}(\xi) \sin(n\pi\xi/L) \, d\xi$$

So much for rods with *fixed* temperatures at the boundaries. What about more realistic-type derivative BCs with *time-varying* right-hand sides? The ideas are similar to the previous problem but a little more complicated.

Transforming Time Varying BCs to Zero BCs

Consider the typical problem

$$\text{PDE} \qquad u_t = \alpha^2 u_{xx} \qquad 0 < x < L \qquad 0 < t < \infty$$

$$(6.6) \qquad \text{BCs} \qquad \begin{cases} u(0,t) = g_1(t) & 0 < t < \infty \\ u_x(L,t) + hu(L,t) = g_2(t) \end{cases}$$

$$\text{IC} \qquad u(x,0) = \phi(x) \qquad 0 \leqslant x \leqslant L$$

To change these nonzero BCs to homogeneous ones, we (after some trial and error) seek a solution of the form

$$(6.7) \qquad u(x,t) = A(t)[1 - x/L] + B(t)[x/L] + U(x,t)$$

where $A(t)$ and $B(t)$ are chosen so that the steady-state part

$$(6.8) \qquad S(x,t) = A(t)[1 - x/L] + B(t)[x/L]$$

satisfies the BCs of the problem. In this way, the transformed problem in $U(x,t)$ will have homogeneous BCs. Substituting $S(x,t)$ into the BCs

$$S(0,t) = g_1(t)$$
$$S_x(L,t) + hS(L,t) = g_2(t)$$

gives us two equations in which we can solve for $A(t)$ and $B(t)$. Doing this, we get

$$(6.9) \qquad A(t) = g_1(t)$$
$$B(t) = \frac{g_1(t) + Lg_2(t)}{1 + Lh}$$

Hence, we have

$$u(x,t) = g_1(t)[1 - x/L] + \frac{g_1(t) + Lg_2(t)}{1 + Lh}[x/L] + U(x,t)$$

and so if we substitute this into the original problem (6.6), we get our transformed problem in $U(x,t)$ (the reader should do this)

$$\text{PDE} \quad U_t = \alpha^2 U_{xx} - S_t \quad \text{(nonhomogeneous PDE)}$$

(6.10) BCs $\begin{cases} U_x(L,t) + hU(L,t) = 0 \\ U(0,t) = 0 \end{cases}$ (homogeneous BCs)

$$\text{IC} \quad U(x,0) = \phi(x) - S(x,0) \quad \text{(new IC—but known)}$$

We now have our new problem with zero BCs (unfortunately, the PDE is nonhomogeneous). We can't solve this problem by separation of variables, but if the reader can wait for a few lessons, we will solve it by integral transforms and eigenfunction expansions.

NOTES

1. Our goal in this lesson was to transform problems with nonhomogeneous BCs into those with zero BCs. In so doing, if the new PDE just happens to be homogeneous, we are fortunate (like the first example) because we can then solve the problem by separation of variables.

    If, on the other hand, the new transformed PDE is nonhomogeneous, then we must solve the new problem by some other method.
2. The most general nonhomogeneous linear BCs

    $$\alpha_1 u_x (0,t) + \beta_1 u(0,t) = g_1(t)$$
    $$\alpha_2 u_x (L,t) + \beta_2 u(L,t) = g_2(t)$$

    can also be transformed into zero BCs in a manner similar to the technique in the second example. Of course, the new PDE would most likely be nonhomogeneous.
3. Some methods of solution do not require the BCs to be homogeneous at all, and, hence, it isn't necessary to make any preliminary transformation. Later, when we study the *Laplace transform*, we will see it isn't necessary to have zero BCs (it's just that it's sometimes easier).
4. For BCs of the form

    $$u(0,t) = g_1(t)$$
    $$u(L,t) = g_2(t)$$

    the method discussed in the second example will give us the transformation

    $$u(x,t) = \{g_1(t) + \frac{x}{L} [g_2(t) - g_1(t)]\} + U(x,t)$$

PROBLEMS

1. Solve the initial-boundary-value problem

$$\text{PDE} \qquad u_t = \alpha^2 u_{xx} \qquad\qquad 0 < x < 1$$

$$\text{BCs} \qquad \begin{cases} u(0,t) = 1 \\ u_x(1,t) + hu(1,t) = 1 \end{cases} \qquad 0 < t < \infty$$

$$\text{IC} \qquad u(x,0) = \sin(\pi x) + x \qquad 0 \leqslant x \leqslant 1$$

by transforming it into homogeneous BCs and then solving the transformed problem. Does the solution agree with your intuition of the problem?

2. Transform

$$\text{PDE} \qquad u_t = u_{xx} \qquad 0 < x < 1$$

$$\text{BCs} \qquad \begin{cases} u(0,t) = 0 \\ u(1,t) = 1 \end{cases} \qquad 0 < t < \infty$$

$$\text{IC} \qquad u(x,0) = x^2 \qquad 0 \leqslant x \leqslant 1$$

to zero BCs and solve the new problem. What will the solution to this problem look like for different values of time? Does the solution agree with your intuition? What is the steady-state solution? What does the transient solution look like?

3. Transform

$$\text{PDE} \qquad u_t = u_{xx} \qquad 0 < x < 1$$

$$\text{BCs} \qquad \begin{cases} u_x(0,t) = 0 \\ u_x(1,t) + hu(1,t) = 1 \end{cases} \qquad 0 < t < \infty$$

$$\text{IC} \qquad u(x,0) = \sin(\pi x) \qquad 0 \leqslant x \leqslant 1$$

into a new problem with zero BCs; is the new PDE homogeneous?

OTHER READING

Analysis and Solution of Partial Differential Equations by R. L. Street. Brooks-Cole, 1973. This excellent text contains an extensive section on transforms of the type we discuss in this lesson and a good section on separation of variables.

Solving More Complicated Problems by Separation of Variables

PURPOSE OF LESSON: To show how more complicated heat-flow problems can be solved by separation of variables. This lesson essentially consists of a worked problem that will give the reader more familiarity with the method. Hopefully, the reader will be able to extrapolate the ideas presented here to solve problems on his or her own.

Eigenvalue problems, known as *Sturm-Liouville problems*, are introduced, and some properties of these general problems are discussed.

The purpose of this lesson is to solve an *initial-boundary-value problem* by the separation of variables method that the reader might have trouble working on his or her own. Hopefully, the reader can extrapolate from this problem to other problems not specifically mentioned in this text.

We start with a one-dimensional heat-flow problem where one of the BCs contains derivatives.

Heat-Flow Problem with Derivative BC

Consider an apparatus

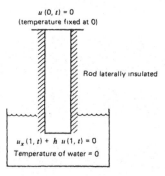

$u(0, t) = 0$
(temperature fixed at 0)

Rod laterally insulated

$u_x(1, t) + h\, u(1, t) = 0$
Temperature of water = 0

FIGURE 7.1 Diagram for the initial-boundary-value problem.

in which we fix the temperature at the top of the rod at $u(0,t) = 0$ and immerse the bottom of the rod in a solution of water fixed at the same temperature of zero (zero refers to some reference temperature). The natural flow of heat (Newton's law of cooling) says that the BC at $x = 1$ is

$$u_x (1,t) = - hu (1,t)$$

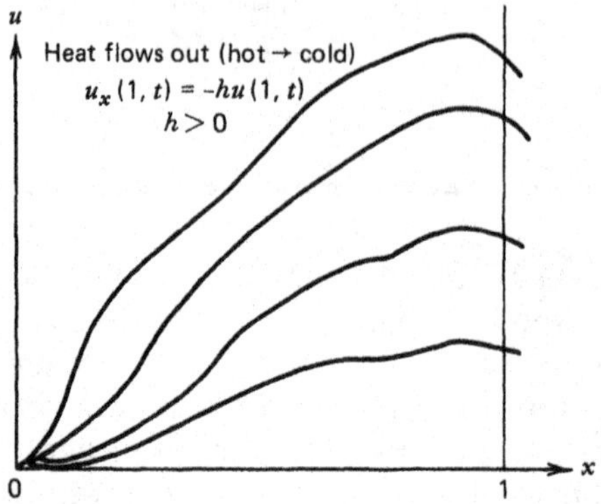

FIGURE 7.2 The nature of curves with BCs $\begin{cases} u(0,t) = 0 \\ u_x(1,t) = -hu(1,t) \end{cases}$

Suppose now the *initial temperature* of the rod is $u(x,0) = x$, but instantaneously thereafter $(t > 0)$, we apply our BCs. To find the ensuing temperature, we must solve the IBVP

PDE $\qquad u_t = \alpha^2 u_{xx} \qquad 0 < x < 1 \qquad 0 < t < \infty$

(7.1) BCs $\qquad \begin{cases} u (0,t) = 0 \\ u_x (1,t) + hu (1, t) \cdot = 0 \end{cases} \qquad$ (homogeneous BCs)

IC $\qquad u (x,0) = x \qquad 0 \leqslant x \leqslant 1$

To apply the separation of variables method, we carry out the following steps:

STEP 1 (Separating the PDE into two ODEs)
Substituting $u(x, t) = X(x)T(t)$ into the PDE gives

$$XT' = \alpha^2 X''T$$

and dividing by $\alpha^2 XT$, we get

$$\frac{T'}{\alpha^2 T} = \frac{X''}{X}$$

Since the left-hand side depends *only* on time and the right-hand side depends only on x (and since x and t are independent), both sides of this equation must be constants. Setting them both equal to μ gives the two ODEs

(7.2)
$$T' - \mu\alpha^2 T = 0$$
$$X'' - \mu X = 0$$

We have now completed the *separation process*.

STEP 2 (Finding the separation constant)

First of all, μ must not be positive or else $T(t)$ will grow exponentially to infinity (which would make $u = XT$ go to infinity—which we can reject on physical grounds).

Secondly, suppose $\mu = 0$. This being the case, we have

$$X'' = 0$$

and thus

$$X(x) = A + Bx$$

But since the BCs of the problem are

$$u\,(0,\,t) = X\,(0)T(t) = 0$$
$$u_x\,(1,t) + h\,u\,(1,t) = X'(1)T(t) + h\,X(1)\,T(t) = 0$$

we could conclude that

$$\begin{array}{c} X\,(0) = 0 \\ X'\,(1) + h\,X(1) = 0 \end{array} \Rightarrow \begin{array}{c} A = 0 \\ B = 0 \end{array}$$

which would mean $u(x,t) = 0$. In other words, $\mu = 0$ gives only $u = 0$; hence we throw it out (we are looking for nonzero solutions).

Finally, if $\mu < 0$, we call $\mu = -\lambda^2$ and write the two ODEs (7.2) as

$$T' + \lambda^2\alpha^2 T = 0$$
$$X'' + \lambda^2 X = 0$$

which gives us solutions

$$T(t) = Ae^{-(\lambda\alpha)^2 t}$$
$$X(x) = B \sin\,(\lambda x) + C \cos\,(\lambda x)$$

Hence, what we have is that *any function*

(7.3) $$u(x, t) = e^{-(\lambda\alpha)^2 t} [A \sin (\lambda x) + B \cos (\lambda x)]$$

for *any* λ and *any* A and B will satisfy the PDE (the reader can verify this calculation on his or her own). What we'd like to do now is find out how many of these functions will satisfy the BCs

(7.4)
$$u (0,t) = 0$$
$$u_x (1,t) + hu (1, t) = 0$$

Substituting the solution (7.3) into the BCs (7.4) gives us conditions on λ, A, and B that must be satisfied; namely,

$$Be^{-(\lambda\alpha)^2 t} = 0 \Rightarrow B = 0$$
$$A\lambda e^{-(\lambda\alpha)^2 t} \cos \lambda + hAe^{-(\lambda\alpha)^2 t} \sin \lambda = 0$$

Performing a little algebra on this last equation gives us our desired condition on λ

$$\tan \lambda = -\lambda/h$$

In other words, to find λ, we must find the intersections of the curves tan λ and −λ/h (Figure 7.3).

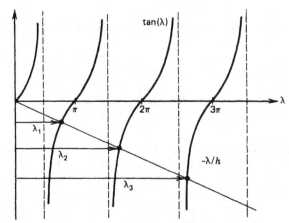

FIGURE 7.3 Graph showing intersections of tan (λ) and −λ/h.

These values $\lambda_1, \lambda_2, \ldots$ can be computed numerically for a given h on a computer and are called the **eigenvalues** of the boundary-value problem

$$X'' + \lambda^2 X = 0$$
(7.5) $$X (0) = 0$$
$$X'(1) + hX (1) = 0$$

In other words, they are the values of λ for which there exists a *nonzero solution*. The eigenvalues λ_n of (7.5), which, in this case, are the roots of $\tan \lambda = -\lambda/h$, have been computed (for $h = 1$) numerically, and the first five values are listed in Table 7.1.

TABLE 7.1   Roots of
$\tan \lambda = -\lambda$

n	λ_n
1	2.02
2	4.91
3	7.98
4	11.08
5	14.20

The solutions of (7.5) corresponding to the eigenvalues λ_n are called the **eigenfunctions** $X_n(x)$, and for this problem, we have

$$X_n(X) = \sin(\lambda_n x)$$

See Figure 7.4.

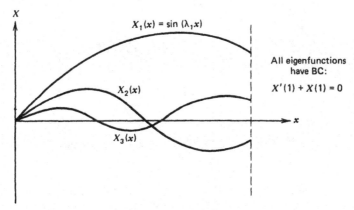

FIGURE 7.4   Eigenfunctions $X_n(x)$ of (7.5) for $h = 1$.

STEP 3   (Finding the fundamental solutions)
We now have an infinite number of functions (fundamental solutions),

$$u_n(x, t) = X_n(x) T_n(t) = e^{-(\lambda_n a)^2 t} \sin(\lambda_n x)$$

each one satisfying the PDE and the BCs. The final step is to add these functions together (the sum will still satisfy the PDE and BCs, since both the PDE and BCs are *linear* and *homogeneous*) in such a way that they agree with the IC when $t = 0$; that is, we sum

$$u(x, t) = \sum_{n=1}^{\infty} a_n X_n (x) T_n (t)$$

$$= \sum_{n=1}^{\infty} a_n e^{-(\lambda_n \alpha)^2 t} \sin (\lambda_n x)$$

so that the IC $u(x,0) = x$ is satisfied. In other words,

(7.6)
$$u(x, 0) = x = \sum_{n=1}^{\infty} a_n \sin (\lambda_n x)$$

This brings us to our final step.

STEP 4 (Expansion of the IC as a sum of eigenfunctions)
To find the constants a_n in the eigenfunction expansion (7.6), we must multiply each side of the equation by $\sin (\lambda_m x)$ and integrate x from 0 to 1; that is;

$$\int_0^1 \xi \sin (\lambda_m \xi) \, d\xi = \sum_{n=1}^{\infty} a_n \int_0^1 \sin (\lambda_n \xi) \sin (\lambda_m \xi) \, d\xi$$

$$= a_m \int_0^1 \sin^2 (\lambda_m \xi) \, d\xi$$

$$= a_m \left(\frac{\lambda_m - \sin \lambda_m \cos \lambda_m}{2\lambda_m} \right)$$

Solving for a_m (we'll change the notation to a_n), we get our desired result

(7.7)
$$a_n = \frac{2\lambda_n}{(\lambda_n - \sin \lambda_n \cos \lambda_n)} \int_0^1 \xi \sin (\lambda_n \xi) \, d\xi$$

In other words, our solution to (7.1) is

(7.8)
$$u (x,t) = \sum_{n=1}^{\infty} a_n e^{-(\lambda_n \alpha)^2 t} \sin (\lambda_n x)$$

where the constants a_n are given by (7.7). In this problem, the first five constants a_n have been computed and are listed in Table 7.2.

54   Diffusion-Type Problems

TABLE 7.2   Coefficients
a_n in (7.8)

n	a_n
1	0.24
· 2	0.22
3	−0.03
4	−0.11
5	−0.09

Hence, the first three terms of the IBVP

$$\text{PDE} \qquad u_t = u_{xx} \qquad 0 < x < 1 \qquad 0 < t < \infty$$

(7.9) $\qquad$ BCs $\qquad \begin{cases} u\,(0,t) = 0 \\ u_x\,(1,t) + u\,(1,t) = 0 \end{cases} \qquad 0 < t < \infty$

$$\text{IC} \qquad u(x, 0) = x \qquad 0 \leqslant x \leqslant 1$$

are

$$u(x, t) = 0.24\ e^{-4t} \sin (2x) + 0.22\ e^{-24t} \sin (4.9x)$$
$$+ 0.03e^{-63.3t} \sin (7.98x) + \ldots$$

The graph of this solution is drawn for various values of time in Figure 7.5. The reader can ask himself or herself if this solution agrees with his or her intuition and whether or not it satisfies the BCs of the problem.

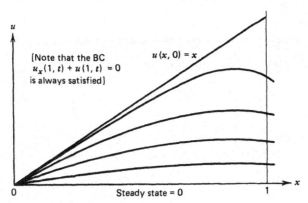

FIGURE 7.5   Solution to (7.8).

NOTES

The eigenvalue problem (7.5) is a special case of the general problem

$$\text{ODE} \qquad [p(x)\, y']' - q(x)y + \lambda\, r(x)y = 0 \qquad 0 < x < 1$$

(7.10)

$$\text{BCs} \qquad \begin{cases} \alpha_1 y\,(0) + \beta_1 y'(0) = 0 \\ \alpha_2 y\,(1) + \beta_2 y'(1) = 0 \end{cases}$$

known as the **Sturm-Liouville problem**. When we solve PDEs by separation of variables with linear homogeneous BCs, the ODE in $X(x)$ along with its BCs will always be some particular Sturm-Liouville problem. We observe that the eigenvalue problem (7.5) is a special case of (7.10).

What Sturm and Liouville proved is that under suitable conditions on the functions $p(x)$, $q(x)$, and $r(x)$, the problem (7.10) has

1. An infinite sequence of eigenvalues

$$\lambda_1 < \lambda_2 < \lambda_3 < \ldots < \lambda_n < \ldots \rightarrow \infty$$

2. Corresponding to *each* eigenvalue λ_n, there is *one* nonzero solution $y_n\,(x)$ [not including other constant multiples of $y_n(x)$].

3. If $y_n(x)$ and $y_m(x)$ are two *different* eigenfunctions (corresponding to $\lambda_n \neq \lambda_m$), then they are *orthogonal* with respect to the *weight function* $r(x)$ on the interval of $[0,1]$; that is, they satisfy

$$\int_0^1 r(x)\, y_n(x)\, y_m(x)\, dx = 0$$

More details of Sturm-Liouville-type problems can be found in references 1 and 2.

PROBLEMS

1. Solve the following heat-flow problem:

$$\text{PDE} \qquad u_t = u_{xx} \qquad 0 < x < 1 \qquad 0 < t < \infty$$

$$\text{BCs} \qquad \begin{cases} u(0,t) = 0 \\ u_x(1,t) = 0 \end{cases} \qquad 0 < t < \infty$$

$$\text{IC} \qquad u(x,0) = x \qquad 0 \leq x \leq 1$$

by separation of variables. Does your solution agree with your intuition? What is the steady-state solution?

2. What are the eigenvalues and eigenfunctions of the Sturm-Louiville problem

$$\text{ODE} \qquad X'' + \lambda X = 0 \qquad 0 < x < 1$$

$$\text{BCs} \qquad \begin{cases} X(0) = 0 \\ X'(1) = 0 \end{cases}$$

What are the functions $p(x)$, $q(x)$, and $r(x)$ in the general Sturm-Liouville problem for this equation?

3. Solve the following problem with insulated boundaries:

$$\text{PDE} \qquad u_t = u_{xx} \qquad 0 < x < 1 \qquad 0 < t < \infty$$

$$\text{BCs} \qquad \begin{cases} u_x(0,t) = 0 \\ u_x(1,t) = 0 \end{cases} \qquad 0 < t < \infty$$

$$\text{IC} \qquad u(x,0) = x \qquad 0 \leqslant x \leqslant 1$$

Does your solution agree with your interpretation of the problem? What is the steady-state solution?; does this make sense?

4. What are the eigenvalues and eigenfunctions of

$$\text{ODE} \qquad X'' + \lambda X = 0 \qquad 0 < x < 1$$

$$\text{BCs} \qquad \begin{cases} X'(0) = 0 \\ X'(1) = 0 \end{cases}$$

OTHER READING

1. *Elementary Differential Equations and Boundary-Value Problems* by W. E. Boyce and R. C. DiPrima. John Wiley & Sons, 1965. Chapter 11. This is an ordinary-differential-equations text that contains an excellent section on the Sturm-Liouville problem, one of the better undergraduate texts in ODEs.

2. *Advanced Engineering Mathematics* by E. Kreyszig. John Wiley & Sons, 1967. This text contains many worked examples of typical problems; very readable.

Transforming Hard Equations into Easier Ones

PURPOSE OF LESSON: To show how one can transform a PDE in $u(x,t)$ into a new (easier) one in a new variable $w(x,t)$. The transformation is generally based on intuition, and in this lesson, the PDEs

$$u_t = \alpha^2 u_{xx} - \beta u$$
$$u_t = \alpha^2 u_{xx} - \nu u_x$$

are transformed into the simple heat equation

$$w_t = \alpha^2 w_{xx}$$

by means of the transformations

$$u(x,t) = e^{-\beta t} w(x,t)$$
$$u(x,t) = e^{\nu[x-\nu t/2]/2\alpha^2} w(x,t)$$

After the transformations are made, the heat equation (the easy one) can be solved for $w(x,t)$, hence,

$$u = e^{-\beta t} w(x,t)$$
$$u = e^{\nu[x-\nu t/2]/2\alpha^2} w(x,t)$$

are the solutions of the original equations (of course, the BCs and the IC must be transformed too).

The reader may get the impression from the last two lessons that the only type of PDE that can be solved by separation of variables is

$$u_t = \alpha^2 u_{xx}$$

It is true the heat equation is the easiest parabolic PDE to solve by separation of variables, but it is in no way the *only* equation we can solve by this technique.

As mentioned earlier, as long as the equation is *linear* and *homogeneous*, we can separate variables. For example, two-dimensional heat flow inside a circle would be described by the equation

$$u_t = \alpha^2 \left[u_{rr} + \frac{1}{r} u_r + \frac{1}{r^2} u_{\theta\theta} \right]$$

and although it has *variable coefficients,* it can still be separated into three ODEs.

This lesson will show the reader that sometimes a PDE doesn't have to be attacked directly but that the original PDE can be *transformed* into an easier one. In this way, the easier problem can be solved (by separation of variables or some other technique). We now present an example that illustrates this technique.

Transforming a Heat-Flow Problem with Lateral Heat Loss into an Insulated Problem

Consider the following problem:

$$\text{PDE} \qquad u_t = \alpha^2 u_{xx} - \beta u \qquad 0 < x < 1 \qquad 0 < t < \infty$$

(8.1)   BCs $\begin{cases} u(0,t) = 0 \\ u(1,t) = 0 \end{cases} \quad 0 < t < \infty$

$$\text{IC} \qquad u(x,0) = \phi(x) \qquad 0 \leq x \leq 1$$

where the term $-\beta u$ represents heat flow across the lateral boundary (Figure 8.1).

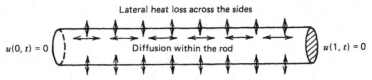

Lateral heat loss across the sides

$u(0, t) = 0$ Diffusion within the rod $u(1, t) = 0$

FIGURE 8.1   Heat flow described by $u_t = \alpha^2 u_{xx} - \beta u$.

The goal of this lesson is to introduce a *new temperature* $w(x,t)$ in place of $u(x,t)$, so that the PDE in w is simpler than the original one

$$u_t = \alpha^2 u_{xx} - \beta u$$

This is a common technique in PDEs, and the transformation is generally based on an intuitive feeling of how the solution of the original PDE behaves. For example, in our problem (8.1), the temperature $u(x,t)$ at any point x_0 is changing as a result of two phenomena

1. *diffusion* of heat within the rod (due to $\alpha^2 u_{xx}$).
2. *heat flow* across the lateral boundary (due to $-\beta u$).

The important point is that if there were *no* diffusion *within* the rod ($\alpha = 0$), then the temperature at each point x_0 would "damp" exponentially to zero according to

$$u(x_0, t) = u(x_0, 0)e^{-\beta t}$$

By means of this observation, we wonder if we can essentially decompose the temperature $u(x,t)$ of problem (8.1) into two factors

(8.2) $$u(x,t) = e^{-\beta t} w(x,t)$$

or

$$\text{Noninsulated temperature} = e^{-\beta t} \quad \text{(insulated temperature)}$$

where $w(x,t)$ would represent the temperature due to diffusion only. Let's see what happens if we substitute this expression into problem (8.1); this is a routine calculation (the reader can do it on his or her own), and we arrive at

$$\text{PDE} \qquad w_t = \alpha^2 w_{xx} \qquad 0 < x < 1 \qquad 0 < t < \infty$$

(8.3) $$\text{BCs} \qquad \begin{cases} w(0,t) = 0 \\ w(1,t) = 0 \end{cases} \qquad 0 < t < \infty$$

$$\text{IC} \qquad w(x,0) = \phi(x) \qquad 0 \leqslant x \leqslant 1$$

This is exactly the same problem we started with except that now the PDE doesn't contain $-\beta u$; so all we have to do to solve (8.1) is solve the transformed problem (8.3) and then multiply the solution $w(x,t)$ by $e^{-\beta t}$. In this case, we have already solved (8.3) previously by the separation of variables method and found

(8.4) $$w(x,t) = \sum_{n=1}^{\infty} a_n e^{-(n\pi\alpha)^2 t} \sin(n\pi x)$$

$$a_n = 2 \int_0^1 \phi(\xi) \sin(n\pi\xi) \, d\xi$$

and, hence, the solution of the original problem (8.1) is

$$u(x,t) = e^{-\beta t} w(x,t)$$

The following example in the notes is solved by this technique.

NOTES

1. To solve the problem

$$u_t \; = u_{xx} - u \qquad 0 < x < 1 \qquad 0 < t < \infty$$
$$u(0,t) = 0$$
$$u(1,t) = 0$$
$$u(x,0) = \sin(\pi x) + 0.5 \sin(3\pi x)$$

by the preceding strategy, we
   (a) neglect the convection term $-u$ for the time being.
   (b) solve the initial-boundary-value problem without the term $-u$ to get

$$u(x,t) \; = \; e^{-\pi^2 t} \sin(\pi x) + 0.5\, e^{-(3\pi)^2 t} \sin(3\pi x)$$

   (c) multiply this solution by the convection factor $e^{-\beta t} = e^{-t}$ to get the solution

$$u(x,t) \; = \; e^{-t} \left[e^{-\pi^2 t} \sin(\pi x) + 0.5\, e^{-(3\pi)^2 t} \sin(3\pi x) \right]$$

2. The *diffusion-convection* equation

$$u_t \; = \; \alpha^2 u_{xx} - v u_x$$

(v is a constant) can also be transformed to

$$w_t \; = \; \alpha^2 w_{xx}$$

In this case, the transformation is

$$u(x,t) \; = \; e^{v(x - vt/2)/2\alpha^2}\, w(x,t)$$

This transformation essentially factors out the part of the solution (exponential factor) that is due to the moving medium. Note that the exponential factor consists of a moving exponential (moving to the right with velocity $v/2$). The reader will get a chance to use this transformation in the problem set.

PROBLEMS

1. Solve the diffusion problem

$$\text{PDE} \qquad u_t = u_{xx} - u_x \qquad 0 < x < 1 \qquad 0 < t < \infty$$

$$\text{BCs} \qquad \begin{cases} u(0,t) = 0 \\ u(1,t) = 0 \end{cases} \qquad 0 < t < \infty$$

$$\text{IC} \qquad u(x,0) = e^{x/2} \qquad 0 \leqslant x \leqslant 1$$

by transforming it into an easier problem. What does the solution look like? We could interpret this problem as describing the concentration $u(x,t)$ in a moving medium (moving from left to right with velocity $v = 1$) where the concentration at the *ends* of the medium are kept at zero (by some filtering device) and the *initial concentration* is $e^{x/2}$. Does your solution agree with this interpretation?

2. Solve the problem

$$\text{PDE} \qquad u_t = u_{xx} - u + x \qquad 0 < x < 1 \qquad 0 < t < \infty$$

$$\text{BCs} \qquad \begin{cases} u(0,t) = 0 \\ u(1,t) = 1 \end{cases} \qquad 0 < t < \infty$$

$$\text{IC} \qquad u(x,0) = 0 \qquad 0 \leqslant x \leqslant 1$$

by
   (a) changing the nonhomogeneous BCs to homogeneous ones.
   (b) transforming into a new equation without the term $-u$.
   (c) solving the resulting problem.
3. Solve

$$\text{PDE} \qquad u_t = u_{xx} - u \qquad 0 < x < 1 \qquad 0 < t < \infty$$

$$\text{BCs} \qquad \begin{cases} u(0,t) = 0 \\ u(1,t) = 0 \end{cases} \qquad 0 < t < \infty$$

$$\text{IC} \qquad u(x,0) = \sin(\pi x) \qquad 0 \leqslant x \leqslant 1$$

directly by separation of variables without making any preliminary transformation. Does your solution agree with the solution you would obtain if the transformation

$$u(x,t) = e^{-t}w(x,t)$$

were made in advance?

OTHER READING

Nonlinear Partial Differential Equations in Engineering by W. F. Ames. Academic Press, 1965. This text discusses many types of transformations for changing old problems into new ones, so that sometimes even nonlinear problems can be transformed into linear ones.

Solving Nonhomogeneous PDEs (Eigenfunction Expansions)

PURPOSE OF LESSON: To show how to solve the initial-boundary-value problem

PDE $\quad u_t = \alpha^2 u_{xx} + f(x,t) \quad 0 < x < 1 \quad 0 < t < \infty$

BCs $\quad \begin{cases} \alpha_1 u_x(0,t) + \beta_1 u(0,t) = 0 \\ \alpha_2 u_x(1,t) + \beta_2 u(1,t) = 0 \end{cases} \quad 0 < t < \infty$

IC $\quad u(x,0) = \phi(x) \quad 0 \leq x \leq 1$

A nonhomogeneous PDE can be solved by finding a series solution of the form

$$u(x,t) = \sum_{n=1}^{\infty} T_n(t)X_n(x)$$

where the $X_n(x)$ are the eigenfunctions we find when solving the associated homogeneous problem

PDE $\quad u_t = \alpha^2 u_{xx} \quad 0 < x < 1 \quad 0 < t < \infty$

BCs $\quad \begin{cases} \alpha_1 u_x(0,t) + \beta_1 u(0,t) = 0 \\ \alpha_2 u_x(1,t) + \beta_2 u(1,t) = 0 \end{cases} \quad 0 < t < \infty$

IC $\quad u(x,0) = \phi(x) \quad 0 \leq x \leq 1$

and $T_n(t)$ are functions that can be found by solving a sequence of ODEs.

In Lesson 6, we discussed how transformations could be made to transform nonhomogeneous BCs into *homogeneous* ones. Unfortunately, the PDE was left nonhomogeneous by this process, and we were left with the problem

$$\text{PDE} \qquad u_t = \alpha^2 u_{xx} + f(x,t) \qquad 0 < x < 1 \qquad 0 < t < \infty$$

(9.1) $\qquad$ BCs $\qquad \begin{cases} \alpha_1 u_x(0,t) + \beta_1 u(0,t) = 0 \\ \alpha_2 u_x(1,t) + \beta_2 u(1,t) = 0 \end{cases} \quad 0 < t < \infty$

$$\text{IC} \qquad u(x,0) = \phi(x) \qquad 0 \leq x \leq 1$$

The purpose of this lesson is to solve this problem by a method that is analogous to the method of *variation of parameters* in ODEs and is known as the **eigenfunction expansion method.**

The idea is quite simple. Inasmuch as the solution of (9.1) with $f(x,t) = 0$ (so-called corresponding homogeneous problem) is given by

$$u(x,t) = \sum_{n=1}^{\infty} a_n e^{-(\lambda_n \alpha)^2 t} X_n(x)$$

where λ_n and $X_n(x)$ are the eigenvalues and eigenfunctions of the Sturm-Liouville problem,

(9.2)
$$\begin{aligned} X'' + \lambda^2 X &= 0 \\ \alpha_1 X'(0) + \beta_1 X(0) &= 0 \\ \alpha_2 X'(1) + \beta_2 X(1) &= 0 \end{aligned}$$

we ask whether the solution of the nonhomogeneous problem (9.1) can be written in the slightly more general form

$$u(x,t) = \sum_{n=1}^{\infty} T_n(t) X_n(x)$$

The reason for this speculation is physically appealing, inasmuch as a source of heat $f(x,t)$ *within* the rod will give rise to a new time component and not the damping factor

$$e^{-(\lambda_n \alpha)^2 t}$$

as was the case when the only input into the problem was the IC.

To show how this method works, we apply it to a simple problem so the details aren't as complicated.

Solution by the Eigenfunction Expansion Method

Consider the nonhomogeneous problem

$$\text{PDE} \qquad u_t = \alpha^2 u_{xx} + f(x,t) \qquad 0 < x < 1 \qquad 0 < t < \infty$$

$$(9.3) \qquad \text{BCs} \qquad \begin{cases} u(0,t) = 0 \\ u(1,t) = 0 \end{cases} \qquad 0 < t < \infty$$

$$\text{IC} \qquad u(x,0) = \phi(x) \qquad 0 \leqslant x \leqslant 1$$

To solve this problem, we divide the procedure into the following steps:

STEP 1 The basic idea in this method is to decompose the heat source $f(x,t)$ into simple components

$$f(x,t) = f_1(t)X_1(x) + f_2(t)X_2(x) + \ldots + f_n(t)X_n(x) + \ldots$$

and find the response $u_n(x,t)$ to *each* of these individual components $f_n(t)X_n(x)$. The solution to our problem is then

$$u(x,t) = \sum_{n=1}^{\infty} u_n(x,t)$$

To determine how to decompose $f(x,t)$ into its component parts $f_n(t)X_n(x)$ is one of the major problems. It turns out that the $X_n(x)$ factors in this problem are the *eigenvectors* of the Sturm-Liouville system we get when solving the *associated homogeneous problem* to (9.3) by separation of variables; that is,

$$(9.4) \qquad \begin{aligned} u_t &= \alpha^2 u_{xx} \qquad \text{(note that } f(x,t) = 0) \\ u(0,t) &= 0 \\ u(1,t) &= 0 \\ u(x,0) &= \phi(x) \end{aligned}$$

in this case, the Sturm-Liouville problem we find when separating variables is

$$\begin{aligned} X'' + \lambda^2 X &= 0 \\ X(0) &= 0 \\ X(1) &= 0 \end{aligned}$$

and, hence, the $X_n(x)$ are

$$X_n(x) = \sin (n\pi x) \qquad n = 1, 2, \ldots$$

Hence, our decomposition of the heat source has the form

$$(9.5) \quad f(x,t) = f_1(t) \sin (\pi x) + f_2(t) \sin (2\pi x) + \ldots + f_n(t) \sin (n\pi x) + \ldots$$

Finally, to find the functions $f_n(t)$, we merely multiply each side of this equation by $\sin (m\pi x)$ and integrate from zero to one (with respect to x); hence, we have

$$\int_0^1 f(x,t) \sin{(m\pi x)}\, dx = \sum_{n=1}^{\infty} f_n(t) \int_0^1 \sin{(m\pi x)} \sin{(n\pi x)}\, dx$$

$$= \frac{1}{2} f_m(t)$$

or (changing m to n)

(9.6) $$f_n(t) = 2 \int_0^1 f(x,t) \sin{(n\pi x)}\, dx$$

This will give us an equation for the coefficients $f_n(t)$ in terms of the heat source $f(x,t)$.

STEP 2 (Find the response $u_n(x,t) = T_n(t)X_n(x)$ to input $f_n(t)X_n(x)$)
We now replace the heat source $f(x,t)$ by its decomposition

$$f(x,t) = \sum_{n=1}^{\infty} f_n(t) \sin{(n\pi x)}$$

and try to find the individual responses

$$u(x,t) = \sum_{n=1}^{\infty} T_n(t) \sin{(n\pi x)}$$

in other words, we seek the functions $T_n(t)$. Knowing these, the answer to our problem is

$$u(x,t) = \sum_{n=1}^{\infty} T_n(t) \sin{(n\pi x)}$$

Substituting $u(x,t) = \sum_{n=1}^{\infty} T_n(t) \sin{(n\pi x)}$ into the system

$$u_t = \alpha^2 u_{xx} + \sum_{n=1}^{\infty} f_n(t) \sin{(n\pi x)}$$
$$u(0,t) = 0$$
$$u(1,t) = 0$$
$$u(x,0) = \phi(x)$$

gives us

$$\sum_{n=1}^{\infty} T'_n(t) \sin(n\pi x) = -\alpha^2 \sum_{n=1}^{\infty} (n\pi)^2 T_n(t) \sin(n\pi x) + \sum_{n=1}^{\infty} f_n(t) \sin(n\pi x)$$

$$\sum_{n=1}^{\infty} T_n(t) \sin 0 = 0 \qquad \text{(says nothing; zero = zero)}$$

(9.7)

$$\sum_{n=1}^{\infty} T_n(t) \sin(n\pi) = 0 \qquad \text{(says nothing; zero = zero)}$$

$$\sum_{n=1}^{\infty} T_n(0) \sin(n\pi x) = \phi(x)$$

Rewriting the PDE and the IC as

$$\text{PDE} \qquad \sum_{n=1}^{\infty} [T'_n + (n\pi\alpha)^2 T_n - f_n(t)] \sin(n\pi x) = 0$$

$$\text{IC} \qquad \sum_{n=1}^{\infty} T_n(0) \sin(n\pi x) = \phi(x)$$

we can see fairly easily that $T_n(t)$ will satisfy the simple initial value problem

(9.8)
$$T'_n + (n\pi\alpha)^2 T_n = f_n(t)$$

$$T_n(0) = 2 \int_0^1 \phi(\xi) \sin(n\pi\xi) \, d\xi = a_n$$

This ODE is one of the easier ones to solve (use an integrating factor) and has the solution

(9.9)
$$T_n(t) = a_n e^{-(n\pi\alpha)^2 t} + \int_0^t e^{-(n\pi\alpha)^2(t-\tau)} f_n(\tau) \, d\tau$$

Hence, the solution of problem (9.3) is

(9.10)
$$u(x,t) = \sum_{n=1}^{\infty} T_n(t) \sin(n\pi x)$$

$$= \sum_{n=1}^{\infty} [a_n e^{-(n\pi\alpha)^2 t} \sin(n\pi x)] + \sum_{n=1}^{\infty} [\sin(n\pi x) \int_0^t e^{-(n\pi\alpha)^2(t-\tau)} f_n(\tau) \, d\tau]$$

Transient part   Steady state
(because of the initial condition)   (because of the right-hand side $f(x,t)$)

We can see from this solution that the *temperature response* is due to *two parts*: the first part that is due to the IC and the second part that is due to the heat source $f(x,t)$. The phrase steady state is not the best phrase to describe the second part, since it doesn't necessarily come to rest (it may approach a periodic steady state, *if* $f(x,t)$ is periodic in t).

This completes the problem. Before stopping, however, we will show how this method can be applied to a specific example.

Solution of a Problem by the Eigenfunction-Expansion Method

Consider the simple problem

$$\text{PDE} \qquad u_t = \alpha^2 u_{xx} + \sin(3\pi x) \qquad 0 < x < 1$$

$$(9.11) \qquad \text{BCs} \quad \begin{cases} u(0,t) = 0 \\ u(1,t) = 0 \end{cases} \quad 0 < t < \infty$$

$$\text{IC} \qquad u(x,0) = \sin(\pi x) \qquad 0 \leq x \leq 1$$

Our goal is to compute the coefficients $T_n(t)$ in the expansion

$$u(x,t) = \sum_{n=1}^{\infty} T_n(t) \sin(n\pi x)$$

(the eigenfunctions $X_n(x)$ are still the same for this problem). If we substitute this expansion in the problem, we will get an ODE for the functions $T_n(t)$ In fact, we will get

$$T_n' + (n\pi\alpha)^2 T_n = f_n(t) = 2\int_0^1 \sin(3\pi x) \sin(n\pi x)\, dx = \begin{cases} 1 & n = 3 \\ 0 & n \neq 3 \end{cases}$$

$$T_n(0) = 2\int_0^1 \sin(\pi\xi) \sin(n\pi\xi)\, d\xi = \begin{cases} 1 & n = 1 \\ 0 & n \neq 1 \end{cases}$$

Writing out these equations for $n = 1, 2, \ldots$, we see

$$(n = 1) \qquad \left. \begin{matrix} T_1' + (\pi\alpha)^2 T_1 = 0 \\ T_1(0) = 1 \end{matrix} \right\} \Rightarrow T_1(t) = e^{-(\pi\alpha)^2 t}$$

$$(n = 2) \qquad \left. \begin{matrix} T_2' + (2\pi\alpha)^2 T_2 = 0 \\ T_2(0) = 0 \end{matrix} \right\} \Rightarrow T_2(t) = 0$$

$$(n = 3) \qquad \left. \begin{matrix} T_3' + (3\pi\alpha)^2 T_3 = 1 \\ T_3(0) = 0 \end{matrix} \right\} \Rightarrow T_3(t) = \frac{1}{(3\pi\alpha)^2}\left[1 - e^{-(3\pi\alpha)^2 t}\right]$$

$$(n \geq 4) \qquad \left. \begin{matrix} T_n' + (n\pi\alpha)^2 T_n = 0 \\ T_n(0) = 0 \end{matrix} \right\} \Rightarrow T_n(t) = 0$$

Hence the solution of our problem is

$$(9.12) \qquad u(x,t) = e^{-(\pi\alpha)^2 t} \sin{(\pi x)} + \frac{1}{(3\pi\alpha)^2} [1 - e^{-(3\pi\alpha)^2 t}] \sin{(3\pi x)}$$

Transient   Steady state
(because of initial conditions)   (because of the right-hand side of the PDE)

NOTES

1. The method of eigenfunction expansion is one of the most powerful for solving nonhomogeneous PDEs. Later, when we study integral transforms, we will see that there are other methods for solving these types of problems.
2. The eigenfunctions $X_n(x)$ in the expansion *change* from problem to problem and depend on the PDE and BCs. The reader should look at problem 4 in the problem set to make sure he or she knows how to find the eigenfunctions $X_n(x)$.
3. If the reader remembers ODE theory, he or she will remember that solutions of equations corresponding to nonhomogeneous terms like

$$P_n(x) \; e^{\alpha x} \begin{cases} \sin{(\beta x)} \\ \cos{(\beta x)} \end{cases}$$

could be found by the method of *undetermined coefficients*. The same is true here. Problem (9.11) could be solved by this method. Any reader interested in this method should consult the reference.

PROBLEMS

1. The solution of the problem

$$\text{PDE} \qquad u_t = u_{xx} + \sin{(3\pi x)} \qquad 0 < x < 1$$

$$\text{BCs} \qquad \begin{cases} u(0,t) = 0 \\ u(1,t) = 0 \end{cases} \qquad 0 < t < \infty$$

$$\text{IC} \qquad u(x,0) = \sin{(\pi x)} \qquad 0 \le x \le 1$$

is given by (9.12). Does this solution agree with your intuition? What does the solution look like?

2. Solve the problem

$$\text{PDE} \qquad u_t = u_{xx} + \sin{(\pi x)} + \sin{(2\pi x)} \qquad 0 < x < 1 \qquad 0 < t < \infty$$

70   Diffusion-Type Problems

$$\text{BCs} \quad \begin{cases} u(0,t) = 0 \\ u(1,t) = 0 \end{cases} \quad 0 < t < \infty$$

$$\text{IC} \quad u(x,0) = 0 \quad 0 \leqslant x \leqslant 1$$

3. Solve the problem

$$\text{PDE} \quad u_t = u_{xx} + \sin(\pi x) \quad 0 < x < 1 \quad 0 < t < \infty$$

$$\text{BCs} \quad \begin{cases} u(0,t) = 0 \\ u(1,t) = 0 \end{cases} \quad 0 < t < \infty$$

$$\text{IC} \quad u(x,0) = 1 \quad 0 \leqslant x \leqslant 1$$

by the method of eigenfunction expansion.

4. Find the solution of

$$\text{PDE} \quad u_t = u_{xx} + \sin(\lambda_1 x) \quad 0 < x < 1 \quad 0 < t < \infty$$

$$\text{BCs} \quad \begin{cases} u(0,t) = 0 \\ u_x(1,t) + u(1,t) = 0 \end{cases} \quad 0 < t < \infty$$

$$\text{IC} \quad u(x,0) = 0 \quad 0 \leqslant x \leqslant 1$$

by the method of eigenfunction expansion where λ_1 is the first root of the equation $\tan \lambda = -\lambda$. What are the eigenfunctions $X_n(x)$ in this problem?

5. Solve the problem

$$\text{PDE} \quad u_t = u_{xx} \quad 0 < x < 1 \quad 0 < t < \infty$$

$$\text{BCs} \quad \begin{cases} u(0,t) = 0 \\ u(1,t) = \cos t \end{cases} \quad 0 < t < \infty$$

$$\text{IC} \quad u(x,0) = x \quad 0 \leqslant x \leqslant 1$$

by
(a) transforming it to one with zero BCs.
(b) solving the resulting problem by expanding it in terms of eigenfunctions.

OTHER READING

Elementary Partial Differential Equations by P. W. Berg and J. L. McGregor. Holden-Day, 1966. One of the more popular texts on PDEs; slightly more advanced than this text; clearly written. An extensive section on nonhomogeneous problems (Chapter 5).

Integral Transforms (Sine and Cosine Transforms)

PURPOSE OF LESSON: To introduce the idea of integral transforms and show how they transform PDEs in n variables into differential equations in $n - 1$ variables.

It is also shown that these transforms can be interpreted as resolving the input of the problem into simple parts (frequency resolution), finding the solution for each subpart, and adding the results.

In summary, integral transforms change differentiation to multiplication, and, hence, certain partial derivatives are changed into algebraic expressions.

The sine and cosine transforms are introduced and are used to solve an infinite-diffusion problem. The solution is interesting in that it involves the complementary-error function.

An integral transformation is merely a transformation that assigns to one function $f(t)$ a new function $F(s)$ by means of a formula like

$$F(s) = \int_A^B K(s,t)f(t) \, dt$$

Note that we *start* with a function of t and *end* with a function of s. The function $K(s,t)$ is called the **kernel of the transformation** and is the major ingredient that distinguishes one transform from another; it is chosen so that the transform has certain desirable properties. The limits of integration A and B also change from transformation to transformation.

The general philosophy behind integral transformations is that they eliminate *partial derivatives* with respect to one of the variables; hence, the new equation has one less variable. For example, if we apply a transform to the PDE

$$u_t = u_{xx}$$

for the purpose of eliminating the time derivative, then we would arrive at an ODE in x. On the other hand, if we had the PDE

$$u_{xx} + u_{yy} + u_{zz} = 0$$

and applied the Fourier transform to the x-variable, then we would eliminate the derivative u_{xx} and would have a new PDE in y and z. We could, of course, apply the Fourier transform again to eliminate one of the other variables (and arrive at an ODE in the last remaining variable). In other words, integral transforms change problems into easier ones. The transformed problem is then solved, and its *inverse* is obtained to find the solution to the original problem; this general strategy is illustrated in Figure 10.1.

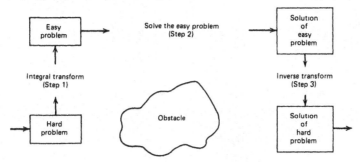

FIGURE 10.1   General philosophy of transforms.

In Figure 10.1, we see that along with every integral transform, there is an *inverse transform* that will reproduce that original function from its transform. The transform and its inverse together form what is called a **transform pair**. Table 10.1 lists several common transform pairs that we will use to solve PDEs.

TABLE 10.1   Some Common Transform Pairs

Transform pairs

1.
$$\mathscr{F}_s[f] = F(\omega) = \frac{2}{\pi} \int_0^\infty f(t) \sin(\omega t)\, dt \qquad \text{(Fourier sine transform)}$$

$$\mathscr{F}_s^{-1}[F] = f(t) = \int_0^\infty F(\omega) \sin(\omega t)\, d\omega \qquad \text{(inverse sine transform)}$$

2.
$$\mathscr{F}_c[f] = F(\omega) = \frac{2}{\pi} \int_0^\infty f(t) \cos(\omega t)\, dt \qquad \text{(Fourier cosine transform)}$$

$$\mathscr{F}_c^{-1}[F] = f(t) = \int_0^\infty F(\omega) \cos(\omega t)\, d\omega \qquad \text{(inverse cosine transform)}$$

3.
$$\mathscr{F}[f] = F(\omega) = \frac{1}{\sqrt{2\pi}} \int_{-\infty}^\infty f(x)\, e^{-i\omega x} dx \qquad \text{(Fourier transform)}$$

$$\mathscr{F}^{-1}[F] = f(x) = \frac{1}{\sqrt{2\pi}} \int_{-\infty}^\infty F(\omega)\, e^{i\omega x} d\omega \qquad \text{(inverse Fourier transform)}$$

4.
$$\begin{cases} \mathcal{F}_s[f] = S_n = \dfrac{2}{L} \displaystyle\int_0^L f(x) \sin(n\pi x/L)\, dx \quad \text{(finite-sine transform)} \\[2em] \mathcal{F}_s^{-1}[F_n] = f(x) = \displaystyle\sum_{n=1}^{\infty} S_n \sin(n\pi x/L) \quad \text{(inverse finite-sine transform)} \end{cases}$$

5.
$$\begin{cases} \mathcal{F}_c[f] = C_n = \dfrac{2}{L} \displaystyle\int_0^L f(x) \cos(n\pi x/L)\, dx \quad \text{(finite-cosine transform)} \\[2em] \mathcal{F}_c^{-1}[F_n] = f(x) = \dfrac{C_0}{2} + \displaystyle\sum_{n=1}^{\infty} C_n \cos(n\pi x/L) \quad \text{(inverse finite-cosine transform)} \end{cases}$$

6.
$$\begin{cases} \mathcal{L}[f] = F(s) = \displaystyle\int_0^{\infty} f(t)\, e^{-st}\, dt \quad \text{(Laplace transform)} \\[2em] \mathcal{L}^{-1}[F] = f(t) = \dfrac{1}{2\pi i} \displaystyle\int_{c-i\infty}^{c+i\infty} F(s) e^{st}\, ds \quad \text{(inverse Laplace transform)} \end{cases}$$

7.
$$\begin{cases} H[f] = F_n(\xi) = \displaystyle\int_0^{\infty} r J_n(\xi r) f(r)\, dr \quad \text{(Hankel transform)} \\[2em] H^{-1}[F_n] = f(r) = \displaystyle\int_0^{\infty} J_n(\xi r) F_n(\xi)\, d\xi \quad \text{(inverse Hankel transform)} \end{cases}$$

Note that in these transforms we have alternative notations. For instance, in the case of the Laplace transform, the notation $\mathcal{L}[f]$ indicates that we are taking the transform of f, whereas the alternative notation $F(s)$ indicates a function of s.

The current lesson does not attempt to study all of these transform pairs— only the sine and cosine transforms (1. and 2.); later, we will study several of the others. Questions about the relationship between the transforms, when to apply them, advantages and disadvantages of each, will be answered as we go along. However, before we begin the study of integral transforms, it will be useful to study what is called the *spectrum of a function* (or the *spectral resolution* of a function).

The Spectrum of a Function

Integral transforms and the spectrum of a function are closely related; in fact, an **integral transformation** can be thought of as a resolution of a function into a certain spectrum of components. How the transform actually resolves the function changes from transform to transform, but the function is being resolved into something nevertheless.

For instance, let's consider the resolution of a periodic funciton $f(x)$ into sines and cosines (Fourier series)*

* Fourier series will be discussed in detail in Chapter 11.

$$f(x) = \sum_{n=0}^{\infty} [A_n \cos(nx) + B_n \sin(nx)]$$

(Figure 10.2).

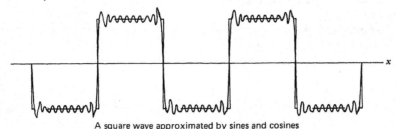

A square wave approximated by sines and cosines

FIGURE 10.2 Expansion of a periodic function into sines and cosines.

Here, the coefficients A_n and B_n represent the amount of the function $f(x)$ made up by $\cos(nx)$ and $\sin(nx)$, respectively, while the square root

$$\sqrt{A_n^2 + B_n^2}$$

(called the **spectrum of the function**) measures the amount of $f(x)$ with frequency n.

For example, if the function $f(x)$ were a simple sum of sines and cosines

$$f(x) = 1 + \sin x + \frac{1}{5} \sin(3x) + \cos x + \frac{1}{2} \cos(2x) + \frac{1}{4} \cos(4x)$$

then its spectrum (discrete) would be as given in Figure 10.3.

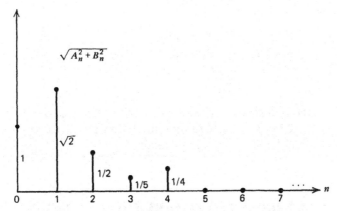

FIGURE 10.3 Discrete spectrum of f(x).

By reading off the values of $\sqrt{A_n^2 + B_n^2}$, we can tell the magnitude of the component in $f(x)$ with frequency n.

Functions that are *periodic* can be resolved into *infinite series* (they have discrete spectrums), whereas functions that are *not periodic* must be resolved into a *continuous spectrum* of values (of course, if a function is defined only on a *finite interval*, we could extend the function outside the interval in a periodic way, so that a Fourier series representation could be obtained for the function inside the interval).

For example, although a nonperiodic function $f(x)$ cannot be represented by an infinite series of sines and cosines, we might be tempted to write it as a *continuous analog* of the Fourier series; that is,

$$f(x) = \int_{-\infty}^{\infty} [C(\omega) \cos (\omega x) + S(\omega) \sin (\omega x)] \, d\omega$$

where the functions $S(\omega)$ and $C(\omega)$ measure the sine and cosine component of $f(x)$ and

$$\sqrt{S^2(\omega) + C^2(\omega)}$$

measures the ω frequency component of $f(x)$ and is called the spectrum (continuous spectrum) of $f(x)$.

With this intuitive explanation of the spectrum of a function, we now get to the nuts and bolts of integral transforms. The first step would be to list a few properties of these transforms that make them work.

Sine and Cosine Transforms of Derivatives

(10.1)

1. $\mathcal{F}_s[f'] = -\omega \, \mathcal{F}_c[f]$ (proved by integration by parts)

2. $\mathcal{F}_s[f''] = \dfrac{2}{\pi} \omega f(0) - \omega^2 \mathcal{F}_s[f]$

3. $\mathcal{F}_c[f'] = \dfrac{-2}{\pi} f(0) + \omega \mathcal{F}_s [f]$

4. $\mathcal{F}_c[f''] = -\dfrac{2}{\pi} f'(0) - \omega^2 \mathcal{F}_c[f]$

Several other sine and cosine transforms and their inverses can be found in the tables at the end of the lessons. We now show how the sine transform can solve an important initial-boundary-value problem.

Solution of an Infinite-Diffusion Problem via the Sine Transform

The problem we are interested in is the *infinite diffusion problem*

$$
\begin{array}{lll}
\text{PDE} & u_t = \alpha^2 u_{xx} & 0 < x < \infty \qquad 0 < t < \infty \\
\text{BC} & u(0, t) = A & 0 < t < \infty \\
\text{IC} & u(x, 0) = 0 & 0 \leqslant x < \infty
\end{array}
$$

(Figure 10.4).

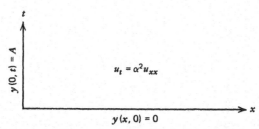

FIGURE 10.4   Diffusion problem in a semi-infinite medium.

STEP 1   To solve this, we break it into three simple steps. First our strategy is to transform the *x-variable* via the Fourier sine transform so that we get an ODE in time. We start by transforming each side of the PDE; in other words

$$
\mathcal{F}_s[u_t] = \alpha^2 \mathcal{F}_s[u_{xx}]
$$

Let's consider each term individually:

$\mathcal{F}_s[u_t]$:   The partial derivative u_t in this problem is what we could call the *off derivative*, since our transform is *with respect to x*. In this case, we can write

$$
\begin{aligned}
\mathcal{F}_s[u_t] &= \frac{2}{\pi} \int_0^\infty u_t(x,t) \sin(\omega x)\, dx \\
&= \frac{\partial}{\partial t}\left[\frac{2}{\pi} \int_0^\infty u(x,t) \sin(\omega x)\, dx\right] \\
&= \frac{d}{dt} \mathcal{F}_s[u] \\
&= \frac{d}{dt} U(t)
\end{aligned}
$$

The fact that we took the derivative *outside* the integral is a property from calculus. Note that u is a function of x and t, whereas its transform

$$
\mathcal{F}_s[u] = U(\omega, t)
$$

is a function of ω and t. The new variable ω will be treated like a parameter in the new problem, and, hence, we call the sine transform a function of t alone; that is,

$$\mathcal{F}_s[u] = U(t)$$

$\mathcal{F}_s[u_{xx}]$:   For this one, we have the identity

$$\mathcal{F}_s[u_{xx}] = \frac{2}{\pi}\,\omega u(0,t) - \omega^2 \mathcal{F}_s[u]$$

$$= \frac{2}{\pi}\,\omega u(0,t) - \omega^2 U(t)$$

$$= \frac{2A\omega}{\pi} - \omega^2 U(t)$$

Note here that when you proved these identities (10.1), you did it for functions of *one* variable $f(x)$. We now have a slight modification, since $u(x,t)$ depends on x and t. You should use the formulas according to which variable is being transformed and treat the others as constants. In this case, the transform is with respect to x, and, hence, t is just carried along as a constant. Also note that the BC $u(0,t) = A$ is used at this point in our operation.

Substituting these expressions into our PDE, we arrive at the ODE

$$\frac{dU}{dt} = \alpha^2 \left[-\omega^2 U(t) + \frac{2A\omega}{\pi} \right]$$

The only thing missing is an IC for $U(t)$; we get this by transforming the IC $u(x,0) = 0$ to get

$$\mathcal{F}_s[u(x,0)] = U(0) = 0$$

This completes the first step in the transform process—we have changed the original problem into an initial-value problem

(10.2)

$$\text{ODE} \quad \frac{dU}{dt} + \omega^2\alpha^2 U = \frac{2A\omega\alpha^2}{\pi}$$

$$\text{IC} \quad U(0) = 0$$

STEP 2   To solve this IVP, we could use a variety of elementary techniques from ordinary differential equations (integrating factor, homogeneous and particular solution); in any case, the solution is

$$U(t) = \frac{2A}{\pi\omega}(1 - e^{-\omega^2\alpha^2 t})$$

We have now found the sine transform for the answer $u(x,t)$. The last step is to find the inverse transform of $U(t)$; that is,

$$u(x,t) = \mathcal{F}_s^{-1}[U]$$

STEP 3 To find the solution, we can either evaluate the inverse transform directly from the integral or else resort to the tables. Using the tables, we see that

$$u(x,t) = A \; erfc \; (x/2\alpha\sqrt{t})$$

where $erfc(x)$, $0 < x < \infty$, is called the **complementary-error function** and is given by

$$erfc(x) = \frac{2}{\sqrt{\pi}} \int_x^\infty e^{-t^2} \, dt$$

See Figure 10.5 for its graph.

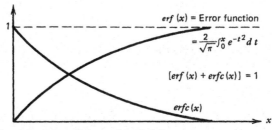

FIGURE 10.5 Graphs of erf(x) and erfc(x).

The exact values of these well-known functions can be found in most tables for physics and chemistry. It should be noted that these integrals cannot be integrated by the usual elementary tricks of calculus.

Interpretation of the Solution

The solution

$$u(x,t) = A \; erfc \; [x/2\alpha\sqrt{t}]$$

makes a lot of sense. For different values of time, we have the graph of a complementary-error function with different scale factors on the x-axis. As time increases, the error function gets pulled out; for a graph of the solution at different values of time, see Figure 10.6.

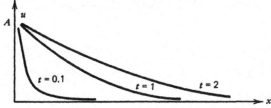

FIGURE 10.6 Solution to semi-infinite rod with fixed temperature A at the end.

PROBLEMS

1. Prove the identities (10.1). What assumptions do you need to assume about the function f?
2. Solve the ordinary-differential equation problem (10.2).
3. Solve by means of the sine *or* cosine transform

$$\begin{array}{lll} \text{PDE} & u_t = \alpha^2 u_{xx} & 0 < x < \infty \\ \text{BC} & u_x(0, t) = 0 & 0 < t < \infty \\ \text{IC} & u(x,0) = H(1 - x) & 0 \leqslant x < \infty \end{array}$$

where $H(x)$ is the *Heaviside function*:

$$H(x) = \begin{cases} 0 & 0 \leqslant x < 1 \\ 1 & 1 \leqslant x \end{cases}$$

In other words, the IC looks like

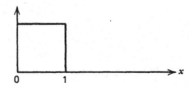

What does the graph of the solution look like for various values of time?

OTHER READING

1. *Operational Mathematics* by R. V. Churchill. McGraw-Hill, 1958. An excellent text covering many of the integral transforms; good problems and many tables.

2. *Tables of Integral Transform* by A. Erdelyi. McGraw-Hill, 1954. One of the most comprehensive tables of integral transform.

3. *Integral Transforms in Mathematical Physics* by C. J. Tranter. Chapman and Hall (Science Paperbacks), 1971. A small, but concise paperback; easy to read with many examples.

The Fourier Series and Transform

PURPOSE OF LESSON: To introduce the Fourier series and to show how it can represent certain periodic functions $f(x)$ by sums of sines and cosines:

$$f(x) = \frac{a_0}{2} + \sum_{n=1}^{\infty} [a_n \cos(n\pi x) + b_n \sin(n\pi x)]$$

In the case of nonperiodic functions on $(-\infty, \infty)$, to show also how the Fourier series is replaced by the *Fourier transform* and how a function $f(x)$ can be represented by a continuous resolution of simple functions. This resolution (the Fourier integral) can be written in the complex form

$$f(x) = \frac{1}{\sqrt{2\pi}} \int_{-\infty}^{\infty} \left[\frac{1}{\sqrt{2\pi}} \int_{-\infty}^{\infty} f(x)e^{-i\xi x}\, dx \right] e^{i\xi x}\, d\xi$$

which gives rise to the Fourier and inverse Fourier transforms

$$\mathcal{F}[f] = F(\xi) = \frac{1}{\sqrt{2\pi}} \int_{-\infty}^{\infty} f(x)e^{-i\xi x}\, dx \quad \text{(Fourier transform)}$$

$$\mathcal{F}^{-1}[F] = f(x) = \frac{1}{\sqrt{2\pi}} \int_{-\infty}^{\infty} F(\xi)e^{i\xi x}\, d\xi \quad \text{(inverse Fourier transform)}$$

The importance of the Fourier series in PDE theory is that periodic functions $f(x)$ defined on $(-\infty, \infty)$ *or* functions defined on *finite intervals* can be represented by infinite series of sines and cosines, and in this way, problems can be resolved into simple ones. For example, the so-called **sawtooth wave**

$$f(x) = x \quad -L < x < L$$
$$f(x + 2L) = f(x) \quad \text{(periodic condition)}$$

shown in Figure 11.1

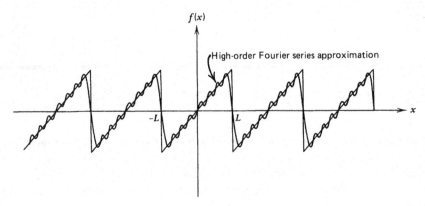

High-order Fourier series approximation

FIGURE 11.1 Sawtooth wave represented by a Fourier series.

can be represented by the Fourier series

(11.1) $$f(x) = \frac{a_0}{2} + \sum_{n=1}^{\infty} [a_n \cos (n\pi x/L) + b_n \sin (n\pi x/L)]$$

where the **Fourier coefficients** a_n *and* b_n *are given by the* **Euler formulas**

(11.2)
$$a_n = \frac{1}{L} \int_{-L}^{L} f(x) \cos (n\pi x/L) \, dx = 0 \qquad n = 0, 1, 2, \ldots$$

$$b_n = \frac{1}{L} \int_{-L}^{L} f(x) \sin (n\pi x/L) \, dx = -(2L/n\pi) (-1)^n \qquad n = 1, 2, \ldots$$

These integrations are routine calculus evaluations. To find Euler's formulas for a_n and b_n, respectively, we multiply each side of equation (11.1) by $\sin (nx)$ or $\cos (nx)$ and integrate the resulting equation from $-L$ to L. The *orthogonality* of the functions $\{\sin (n\pi x/L)\}$ and $\{\cos (n\pi x/L)\}$ allows us to solve for the coefficients a_n and b_n; see problem 6. For the sawtooth wave, the Fourier representation is given by

(11.3) $$f(x) = \frac{2L}{\pi} \left[\sin (\pi x/L) - \frac{1}{2} \sin (2\pi x/L) + \frac{1}{3} \sin (3\pi x/L) - \ldots \right]$$

where each term (called **harmonic**) has a larger frequency than the previous term, and all frequencies are *multiples* of a fundamental frequency that has the same period as the function $f(x)$

One of the drawbacks of the Fourier series is that in order for a function to have a Fourier series representation, the function must be periodic. Of course, if we want to expand a function (say, $f(x) = x$ for $0 \leq x \leq 1$) defined on a *finite interval*, we could use expansion (11.1). The fact that the Fourier series is periodic *outside* the interval [0,1] doesn't concern us, since we're only interested in the function *inside* the interval. As a matter of fact, we can represent a function inside an interval with many different types of Fourier series by considering different types of extensions outside the interval (some converge faster than others).

The reader shouldn't get the idea that every periodic function can be represented by a Fourier series expansion. What we do know is that if a function $f(x)$ can be represented by a Fourier series (11.1), then the coefficients a_n and b_n are given by the *Euler formulas* (11.2). What's more, even if a function $f(x)$ can be represented by a Fourier series, it isn't always true that the *derivative* $f'(x)$ can be found by differentiating the series term by term. In fact, we can easily see that the derivative of $f(x) = x$ (the sawtooth function) cannot be found by differentiating each term of the Fourier series (11.3). Indeed, the differentiated series will not even converge for any x (the reader can verify this himself or herself).

The *exact conditions* that insure that a function $f(x)$ will have a Fourier series representation and that the representation can be differentiated term by term are found in the recommended reading for this lesson. For our purposes, we are content to know the important result of P. G. Dirichlet's (Deer-ish-lay) theorem, which states

---

Dirichlet's theorem (sufficient conditions for a function to have a Fourier series representation):

If $f(x)$ is a bounded periodic function that contains a finite number of maximum points, minimum points, and points of discontinuity in each period, then the Fourier series of $f(x)$ converges to $f(x)$ at each point x where $f(x)$ is continuous and to the *average* of the left- and right-hand limits of $f(x)$ at those points where $f(x)$ is discontinuous.

For example, in Figure 11.1, the Fourier series converges to the function $f(x)$ for all except $x = \pm L, \pm 3L, \ldots$ (points of discontinuity), in which case it converges to zero (the average of $+L$ and $-L$).

We are now almost ready to introduce the *Fourier transform*. Before we do, however, it will be useful to introduce the idea of the *frequency spectrum* of a periodic function.

Discrete Frequency Spectrum of a Periodic Function

For periodic functions, we can interpret the Fourier series as the replacement of a periodic function $f(x)$ by a sequence $\{c_n\}$ of numbers

$$c_n = \sqrt{a_n^2 + b_n^2} \qquad n = 0, 1, 2, \ldots.$$

where the numbers c_n can be taken as measuring the contributions of the various frequency components of the function $f(x)$. For example, the sawtooth wave $f(x)$ has the Fourier series representation

$$f(x) = \frac{2L}{\pi} \left[\sin (\pi x/L) - \frac{1}{2} \sin (2\pi x/L) + \frac{1}{3} \sin (3\pi x/L) - \dots \right]$$

and, hence, the frequency spectrum $\{c_n\}$ is $c_n = 2L/n\pi$ for $n = 1, 2, \dots$ and $c_0 = \left| \frac{a_0}{2} \right| = 0$ (Figure 11.2).

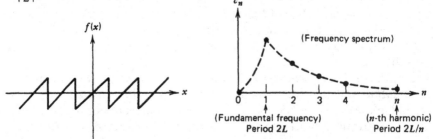

FIGURE 11.2   Discrete frequency spectrum of the sawtooth wave.

The sequence $\{c_n\}$ is somewhat similar to the decomposition of white light into the frequency spectrum of colors obtained with a spectroscope.

We now introduce the Fourier transform.

The Fourier Transform

The major difficulty with Fourier series representation is that nonperiodic functions defined on $(-\infty, \infty)$ cannot be represented. It is possible, however, to find an analogous representation for some of these functions. Without going through the details of the proof, we can show that the Fourier series representation

$$f(x) = \frac{a_0}{2} + \sum_{n=1}^{\infty} [a_n \cos (n\pi x/L) + b_n \sin (n\pi x/L)]$$

changes to the *Fourier integral* representation (continuous frequency resolution)

(11.4) $$f(x) = \int_0^\infty a(\xi) \cos (\xi x) \, d\xi + \int_0^\infty b(\xi) \sin (\xi x) \, d\xi$$

where

(11.5)

$$a(\xi) = \frac{1}{\pi} \int_{-\infty}^{\infty} f(x) \cos (\xi x) \, dx$$

$$b(\xi) = \frac{1}{\pi} \int_{-\infty}^{\infty} f(x) \sin (\xi x) \, dx$$

for nonperiodic functions defined on $(-\infty, \infty)$. Here, we see that the Fourier integral representation has resolved the function $f(x)$ into *all* frequencies $0 < \xi < \infty$ (and not just multiples of *one basic frequency*, as with periodic functions). As we did in the Fourier series, we define the **frequency spectrum**

$$C(\xi) = \sqrt{a^2(\xi) + b^2(\xi)}$$

which measures the composition of the function $f(x)$ in terms of its frequencies. Some examples of functions $f(x)$ and their frequency spectrums are given in Figure 11.3.

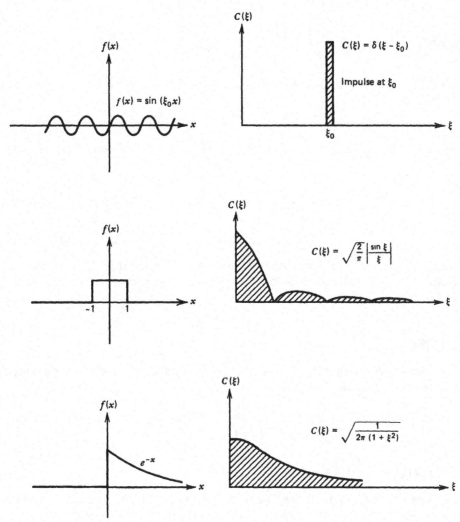

FIGURE 11.3 Frequency spectra for various functions.

Note that functions $f(x)$ that have sharp corners give rise to frequency spectra with large frequencies, since sharp corners require high-frequency components to represent them. On the other hand, the simple periodic function $f(x) = \sin(\xi_0 x)$ obviously has a frequency spectrum that is zero everywhere except at $\xi = \xi_0$.

We are now in a position to define what is generally known as the *exponential Fourier transform* (Equations 11.5 are known as the **Fourier sine and cosine transforms**). By use of Euler's (Oy-ler) equation

$$e^{i\theta} = \cos\theta + i\sin\theta$$

we can rewrite equation (11.4) after a little effort as

$$(11.6) \qquad f(x) = \frac{1}{\sqrt{2\pi}} \int_{-\infty}^{\infty} \left[\frac{1}{\sqrt{2\pi}} \int_{-\infty}^{\infty} f(x)e^{-i\xi x}\, dx \right] e^{i\xi x}\, d\xi$$

which is known as the **Fourier integral representation**. From this, we can write the two equations

$$(11.7)$$

$$\mathscr{F}[f] \equiv F(\xi) = \frac{1}{\sqrt{2\pi}} \int_{-\infty}^{\infty} f(x)e^{-i\xi x}\, dx \qquad \text{(Fourier transform)}$$

$$\mathscr{F}^{-1}[F] \equiv f(x) = \frac{1}{\sqrt{2\pi}} \int_{-\infty}^{\infty} F(\xi)e^{i\xi x}\, d\xi \qquad \text{(inverse Fourier transform)}$$

which are the Fourier and inverse Fourier transforms. Properties of this transform pair will be discussed in the next lesson along with problems using these transforms.

NOTES

1. The Fourier transform $F(\xi)$ of $f(x)$ can be a *complex function*; for example, the Fourier transform of

$$f(x) = \begin{cases} 0 & x \leqslant 0 \\ e^{-x} & x > 0 \end{cases}$$

is $F(\xi) = \dfrac{1}{\sqrt{2\pi}} \dfrac{1 - i\xi}{1 + \xi^2}$

2. The absolute value of the Fourier transform $F(\xi)$ is the frequency spectrum of $f(x)$. For example, in note 1, the frequency spectrum of $f(x)$ is

$$|F(\xi)| = \sqrt{\frac{1}{2\pi\,(1 + \xi^2)}}$$

(the reader should be able to find the magnitude of a complex number).

3. Not all functions have Fourier transforms [the integral (11.7) may not exist]; in fact, $f(x) = c$, $\sin x$, e^x, x^2, do *not* have Fourier transforms. Only functions that go to zero sufficiently fast as $|x| \to \infty$ have transforms. In applications, we apply the Fourier transform to temperature functions, wave functions, and other physical phenomena that go to zero as $|x| \to \infty$.

PROBLEMS

1. What is the Fourier series expansion of the square sine wave

$$f(x) \quad = \begin{cases} -1 & -1 < x < 0 \\ 1 & 0 \leqslant x < 1 \end{cases}$$

$$f(x + 2) = f(x) \qquad \text{(periodic condition)}$$

Graph the first 2, 3, 4 terms of the series to see how it is converging to $f(x)$. Also graph the frequency spectrum of $f(x)$.

2. Show that if we differentiate the Fourier series expansion (11.3) of the sawtooth wave term by term, we arrive at an infinite series that clearly does not represent the derivative of the sawtooth curve.

3. Graph the frequency spectrum of the following periodic functions:
   (a) $f(x) = \sin x$
   (b) $f(x) = \sin x + \cos 2x$
   (c) $f(x) = \sin x + \cos x + 0.5 \sin 3x$

4. What is the Fourier transform $F(\xi)$ and the frequency spectrum $C(\xi) = |F(\xi)|$ of the function

$$f(x) = \begin{cases} 1 & -1 < x < 1 \\ 0 & \text{elsewhere} \end{cases}$$

5. Show that the absolute value of the function $F(\xi) = 1/(1 + i\xi)$ is $|F(\xi)| = \sqrt{1/(1 + \xi^2)}$.
   HINT   First multiply the numerator and denominator by $1 - i\xi$ to get rid of the complex number i in the denominator.

6. Verify the *orthogonality* properties of sines and cosines on the interval $[-L, L]$

$$\int_{-L}^{L} \sin\,(m\pi x/L)\,\sin\,(n\pi x/L)\,dx = \begin{cases} 0 & m \neq n \\ L & m = n \end{cases}$$

$$\int_{-L}^{L} \cos{(m\pi x/L)} \cos{(n\pi x/L)} \, dx = \begin{cases} 0 & m \neq n \\ L & m = n \end{cases}$$

$$\int_{-L}^{L} \sin{(m\pi x/L)} \cos{(n\pi x/L)} \, dx = 0 \qquad \text{all } m, n = 1, 2, 3, \ldots$$

OTHER READING

Partial Differential Equations of Mathematical Physics by Tyn Myint-U. Elsevier, 1973. A well-written text with a fairly extensive section on the Fourier series and transform (Chapters 5, 11). Most of the important questions dealing with whether a function actually has a Fourier series or integral representation, whether the representation can be differentiated term by term or under the integral to get the derivative of the function, and so forth, are answered in these chapters.

The Fourier Transform and Its Application to PDEs

PURPOSE OF LESSON: To illustrate several useful properties of the Fourier transform and to show how these properties can be used to solve PDEs. In particular, it is shown how the Fourier transform changes *differentiation* to *multiplication*, so differential equations change into algebraic equations. Also, the idea of the *infinite convolution* is introduced.

The Fourier transform of the function $f(x)$ for $-\infty < x < \infty$ is given by the formula

$$(12.1) \quad \mathscr{F}[f] = F(\xi) = \frac{1}{\sqrt{2\pi}} \int_{-\infty}^{\infty} f(x)\, e^{-i\xi x}\, dx$$

That is, we start with a function $f(x)$ defined on the real x-axis, substitute it into equation (12.1), and arrive at the new function $F(\xi)$ for $-\infty < \xi < \infty$. For example, Table 12.1 lists some common Fourier transforms.

TABLE 12.1   Some Common Fourier Transforms

Function $f(x)$	Fourier Transform $F(\xi)$
1. $f(x) = \begin{cases} e^{-x} & x \geq 0 \\ -e^{x} & x < 0 \end{cases}$	$F(\xi) = -i\sqrt{\dfrac{2}{\pi}} \dfrac{\xi}{1 + \xi^2}$ (complex function)
2. $f(x) = \begin{cases} 1 & -1 < x < 1 \\ 0 & \text{elsewhere} \end{cases}$	$F(\xi) = \sqrt{\dfrac{2}{\pi}} \dfrac{\sin \xi}{\xi}$ (real function)
3. $f(x) = e^{-x^2}$	$F(\xi) = \dfrac{1}{\sqrt{2}} e^{-(\xi/2)^2}$ (real function)

The reader can refer to the tables in the appendix for additional transforms. We can see from the examples that the transformed function $F(\xi)$ may or may not be a complex-valued function of ξ. In the first example, the transformed function $F(\xi)$ contains the complex number i, so we call it a **complex-valued function** of the *real variable* ξ (ξ ranges from $-\infty$ to ∞). In other words, the argument ξ is real, but the value of the function is complex.

The usefulness of the Fourier transform (as with most integral transforms) comes from the fact that it changes the operation of differentiation into multiplication; that is, differential equations are changed into algebraic equations. There are also a host of other properties that make the Fourier transform a useful operational tool; we list a few of the more important ones.

Useful Properties of the Fourier Transform

Property 1 (Fourier Transform Pair)

The Fourier transform of $f(x)$, $-\infty < x < \infty$, produces a new function $F(\xi)$ defined by the formula

$$\mathcal{F}[f] = F(\xi) = \frac{1}{\sqrt{2\pi}} \int_{-\infty}^{\infty} f(x)e^{-i\xi x}\, dx$$

and the *inverse* Fourier transform of $F(\xi)$, $-\infty < \xi < \infty$ will reproduce the original function $f(x)$ according to

$$\mathcal{F}^{-1}[F] = f(x) = \frac{1}{\sqrt{2\pi}} \int_{-\infty}^{\infty} F(\xi)e^{i\xi x}\, d\xi$$

For example,

$$e^{-|x|} \xrightarrow{\mathcal{F}} \sqrt{\frac{2}{\pi}} \frac{1}{1 + \xi^2} \xrightarrow{\mathcal{F}^{-1}} e^{-|x|}$$

See Figure 12.1.

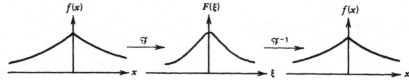

FIGURE 12.1 Graph of a function and its transform.

Property 2 (Linear Transformation)

The Fourier transform is a linear transformation; that is

$$\mathcal{F}[af + bg] = a\mathcal{F}[f] + b\mathcal{F}[g]$$

This is easy to see. The reader can spend a few minutes to verify this property, which is used over and over again. For example, the Fourier transform of the expression

$$\frac{1}{x^2 + 1} + 3e^{-x^2}$$

would be

$$\mathcal{F}[\frac{1}{x^2 + 1}] + 3\mathcal{F}[e^{-x^2}]$$

Property 3 (Transformation of Partial Derivatives)

When we discuss how derivatives transform, we must distinguish partial derivatives with respect to various variables. For instance, if the Fourier transform transforms the x-variable (the variable of integration in the transform) and *if* the function being transformed is a partial derivative of a function $u(x,t)$ *with respect to* x, then the **rules of transformation are**

$$\mathcal{F}[u_x] = \frac{1}{\sqrt{2\pi}} \int_{-\infty}^{\infty} u_x(x,t)e^{-i\xi x} dx = i\xi\mathcal{F}[u]$$

$$\mathcal{F}[u_{xx}] = \frac{1}{\sqrt{2\pi}} \int_{-\infty}^{\infty} u_{xx}(x,t)e^{-i\xi x} dx = -\xi^2\mathcal{F}[u]$$

On the other hand, if we transform the partial derivative $u_t(x,t)$ (and if the variable of integration in the transform is x), then the transform is given by

$$\mathcal{F}[u_t] = \frac{1}{\sqrt{2\pi}} \int_{-\infty}^{\infty} u_t(x,t) e^{-i\xi x} dx = \frac{\partial}{\partial t} \mathcal{F}[u]$$

$$\mathcal{F}[u_{tt}] = \frac{1}{\sqrt{2\pi}} \int_{-\infty}^{\infty} u_{tt}(x,t)e^{-i\xi x} dx = \frac{\partial^2}{\partial t^2} \mathcal{F}[u]$$

Property 4 (Convolution Property)

Every integral transform has what is called a *convolution* property. The general idea is that the transform of a product of two functions $f(x)g(x)$ is *not* the product of the individual transforms; that is,

$$\mathcal{F}[f(x)g(x)] \neq \mathcal{F}[f]\mathcal{F}[g]$$

However, in transform theory there is something called the convolution $f * g$ of two functions that more or less plays the role of the product. What is true about this convolution $f * g$ is that

(12.2)
$$\mathcal{F}[f * g] = \mathcal{F}[f]\mathcal{F}[g]$$

So what is this mysterious convolution $f * g$? It's given by the formula

(12.3)
$$(f * g)(x) = \frac{1}{\sqrt{2\pi}} \int_{-\infty}^{\infty} f(x - \xi)g(\xi) \, d\xi$$

and it can be shown without too much trouble that (12.2) holds. We see from the definition of the convolution that given two functions $f(x)$ and $g(x)$, the convolution $(f * g)(x)$ is a new function.

Example of a Convolution of Two Functions

Given the two functions

$$f(x) = x$$
$$g(x) = e^{-x^2}$$

the convolution is given by

$$(f * g)(x) = \frac{1}{\sqrt{2\pi}} \int_{-\infty}^{\infty} (x - \xi)e^{-\xi^2} \, d\xi$$
$$= x/\sqrt{2} \qquad \text{(a new function)}$$

We have used the formula

$$\int_{-\infty}^{\infty} e^{-\xi^2} \, d\xi = \sqrt{\pi}$$

to arrive at this value.

The importance of the convolution (12.3) in applications is due to the fact that quite often, the final step in solving a PDE boils down to finding the inverse transform of some expression that we can interpret as the product of two transforms $\mathcal{F}[f]\mathcal{F}[g]$; that is, we must find

(12.4)
$$\mathcal{F}^{-1}\{\mathcal{F}[f]\mathcal{F}[g]\}$$

By taking the inverse of each side of (12.2), we arrive at the result

(12.5)
$$f * g = \mathcal{F}^{-1}\{\mathcal{F}[f]\mathcal{F}[g]\}$$

Hence, to find (12.4), all we have to do is find the inverse transform of *each* factor to get *f* and *g* and *then* compute their convolution. We are now in a position to work an important problem in PDE theory.

Solution of an Initial-Value Problem

Consider the heat flow in an *infinite* rod where the initial temperature is $u(x,0) = \phi(x)$. In other words, we look for the solution to the *initial-value problem* (IVP), sometimes called a *Cauchy problem*

(12.6)   PDE $u_t = \alpha^2 u_{xx}$ $-\infty < x < \infty$ $0 < t < \infty$
         IC $u(x,0) = \phi(x)$ $-\infty < x < \infty$

There are three basic steps in solving this problem.

STEP 1   (Transforming the problem)
Since the space variable *x* ranges from $-\infty$ to ∞, we take the Fourier transform of the PDE and IC with respect to this variable *x* (the variable of integration in the transform is *x*). Doing this, we get

$$\mathcal{F}[u_t] = \alpha^2 \mathcal{F}[u_{xx}]$$
$$\mathcal{F}[u(x,0)] = \mathcal{F}[\phi(x)]$$

and using the properties of the Fourier transform, we have

(12.7)
$$\frac{dU(t)}{dt} = -\alpha^2 \xi^2 U(t)$$

$$U(0) = \Phi(\xi) \quad (\Phi \text{ is the Fourier transform of } \phi)$$

where $U(t) = \mathcal{F}[u(x,t)]$. The reader should note here that the function *U* actually depends on *both* *t* and the new transformed variable ξ, but, for simplicity, since ξ is a constant insofar as the differential equation (12.7) is concerned, we will drop the notation and just call $U = U(t)$.

STEP 2   (Solving the transformed problem)

Remember the new variable ξ is nothing more than a constant in this differential equation, so the solution to this problem is

(12.8) $U(t) = \Phi(\xi)\, e^{-\alpha^2 \xi^2 t}$

STEP 3   (Finding the inverse transform)

To find the solution $u(x,t)$, we merely compute

$$u(x,t) = \mathcal{F}^{-1}[U(\xi,t)]$$
$$= \mathcal{F}^{-1}[\Phi(\xi)e^{-\alpha^2\xi^2 t}]$$

Here is where the convolution theorem (12.5) comes to the rescue. Using this property, we can write

$$u(x,t) = \mathcal{F}^{-1}[\Phi(\xi)e^{-\alpha^2\xi^2 t}]$$
$$= \mathcal{F}^{-1}[\Phi(\xi)] * \mathcal{F}^{-1}[e^{-\alpha^2\xi^2 t}]$$

(12.9)
$$= \phi(x) * \left[\frac{1}{\alpha\sqrt{2t}} e^{-(x^2/4\alpha^2 t)}\right] \qquad \text{(using tables)}$$

$$= \frac{1}{2\alpha\sqrt{\pi t}} \int_{-\infty}^{\infty} \phi(\xi)\, e^{-(x-\xi)^2/4\alpha^2 t}\, d\xi$$

We're done; equation (12.9) is the solution to our problem.

Before stopping, however, let's analyze this result. Note that the integrand is made up of two terms

1. The initial temperature $\phi(x)$

2. The function $G(x,t) = \dfrac{1}{2\alpha\sqrt{\pi t}} e^{-(x-\xi)^2/4\alpha^2 t}$ (which is called **Green's function** or the **impulse-response function**)

It can be shown that this impulse-response function $G(x,t)$ is the *temperature response* to an initial temperature impulse at $x = \xi$. In other words, $G(x,t)$ is the temperature along the rod at time t due to an initial *unit* impulse of heat at $x = \xi$ (Figure 12.2).

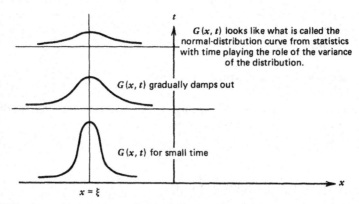

FIGURE 12.2   Impulse response $G(x,t)$ from a temperature impulse at $x = \xi$.

Hence, the interpretation of solution (12.9) is that the initial temperature $u(x,0)$ $= \phi(x)$ is *decomposed* into a continuum of impulses of magnitude $\Phi(\xi)$ (at each point $x = \xi$) and the resulting temperature $\Phi(\xi)G(x,t)$ is found. These resulting temperatures are then added (integrated) to obtain solution (12.9). Later, we will see that this general idea is known as **superposition**.

From a practical point of view, solution (12.9) can often be integrated for some particular initial temperature $\phi(x)$. If this integration cannot be carried out *analytically*, the solution can be found at any point (x,t) by *numerically* integrating the integral.

NOTES

The major drawback of the Fourier transform is that all functions can not be transformed; for example, even simple functions like

$$f(x) = \text{constant}$$
$$f(x) = e^x$$
$$f(x) = \sin x$$

cannot be transformed, since the integral

$$\mathcal{F}[f] = \frac{1}{\sqrt{2\pi}} \int_{-\infty}^{\infty} f(x)e^{-i\xi x} \, dx$$

does not exist. Only functions that damp to zero sufficiently fast as $|x| \to \infty$ have transforms. Also, the Fourier transform could not be used to transform the time variable in the previous initial value problem, since $0 < t < \infty$.

PROBLEMS

1. Find the Fourier transform of

$$f(x) = \begin{cases} 0 & x \le 0 \\ e^{-x} & 0 < x \end{cases}$$

    Check your answer by using the tables in the appendix.
2. Verify that the Fourier and inverse Fourier transforms are linear transforms.
3. Solve the initial-value problem

$$\begin{array}{lll} \text{PDE} & u_t = \alpha^2 u_{xx} & -\infty < x < \infty \\ \text{IC} & u(x,0) = e^{-x^2} & -\infty < x < \infty \end{array}$$

    by using the Fourier transform.
4. Verify the properties

$$\mathcal{F}[u_x] = i\xi\, \mathcal{F}[u]$$
$$\mathcal{F}[u_{xx}] = -\xi^2\, \mathcal{F}[u]$$

HINT   Use integration by parts.

5.   Verify that the convolution of two functions f and g can be written as either

$$(f * g)(x) = \frac{1}{\sqrt{2\pi}} \int_{-\infty}^{\infty} f(x - \xi) g(\xi)\, d\xi$$

or

$$(f * g)(x) = \frac{1}{\sqrt{2\pi}} \int_{-\infty}^{\infty} f(\xi) g(x - \xi)\, d\xi$$

OTHER READING

Fourier Series and Orthogonal Functions by H. Davis. Allyn and Bacon, 1963; Dover, 1989. An excellent book gives the reader an intuitive as well as rigorous viewpoint of Fourier series and transforms.

The Laplace Transform

PURPOSE OF LESSON: To introduce the important transform pair

$$\mathscr{L}[f] = F(s) = \int_0^\infty f(t)e^{-st}\, dt \qquad \text{(Laplace transform)}$$

$$\mathscr{L}^{-1}[F] = f(t) = \frac{1}{2\pi i} \int_{c-i\infty}^{c+i\infty} F(s)\, e^{st}\, ds \qquad \text{(inverse Laplace transform)}$$

and illustrate useful properties. The Laplace transform has an advantage over the Fourier transform because it contains the damping factor e^{-st} that allow us to transform a wider class of functions. Inasmuch as the transform operates on functions defined on $[0,\infty)$, it is mostly applied to the time variable t.

After discussing some useful properties of the Laplace transform, we will solve an important problem in PDE theory.

Of all the integral transforms we will study in this book, the Laplace transform

$$(13.1) \qquad\qquad \mathscr{L}[f] = \int_0^\infty f(t)e^{-st}\, dt$$

is probably the only one the reader has seen before, since it is a very powerful tool for transforming initial-value problems in ODE into algebraic equations. Not only is the Laplace transform useful in transforming ODEs into algebraic equations, but now we will use the Laplace transform to transform PDEs into ODEs. In particular, we will attempt to apply the Laplace transform to any variable x, y, z, t, ... that ranges from 0 to ∞ (although it will generally be applied to time). The major difference in applying the Laplace transform to PDEs in contrast to ODEs is that now when the original PDE is transformed, the new resulting equation will be either a new PDE with one less independent variable or else an ODE in one variable. We must then decide how to solve this new problem (maybe by *another* transform, by separation of variables, and so on). Before actually solving a very interesting problem, we enumerate some useful properties of this transform.

Properties of the Laplace Transform

Property 1 (Transform Pair)

The Laplace transform and its inverse are given by

(13.2)
$$\mathcal{L}[f] = F(s) = \int_0^\infty f(t)e^{-st}\, dt \qquad \text{(Laplace transform)}$$

$$\mathcal{L}^{-1}[F] = f(t) = \frac{1}{2\pi i}\int_{c-i\infty}^{c+i\infty} F(s)e^{st}\, ds \qquad \text{(inverse Laplace transform)}$$

The Laplace transform has one major advantage over the Fourier transform in that the damping factor e^{-st} in the integrand allows us to transform a wider class of functions (the factor $e^{i\xi x}$ in the Fourier transform doesn't do any damping, since its absolute value is one). In fact, the exact conditions that insure that a function $f(t)$ has a Laplace transform are given by the following theorem:

Sufficient Conditions to Insure the Existence of a Laplace Transform

If
1. f is piecewise continuous on the interval $0 \leqslant t \leqslant A$ for any positive A
2. we can find constants M and a such that $|f(t)| \leqslant Me^{at}$ for all values of t greater than some number T

then

the Laplace transform $\mathcal{L}[f(t)] = F(s) = \int_0^\infty f(t)e^{-st}\, dt$ exists for $s > a$

We now list a few functions that have Laplace transforms and graph them on the s-axis.

1. $f(t) = 1 \qquad 0 < t < \infty$
   (pick $M = 1 \qquad a = 0$)

   $$F(s) = \int_0^\infty e^{-st}\, dt = \frac{1}{s}$$

   (see Figure 13.1a)

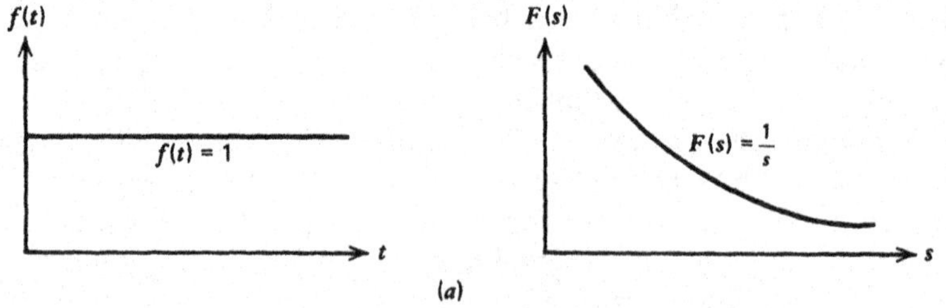

(a)

FIGURE 13.1a–13.1c Graphs of a few Laplace transforms.

2. $f(t) = e^{2t}$ $0 < t < \infty$
   (pick $M = 1$ $a = 2$)

$$F(s) = \frac{1}{s - 2} \quad s > 2$$

(see Figure 13.1*b*)

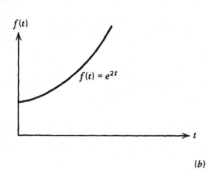
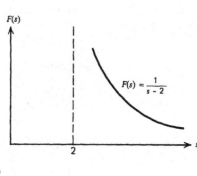

(b)

3. $f(t) = \sin(\omega t)$
   (pick $M = 1$ $a = 0$)

$$F(s) = \frac{\omega}{s^2 + \omega^2}$$

(see Figure 13.1*c*)

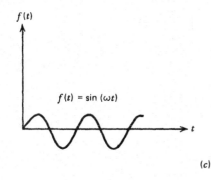
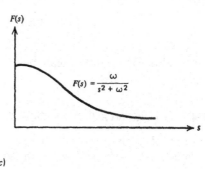

(c)

4. $f(t) = e^{t^2}$ (doesn't have a Laplace transform)

In the definition of the Laplace transform, the variable s is taken to be a real variable $0 < s < \infty$. It is possible, however (often desirable), to extend this definition to *complex* values of s and, in fact, to evaluate the *inverse Laplace transform*

$$\mathcal{L}^{-1}[F] = \frac{1}{2\pi i} \int_{c - i\infty}^{c + i\infty} F(s)e^{st} \, ds$$

We must often resort to *contour integration* in the complex plane and the theory of residues. We won't bother ourselves with this topic here but will use the tables in the appendix for finding inverse transforms.

Property 2 (Transforms of Partial Derivatives)

Suppose we have a function $u(x,t)$ of two variables and wish to transform various partial derivatives $u_t, u_{tt}, u_x, u_{xx}, \ldots$. Since the Laplace transform transforms the *t-variable* (variable of integration), the rules of transformation for partial derivatives are

$$\mathcal{L}[u_t] = \int_0^\infty u_t(x,t)e^{-st}\, dt = sU(x,s) - u(x,0)$$

$$\mathcal{L}(u_{tt}) = \int_0^\infty u_{tt}(x,t)e^{-st}\, dt = s^2 U(x,s) - su(x,0) - u_t(x,0)$$

$$\mathcal{L}[u_x] = \int_0^\infty u_x(x,t)e^{-st}\, dt = \frac{\partial U}{\partial x}(x,s)$$

$$\mathcal{L}[u_{xx}] = \int_0^\infty u_{xx}(x,s)e^{-st}\, dt = \frac{\partial^2 U}{\partial x^2}(x,s)$$

where $U(x,s) = \mathcal{L}[u(x,t)]$. The transform rules for u_x and u_{xx} are a result of a basic rule in calculus

$$\frac{d}{dx}\int_a^b f(x,y)\, dy = \int_a^b \frac{\partial f}{\partial x}(x,y)\, dy$$

while the rules for u_t and u_{tt} can be proven by using the integration by parts formula.

Property 3 (Convolution Property)

Convolution plays the same role here as it did in the Fourier transform, but now the convolution is defined slightly differently.

Definition of the Finite Convolution

The **finite convolution** of two functions f and g is defined by

$$(f * g)(t) = \int_0^t f(\tau)g(t - \tau)\, d\tau$$

$$= \int_0^t f(t - \tau)g(\tau)\, d\tau$$

(these two integrals are the same). In other words, in the *finite* convolution, we integrate from 0 to t instead of from $-\infty$ to ∞, as we did in the *infinite* convolution. An example of the finite convolution of two functions

$$f(t) = t$$
$$g(t) = t$$

would be

$$(f * g)(t) = \int_0^t \tau(t - \tau) \, d\tau = t^3/6$$

As in the case of the infinite convolution, the important property of this new convolution is that

(13.4) $$\mathscr{L}[f * g] = \mathscr{L}[f] \, \mathscr{L}[g]$$

or the equivalent formula

(13.5) $$\mathscr{L}^{-1}\{\mathscr{L}[f] \, \mathscr{L}[g]\} = f * g$$

This property will allow us to find the inverse Laplace transform of a product of two functions (which we interpret as $\mathscr{L}[f]\mathscr{L}[g]$) by finding the inverses of each factor $\mathscr{L}[f]$ and $\mathscr{L}[g]$ to get f and g and then finding their convolution. For example

$$\mathscr{L}^{-1}\left[\frac{1}{s} \cdot \frac{1}{s^2 + 1}\right] = \int_0^t \sin \tau \, d\tau = 1 - \cos t$$

$$F(s) = \frac{1}{s} \xrightarrow{\mathscr{L}^{-1}} f(t) = 1 \qquad G(s) \xrightarrow{\mathscr{L}^{-1}} g(t) = \sin t$$

We are now ready to work an important initial-boundary-value problem.

Heat Conduction In a Semi Infinite Medium

Consider a large (deep) container of liquid that is insulated on the sides. Suppose the liquid has an initial temperature of u_0 and that the temperature of the air above the liquid is zero (some reference temperature). Our goal is to find the temperature of the liquid at various depths of the container at different values of time. To do so, we must solve the problem

(13.6)

PDE	$u_t = u_{xx}$	$0 < x < \infty$	$0 < t < \infty$
BC	$u_x(0,t) - u(0,t) = 0$	$0 < t < \infty$	
IC	$u(x,0) = u_0$	$0 < x < \infty$	

See Figure 13.2.

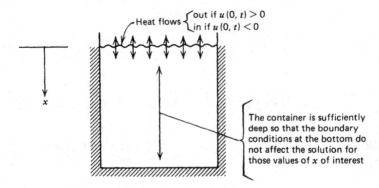

Heat flows {out if $u(0, t) > 0$ / in if $u(0, t) < 0$

The container is sufficiently deep so that the boundary conditions at the bottom do not affect the solution for those values of x of interest

FIGURE 13.2   Diagram illustrating the heat-flow problem.

To solve this problem, we transform the *time variable* t by means of the Laplace transform (conceivably, we could also transform x by means of the Laplace transform, since x also ranges from 0 to ∞). Transforming our problem, we arrive at an ODE in x

(13.7)
$$\text{ODE} \quad sU(x) - u_0 = \frac{d^2U}{dx^2}(x) \quad 0 < x < \infty$$
$$\text{BC} \quad \frac{dU}{dx}(0) = U(0)$$

(we transform the PDE and the BC—not the IC). This is a second-order ODE with one BC at $x = 0$ [for physical reasons, we *really* have a second, implied BC that says $U(x)$ is bounded]. Note that we have dropped the s-notation in $U(x,s)$ in favor of the simpler notation $U(x)$, since the differential equation in (13.7) depends only on x.

To solve (13.7), we first find the general solution (homogeneous + a particular solution), which is

$$U(x) = c_1 e^{\sqrt{s}\, x} + c_2 e^{-\sqrt{s}\, x} + \frac{u_0}{s}$$

Substituting this expression into the BCs of (13.7) allows us to find the constants c_1 and c_2 (first note that $c_1 = 0$ or else the temperature will go to infinity as x gets large). Finding c_2 from the BC at $x = 0$ gives us the answer for $U(x)$

(13.8)
$$U(x) = - u_0 \left\{ \frac{e^{-\sqrt{s}\, x}}{s\,(\sqrt{s} + 1)} \right\} + \frac{u_0}{s}$$

Now for the last step. To find the temperature $u(x,t)$, we compute

$$u(x,t) = \mathcal{L}^{-1}[U(x,s)]$$

[we now put back s in $U(x,s)$]. To find this inverse transform, we must resort to the tables of inverse Laplace transforms in the appendix; they will give us

(13.9) $u(x,t) = u_0 - u_0 \left[erfc(x/2\sqrt{t}) - erfc\left(\sqrt{t} + x/2\sqrt{t}\right) e^{(x+t)} \right]$

where

$$erfc(x) = \frac{2}{\sqrt{\pi}} \int_x^\infty e^{-\xi^2}\, d\xi$$

is the *complementary-error function* whose graph is given in Figure 13.3.

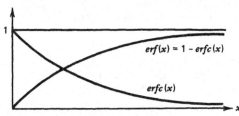

FIGURE 13.3   Graphs of the error (*erf*) and complementary-error (*erfc*) functions.

If we spend a little time analyzing this equation and graphing it by means of a computer with a plotter attachment, we will see that it looks like Figure 13.4.

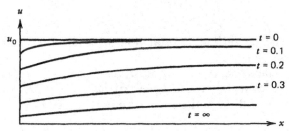

FIGURE 13.4   Temperatures inside the infinite medium for different values of time.

NOTES

1. The Laplace transform can also be applied to problems where the PDE is nonhomogeneous (in separation of variables, the equation had to be homogeneous), but the Laplace transform will generally work only if the equation has constant coefficients (in separation of variables, we could have variable coefficients). The following table lists the types of problems the two methods will handle.

TABLE 13.2 Comparison of Laplace Transform and Separation of Variables

	Method	
	Laplace Transform	Separation of Variables
Nonhomogeneous PDE	yes	no
Nonhomogeneous BC	yes	no
Variable coefficients	no	yes
Nonlinear equations	no	no

2. The Hankel and Mellin transforms are also used to solve IBVPs and BVPs but differ from the Laplace transform in one regard. The Laplace transform converts derivatives to multiplication operations by means of a formula like

$$\mathscr{L}[y'] = s\mathscr{L}[y] - y(0)$$

whereas the Hankel and Mellin transforms convert *differential operators* to multiplication; for example, the Hankel transform

$$H[y] = \int_0^\infty rJ_0(\xi r)y(r) \, dr$$

transforms the differential operator

$$H[y''(r) + \frac{1}{r}y'(r)] = -\xi^2 H[y]$$

In this way, specific differential equations with variable coefficients (Bessel's equation) can be solved.

3. The Laplace transform (which transforms t) can be interpreted as projecting the xt-plane onto the x-axis, and the original BCs, PDE, and IC are transformed into a new differential equation and BCs. See the following diagram.

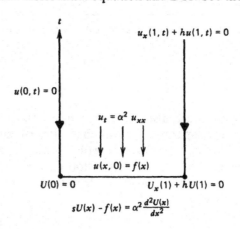

$u_x(1, t) + hu(1, t) = 0$

$u(0, t) = 0$

$u_t = \alpha^2 u_{xx}$

$u(x, 0) = f(x)$

$U(0) = 0$

$U_x(1) + hU(1) = 0$

$sU(x) - f(x) = \alpha^2 \dfrac{d^2 U(x)}{dx^2}$

PROBLEMS

1. Verify the following formula for the transform of the partial derivative u_t

$$\mathcal{L}[u_t(x,t)] = sU(x,s) - u(x,0)$$

2. Solve the following initial-value problem by means of the Laplace transform

$$\text{PDE} \quad u_t = \alpha^2 u_{xx} \quad -\infty < x < \infty \quad 0 < t < \infty$$
$$\text{IC} \quad u(x,0) = \sin x \quad -\infty < x < \infty$$

3. Solve the problem

$$\text{PDE} \quad u_t = u_{xx} \quad 0 < x < \infty \quad 0 < t < \infty$$
$$\text{BC} \quad u(0,t) = \sin t \quad 0 < t < \infty$$
$$\text{IC} \quad u(x,0) = 0 \quad 0 \leqslant x < \infty$$

by means of the Laplace transform (transform t). What is the physical interpretation of this problem?

4. Solve the boundary-value problem

$$\text{ODE} \quad \frac{d^2U}{dx^2} - sU = A \quad 0 < x < 1$$

$$\text{BCs} \quad \begin{cases} \dfrac{dU}{dx}(0) = 0 \\[2mm] U(1) = 0 \end{cases}$$

OTHER READING

1. *A First Course in PDE* by H. Weinberger. Ginn and Co., 1965. This text contains an extensive section on contour integration, which is the tool used for evaluating the inverse Laplace transform.

2. Almost any beginning text in ODE will contain a chapter on the Laplace transform.

Duhamel's Principle

PURPOSE OF LESSON: To show how the Laplace transform can bring out interesting underlying phenomena concerning solutions of differential equations, in particular, by algebraically manipulating the Laplace transform of the solution of a PDE, we discover an interesting idea known as *Duhamel's principle*. This principle has interpretations in ODE, but we will illustrate how it works in the context of a specific initial-boundary-value problem.

In addition to providing a powerful tool for solving PDEs, the Laplace transform also provides insight into the nature of solutions to physical problems. With the help of the Laplace transform, we illustrate a very important and interesting concept known as *Duhamel's principle* in this lesson. Before getting to this principle, however, let's discuss a problem that occurs frequently in engineering.

Heat Flow within a Rod with Temperature Fixed on the Boundaries

Quite often, it is important to find the temperature inside a medium due to *time-varying boundary conditions*. For example, consider an insulated rod with temperature specified as $f(t)$ on the right end

$$\text{PDE} \quad u_t = u_{xx} \quad 0 < x < 1 \quad 0 < t < \infty$$

(14.1) $\quad$ BCs $\quad \begin{cases} u(0,t) = 0 \\ u(1,t) = f(t) \end{cases} \quad 0 < t < \infty$

$$\text{IC} \quad u(x,0) = 0 \quad 0 \leq x \leq 1$$

See Figure 14.1.

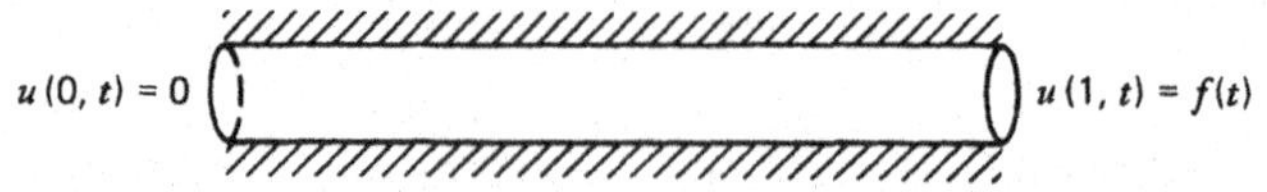

$u(0, t) = 0$ $\qquad\qquad\qquad\qquad\qquad\qquad\qquad$ $u(1, t) = f(t)$

Initial temperature = 0

FIGURE 14.1 Time-varying boundary conditions.

We may think that the solution to problem (14.1) can be easily found once we know the solution to the simpler version (constant temperature on the boundaries)

$$\text{PDE} \qquad w_t = w_{xx} \qquad 0 < x < 1 \qquad 0 < t < \infty$$

(14.2) $\qquad$ BCs $\qquad \begin{cases} w(0,t) = 0 \\ w(1,t) = 1 \end{cases} \qquad 0 < t < \infty$

$$\text{IC} \qquad w(x,0) = 0 \qquad 0 \leqslant x \leqslant 1$$

In fact, if we solve problems (14.1) and (14.2) side by side by the Laplace transform, we will see a striking result (Duhamel's principle) that will give us the solution to (14.1) *in terms of the solution of (14.2)*.

So, solving (14.1) and (14.2) at the same time, we have

Easy Problem (14.2) (constant BC)	**Hard Problem** (14.1) (time-varying BC)
Transform problem (14.2) by the Laplace transform	Transform problem (14.1) by the Laplace transform
$\dfrac{d^2 W}{dx^2} - sW(x) = 0$ $W(0) = 0$ $W(1) = 1/s$	$\dfrac{d^2 U}{dx^2} - sU(x) = 0$ $U(0) = 0$ $U(1) = F(s)$
Solve the ODE	Solve the ODE
$W(x,s) = \dfrac{1}{s}\left[\dfrac{\sinh(x\sqrt{s})}{\sinh(\sqrt{s})}\right]$	$U(x,s) = F(s)\left[\dfrac{\sinh(x\sqrt{s})}{\sinh(\sqrt{s})}\right]$
Find the inverse transform	Don't invert yet—first multiply and divide by s
$w(x,t) =$ $x + \dfrac{2}{\pi}\displaystyle\sum_{n=1}^{\infty} \dfrac{(-1)^n}{n} e^{-(n\pi)^2 t} \sin(n\pi x)$ (solution to the constant BC problem)	$U(x,s) = F(s)\left\{ s\left[\dfrac{\sinh(x\sqrt{s})}{s\,\sinh(\sqrt{s})}\right]\right\}$

Easy Problem (14.2) (Cont.)	*Hard Problem* (14.1) (Cont.)
(constant BC)	(time-varying BC)

Using the relationship

$$\mathcal{L}[w_t] = sW - w(x,0)$$

we have

$$\downarrow$$

$$U(x,s) = F(s) \mathcal{L}[w_t]$$

Hence

$$
\begin{aligned}
u(x,t) &= \mathcal{L}^{-1}\{F(s)\mathcal{L}[w_t]\} \\
&= \mathcal{L}^{-1}[F(s)] \cdot [w_t] \\
&= f(t) \cdot w_t(t) \\
&= \int_0^t f(\tau)\, w_\tau(x,t-\tau)\, d\tau
\end{aligned}
$$

(or by integration by parts)

$$
= \int_0^t w(x,t-\tau)\, f'(\tau)\, d\tau + f(0)w(x,t)
$$

(solution to the time-varying problem in terms of the solution of the constant BC problem)

In other words, we have found the solution $u(x,t)$ to the *time-varying* problem in terms of the solution to the easy problem (constant BCs); that is,

(14.3)
$$
\begin{aligned}
u(x,t) &= \int_0^t w_t(x,t-\tau)f(\tau)\, d\tau \\
&= \int_0^t w(x,t-\tau)f'(\tau)\, d\tau + f(0)w(x,t)
\end{aligned}
$$

Equations (14.3) are known as **Duhamel's principle.** We can now take the solution $w(x,t)$ to the constant BC problem

$$
w(x,t) = x + \frac{2}{\pi}\sum_{n=1}^{\infty}\frac{(-1)^n}{n}e^{-(n\pi)^2 t}\sin(n\pi x)
$$

and substitute it into equation (14.3) to obtain the solution to the time-varying problem [we must use the second equation in (14.3), since if we differentiate the infinite series representation for $w(x,t)$ term by term (with respect to t), it results in a *divergent* series].

There is another aspect of Duhamel's formulas that makes them very useful.

The Importance of Duhamel's Principle

In the problem just discussed, we were able to solve the easy problem with constant BCs, so we used Duhamel's formulas (14.3) to obtain the solution to the time-varying BCs. Quite often, however, even the easy problem (constant BCs) cannot be solved analytically. What we can do, however, is observe the solution *experimentally*; in other words, we can rig a device that has constant BCs and experimentally measure the response. We can then use Duhamel's principle to find the solution for *any* time-varying BC. In fact, we have only to observe the response $w(x,t)$ to the constant BC problem *once*. When we have this data, we can then solve the problem with *arbitrary* BC $f(t)$ by substituting into Duhamel's formulas (14.3).

NOTES

There is another interesting version of Duhamel's principle that gives the answer to the problem

$$\text{PDE} \quad u_t = u_{xx} \quad 0 < x < 1 \quad 0 < t < \infty$$

$$(14.4) \quad \text{BCs} \quad \begin{cases} u(0,t) = 0 \\ u(1,t) = f(t) \end{cases} \quad 0 < t < \infty$$

$$\text{IC} \quad u(x,0) = 0 \quad 0 \leqslant x \leqslant 1$$

in terms of the solution $w(x,t)$ of the *alternative simple problem*

$$\text{PDE} \quad w_t = w_{xx} \quad 0 < x < 1 \quad 0 < t < \infty$$

$$(14.5) \quad \text{BCs} \quad \begin{cases} w(0,t) = 0 \\ w(1,t) = \delta(t) \end{cases} \quad \text{(temperature impulse at } t = 0$$

$$\text{IC} \quad w(x,0) = 0 \quad 0 \leqslant x \leqslant 1$$

Knowing this formula, which is

$$(14.6) \quad u(x,t) = \int_0^t w(x,t - \tau)f(\tau) \, d\tau$$

allows us to find the temperature response $u(x,t)$ to an arbitrary boundary temperature $f(t)$ once we have carried out an experiment to determine the temperature response $w(x,t)$ from an impulse temperature.

PROBLEMS

1. Prove the Duhamel principle (14.6) by transforming both problems (14.4) and (14.5) and using an argument similar to the one for finding (14.3) in the lesson.
    HINT   The Laplace transform of the impulse function $\delta(t)$ is $\mathcal{L}[\delta(t)] = 1$.

2. Show that the partial derivative w_t of

$$w(x,t) = x + \frac{2}{\pi} \sum_{n=1}^{\infty} \frac{(-1)^n}{n} e^{-(n\pi)^2 t} \sin (n\pi x)$$

diverges for all x if we differentiate the series term by term.

3. Suppose we have a metal rod (laterally insulated) and we supply an *initial impulse of heat* at the right-hand side (the left-hand side is fixed at zero). Suppose the initial temperature of the rod is zero (some reference temperature) and the temperature at the midpoint $x = 0.5$ is measured at various values of time, so that we have the following table:

Values of Time	Midpoint Temperature
t_1	w_1
$t_2 = 2t_1$	w_2
$t_3 = 3t_1$	w_3
.	.
.	.
.	.
$t_n = nt_1$	w_n

Using this data, how could we approximate the temperature response at the point $u(0.5, t_n)$ due to the BCs

(a) $u(1,t) = \sin t$
(b) $u(1,t) = f(t)$ (arbitrary $f(t)$)

4. Using Duhamel's principle, what is the solution of the IBVP

$$\text{PDE} \qquad u_t = \alpha^2 u_{xx} \qquad 0 < x < 1 \qquad 0 < t < \infty$$

$$\text{BCs} \qquad \begin{cases} u(0,t) = 0 \\ u(1,t) = \sin t \end{cases} \qquad 0 < t < \infty$$

$$\text{IC} \qquad u(x,0) = 0 \qquad 0 \leqslant x \leqslant 1$$

OTHER READING

1. *Differential Equations* by C. Wylie. McGraw-Hill, 1979. Duhamel's principle is discussed in conjunction with problems in ODE in Chapter 6 of this text.

2. *Equations of Mathematical Physics* by A. N. Tikhonov and A. A. Samarskii. Macmillan, 1963; Dover, 1990. An excellent source of all kinds of applied problems; the Duhamel principle is discussed on page 261.

The Convection Term u_x in the Diffusion Problems

PURPOSE OF LESSON: To show how the term u_x in the diffusion equation

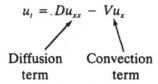

$$u_t = . Du_{xx} - Vu_x$$

Diffusion   Convection
term   term

represents the phenomenon of convection. Phenomena described by this convection-diffusion equation exhibit both diffusion and convection properties and are common in many situations. How much diffusion and convection takes place depends on the relative size of the two coefficients D and V.

Inasmuch as the convection of a substance represents material moving with the medium, it is possible to pick a moving coordinate system that moves with the medium. In this way, the convection term is eliminated and the equation can be solved in terms of the moving coordinate and then transformed back into the original variable x.

So far, we have been concerned with heat flow (or diffusion of some kind) in a one-dimensional domain. Suppose now we consider the problem of finding the *concentration* of a substance upwards from the surface of the earth where the substance both diffuses through the air and is *carried upward* (convected) by moving currents (moving with velocity V). Clearly, it is possible for the convection of the substance to contribute more of a movement in the substance than the diffusion itself. (It would depend on the relative size of the diffusion coefficient and the velocity of the air.) **Diffusion** is mixing the substance through the air, while **convection** is the movement of the substance *by means* of the air (the movement of the medium); in any case, it is our purpose here to solve the diffusion-convection equation

$$u_t = Du_{xx} - Vu_x$$

and to show how it is derived.

To verify this equation for a concentration $u(x,t)$ of a substance, we use *two basic facts*

1. *Flux due to convection*

   The flux of material (from left to right) across a point due to *convection* is given by $Vu(x,t)$, where V is the velocity of the medium (cm/sec) and $u(x,t)$ is the linear concentration (g/cm) (Figure 15.1).

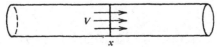

FIGURE 15.1   Amount of material across x (per second) due to convection is $Vu(x,t)$.

2. *Flux due to diffusion*

   The *flux* of material (from left to right) across a point x due to *diffusion* is given by $-Du_x(x,t)$, where D is the diffusion coefficient.

If we substitute these two expressions into the *conservation equation* in Lesson 4, we can prove that the basic PDE is

$$u_t = Du_{xx} - Vu_x$$

To get an idea of what solutions look like or how they behave with the convection term included, let's first work a problem that is pure convection (the diffusion term is zero). A typical problem would be dumping a substance into a clean river (moving with velocity V) and observing the concentration of the substance downstream. For example, if x measures the distance downstream from where the substance is added and if the substance *does not diffuse* with the running water, then the concentration of the substance $u(x,t)$ can be found by solving the following mathematical model:

PDE $\quad u_t = -Vu_x \quad\quad 0 < x < \infty \quad\quad 0 < t < \infty$

(15.1) $\quad$ BC $\quad u(0,t) = P \quad \longleftarrow$ Constant input of the substance

$\quad\quad\quad$ IC $\quad u(x,0) = 0 \quad \longleftarrow$ Initially a clean river

This problem is illustrated in Figure 15.2.

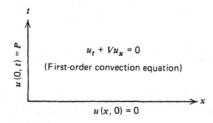

FIGURE 15.2   Pure convection problem.

Before solving this problem, we should think a little about what the solution should be. It's obvious that the pollutant (substance in the river) will initially be zero, but once it is added at a constant rate at $x = 0$, it will move downstream with velocity V. To see this mathematically, let's solve (15.1). Since it is a linear PDE with linear BCs, we should think in terms of separation of variables and integral transforms; however, since the x-variable is unbounded, separation of variables is out. Let's use the Laplace transform on t.

Laplace Transform Solution to the Convection Problem

The convection problem (15.1) can be replaced by

$$sU(x) = -V\frac{dU}{dx} \quad 0 < x < \infty$$
$$U(0) = \frac{P}{s}$$

by using the Laplace transform

$$U(x) = \int_0^\infty u(x,t)e^{-st}\,dt$$

Solving this very simple initial-value problem, we get

$$U(x) = \frac{P}{s}e^{-(sx/V)}$$

Looking up the inverse transform in the tables, we see

$$u(x,t) = \mathcal{L}^{-1}[U] = PH(t - x/V)$$

where $H(\xi)$ is the Heaviside function (step function)

$$H(\xi) = \begin{cases} 0 & \xi < 0 \\ 1 & \xi \geq 0 \end{cases}$$

Hence, the solution of our problem is just

$$u(x,t) = \begin{cases} 0 & t < x/V \\ P & t \geq x/V \end{cases}$$

This was pretty simple; certainly, it isn't any more complicated than dumping something on a conveyor belt and watching it move along. It does, however, become more interesting when the solute (pollutant) *diffuses* with the medium.

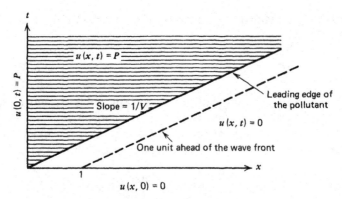

FIGURE 15.3 Pure convection wave.

To see what happens when a moving wave diffuses, we solve the following problem:

(15.2)   PDE $u_t = Du_{xx} - Vu_x$ $-\infty < x < \infty$
            IC $u(x,0) = 1 - H(x)$ $-\infty < x < \infty$

where, as usual, $H(x)$ is the Heaviside function. The initial concentration is shown in Figure 15.4.

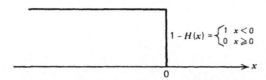

$$1 - H(x) = \begin{cases} 1 & x < 0 \\ 0 & x \geqslant 0 \end{cases}$$

FIGURE 15.4 Initial condition for the diffusion-convection equation.

Note that in the new problem (15.2), we have moved the boundary to $-\infty$ (we now have an *initial-value problem*), so that it doesn't confuse the real issue of measuring the relative effects of D versus V (just a technicality). To solve (15.2), we could use the Laplace transform on t or the Fourier transform on x; however, in this case, there is another alternative that is very interesting. That is, instead of measuring the concentration $u(x,t)$ as a function of x, we introduce a new coordinate ξ, which moves along the x-axis with velocity V. In other words, instead of placing our coordinate system along the bank of the river (so to speak), we now place our coordinate system so that it moves with the *wave front* (of course, now when we have diffusion in addition to convection, we won't have a sharp wave front). Mathematically this says that we change our space coordinate x to $\xi = x - Vt$. It's now clear that

when $\xi = 0$ we are on the wave front
when $\xi = 1$ we are one unit ahead of the front
when $\xi = -1$ we are one unit behind the front

So our goal is to transform the initial-value problem (IVP)

$$\text{PDE} \quad u_t = Du_{xx} - Vu_x \quad -\infty < x < \infty$$
$$\text{IC} \quad u(x,0) = 1 - H(x) \quad -\infty < x < \infty$$

into a *new one* in the moving coordinate system, solve it, and then transform back to get the solution in terms of the original coordinates (x,t). To begin, we make what is called a *change of variables* (change of *independent* variables). In place of the old coordinates (x,t), we introduce new ones (ξ,τ)

$$\xi = x - Vt$$
$$\tau = t$$

The reader should note that τ is the same as t, but it is less confusing if we give it a new name. To rewrite the PDE in terms of (ξ,τ), we use the chain rule

$$u_t = u_\xi \xi_t + u_\tau \tau_t = -Vu_\xi + u_\tau$$
$$u_x = u_\xi \xi_x = u_\xi$$
$$u_{xx} = (u_\xi)_x = u_{\xi\xi}\xi_x = u_{\xi\xi}$$

Using *functional diagrams*, as in Figure 15.5, makes these chain-rule arguments clearer.

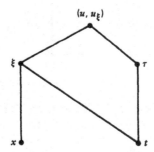

(u, u_ξ)

ξ

τ

x

t

FIGURE 15.5 Diagram illustrating functional dependence of variables.

The diagram in Figure 15.5 is useful for computing the partial derivatives of u, u_ξ with respect to x and t, since it shows exactly how u and u_ξ depend, in general, on ξ and τ and that ξ depends, in turn, on both x and t. The variable τ, on the other hand, depends only on t.

So much for the transformation. We now substitute our computed u_t, u_x, and u_{xx} into the PDE to get

$$-Vu_\xi + u_\tau = Du_{\xi\xi} - Vu_\xi$$

or

$$u_\tau = Du_{\xi\xi}$$

Hence, our new IVP in terms of ξ and τ is

$$\text{PDE} \quad u_\tau = Du_{\xi\xi} \quad -\infty < \xi < \infty$$
$$\text{IC} \quad u(\xi,0) = 1 - H(\xi) \quad -\infty < \xi < \infty$$

(Note that $\xi = x$ when $t = 0$, so our ICs are both the same.) This problem has already been solved in Lesson 12 by the Fourier transform and has the solution

$$u(\xi,\tau) = \frac{1}{2\sqrt{D\pi\tau}} \int_{-\infty}^{\infty} \phi(\beta)e^{-(\xi-\beta)^2/4D\tau}\, d\beta$$

where $\phi(\beta)$ is the initial condition. Hence, in our case, we have

$$u(\xi,\tau) = \frac{1}{2\sqrt{D\pi\tau}} \int_{-\infty}^{0} e^{-(\xi-\beta)^2/4D\tau}\, d\beta$$

By letting

$$\overline{\beta} = \frac{\xi-\beta}{2\sqrt{D\tau}} \qquad d\overline{\beta} = \frac{-1}{2\sqrt{D\tau}}\, d\beta$$

we get the interesting result

$$(15.3) \quad u(\xi,\tau) = \frac{1}{2}\left[\frac{2}{\sqrt{\pi}} \int_{\frac{\xi}{2\sqrt{D\tau}}}^{\infty} e^{-\overline{\beta}^2}\, d\overline{\beta} \right]$$

$$= \begin{cases} \frac{1}{2}\left[1 + erf\left(\frac{-\xi}{2\sqrt{D\tau}}\right)\right] & \xi < 0 \\[2ex] \frac{1}{2}\, erfc\left(\frac{\xi}{2\sqrt{D\tau}}\right) & \xi \geq 0 \end{cases}$$

The graph of this function is plotted for various values of t in Figure 15.6.

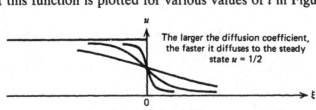

The larger the diffusion coefficient, the faster it diffuses to the steady state $u = 1/2$

FIGURE 15.6 Simple diffusion from high to low concentrations.

Finally, to get the solution of our problem in terms of the coordinates x and t, we substitute

$$\xi = x - Vt$$
$$\tau = t$$

into equation (15.3) to get

(15.4)
$$u(x,t) = \begin{cases} \dfrac{1}{2}\left[1 + erf\left(\dfrac{Vt - x}{2\sqrt{D\tau}}\right)\right] & Vt > x \\[4mm] erfc\left(\dfrac{x - Vt}{2\sqrt{D\tau}}\right) & Vt \leq x \end{cases}$$

This is the solution of our diffusion-convection problem (15.2), and it is really very easy to interpret; it's just a moving version of Figure 15.6. In other words, depending on the relative size of D (diffusion coefficient) and V (velocity of the stream), the solution moves to the right with velocity V while, at the same time, the leading edge is diffusing at a rate defined by D (Figure 15.7 shows the break up of the leading edge).

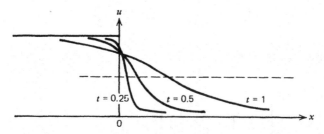

FIGURE 15.7   Diffusion-convection solution moving and diffusing at the same time.

NOTES

Changing coordinates is a very important technique in PDEs. By looking at physical systems with different coordinates, the equations are sometimes simplified.

PROBLEMS

1. Solve the initial-value problem

$$u_t = u_{xx} - 2u_x \qquad -\infty < x < \infty \qquad 0 < t < \infty$$
$$u(x,0) = \sin x \qquad -\infty < x < \infty$$

2. Solve the following diffusion-convection problem by making a transformation as shown in Lesson 8:

$$u_t = u_{xx} - 2u_x \qquad -\infty < x < \infty \qquad 0 < t < \infty$$
$$u(x,0) = e^x \sin x \qquad -\infty < x < \infty$$

3. What is the solution of the following *convection* problem:

PDE $\quad u_t = -2u_x \qquad -\infty < x < \infty \qquad 0 < t < \infty$

IC $\quad u(x,0) = e^{-x^2}$

Check your answer.
4. Solve

$$u_t = Du_{xx} - Vu_x \qquad -\infty < x < \infty \qquad 0 < t < \infty$$
$$u(x,0) = e^{-x^2} \qquad -\infty < x < \infty$$

Does the solution check? What does the solution look like for various values of time?

HINT   Note that our transformation to moving coordinates allows us to essentially neglect the term Vu_x in the PDE. After solving the *new* problem,

PDE $\quad u_\tau = Du_{\xi\xi} \qquad -\infty < \xi < \infty \qquad 0 < \tau < \infty$

IC $\quad u(\xi,0) = e^{-\xi^2} \qquad -\infty < \xi < \infty$

we merely set $\xi = x - Vt$ and $\tau = t$. In this particular problem, it is possible to *evaluate* the integral

$$u(\xi,\tau) = \frac{1}{2\sqrt{D\pi\tau}} \int_{-\infty}^{\infty} e^{-\beta^2} e^{-(\xi-\beta)^2/4D\tau} \, d\beta$$

This is the Fourier transform solution from Lesson 12. It may be more convenient for the reader to rewrite this integrand and then look it up in a table of integrals.

PART 3

Hyperbolic-Type Problems

The One-Dimensional Wave Equation (Hyperbolic Equations)

PURPOSE OF LESSON: To introduce the one-dimensional wave equation

$$u_{tt} = \alpha^2 u_{xx}$$

and show how it describes the motion of a vibrating string. It is also shown how this equation is derived as a result of Newton's equations of motion. In addition, a few other variations of the wave equation, such as

$$u_{tt} = \alpha^2 u_{xx} + F(x,t)$$
$$u_{tt} = \alpha^2 u_{xx} - \beta u_t - \lambda u + F(x,t)$$

are discussed.

So far, we have been concerned with physical phenomenon described by one-dimensional parabolic equations (diffusion problems). We will now begin to study the second major class of PDEs, hyperbolic equations. We start by studying the *one-dimensional wave equation*, which describes (among other things) the *transverse vibrations* of a string.

Vibrating-String Problem

We consider the small vibrations of a string that is fastened at each end. We assume the string is stretched tightly, made of a homogeneous material, unaffected by gravity, and that the vibrations take place in a plane (Figure 16.1).

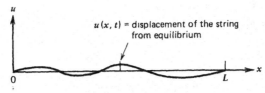

FIGURE 16.1   Transverse vibrations of a string.

To mathematically describe the vibrations of this string, we consider all the forces acting on a small section of the string (Figure 16.2).

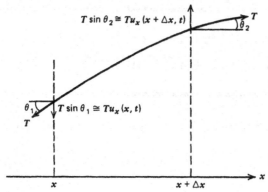

FIGURE 16.2 Small segment $(x, x + \Delta x)$ of the vibrating string.

Essentially, the wave equation is nothing more than Newton's equation of motion applied to the string (the change in momentum mu_{tt} of a small string segment is equal to the applied forces). Looking at Figure 16.1, we can imagine several forces acting on the string in the direction *perpendicular* to the x-axis.

The most important forces are

1. *Net force due to the tension of the string $(\alpha^2 u_{xx})$*
   The tension component has a net transverse force on the string segment of

$$\text{Tension component} = T \sin \theta_2 - T \sin \theta_1$$
$$\cong T\left[u_x(x + \Delta x, t) - u_x(x, t)\right]$$

2. *External force $F(x,t)$*
   An external force $F(x,t)$ may be applied along the string at any value of x and t; some examples would be,
   (a) gravity $F(x,t) = -mg$
   (b) impulses along the string at different values of time
   (c) in the two-dimensional wave equation (which we will study later) that describes the vibrating drumhead, a force could be applied by sound waves impinging on the surface of the membrane.
3. *Frictional force against the string $(-\beta u_t)$*
   If the string is vibrating in a medium that offers a resistance to the string's velocity u_t, then this resistance force is $-\beta u_t$
4. *Restoring force $(-\gamma u)$*
   This is a force that is directed opposite to the displacement of the string. If the displacement u is positive (above the x-axis), then the force is negative (downward)

If we now apply Newton's equation of motion

$$mu_{tt} = \text{applied forces to the segment } (x, x + \Delta x)$$

to the small segment of string, we have

$$\Delta x \rho u_{tt} = T[u_x(x + \Delta x, t) - u_x(x, t)] + \Delta x F(x, t) - \Delta x \beta u_t(x, t) - \Delta x \gamma u(x, t)$$

where ρ is the density of the string. By dividing each side of the equation by Δx and letting $\Delta x \to 0$, we have the well-known **telephone equation**

(16.1) $$u_{tt} = \alpha^2 u_{xx} - \beta u_t - \gamma u + F(x, t)$$

which is the equation we wanted to derive [note that β, γ, and $F(x, t)$ in equation (16.1) should be divided by ρ, but we will relabel them as β, γ, and F for simplicity]. We will now present an intuitive interpretation of the simple wave equation.

(16.2) $$u_{tt} = \alpha^2 u_{xx}$$

Intuitive Interpretation of the Wave Equation

The reader may ask why we would expect an equation like (16.2) to describe something like the vibration of a violin string. To answer this, we must understand that the expression u_{tt} represents the vertical acceleration of the string at a point x. Hence, equation (16.2) can be interpreted as saying that the acceleration of each point of the string is due to the tension in the string and that the larger the *concavity* u_{xx}, the stronger the force (the proportionality constant is $\alpha^2 = T/\rho$). See Figure 16.3.

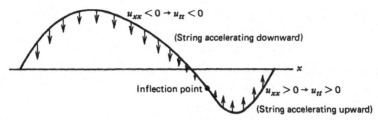

FIGURE 16.3 Interpretation of the wave equation $u_{tt} = \alpha^2 u_{xx}$.

NOTES

1. The wave equation $u_{tt} = \alpha^2 u_{xx}$ also describes *longitudinal* and *torsional* vibrations of a rod. In these vibrations, the displacement is *parallel* to the rod (like hitting the end of the rod with a hammer), and $u(x, t)$ stands for the longitudinal displacement of a point on the rod from its equilibrium

position. The PDE can be derived in more or less the same way as in the transverse-vibrations case and it is given by

$$u_{tt} = ku_{xx}$$

where k is a physical parameter known as Young's modulus (which is a measure of the elasticity of the rod). Materials that have large Young's moduli vibrate rapidly; *sound waves* also constitute longitudinal waves.

2. If the vibrating string had a *variable density* $\rho(x)$, then the wave equation would be:

$$u_{tt} = \frac{\partial}{\partial x}[\alpha^2(x)u_x]$$

In other words, the PDE would have variable coefficients.

3. Since the wave equation $u_{tt} = \alpha^2 u_{xx}$ contains a second-order time derivative u_{tt}, it requires *two initial conditions*

$$
\begin{aligned}
u(x,0) &= f(x) && \text{(initial position of the string)} \\
u_t(x,0) &= g(x) && \text{(initial velocity of the string)}
\end{aligned}
$$

in order to uniquely define the solution for $t > 0$. This is in contrast to the heat equation, where only one IC was required.

4. Another situation that can be described by the one-dimensional wave equation is an electric current along a wire. With the help of Kirchhoff's laws, we can arrive at a system of two first-order PDEs

(16.3)
$$
\begin{aligned}
i_x + Cv_t + Gv &= 0 \\
v_x + Li_t + Ri &= 0
\end{aligned}
$$

where
  x = location along the wire
  t = time
  i = current along the wire
  $v\,(x,t)$ = potential along the wire
  C = capacitance/unit length of wire
  G = leakage conductance/unit length of wire
  R = resistance/unit length of wire
  L = self-inductance/unit length of wire

Equations (16.3) are the well-known **transmission-line equations** and are not generally left in this form, but are further manipulated. This is, by differentiating the first equation with respect to x and the second with respect to t, multiplying by C, and then subtracting, we have

$$i_{xx} + Gv_x - CLi_{tt} - CRi_t = 0$$

Now, by using the second equation in (16.3)

$$v_x = -Li_t - Ri$$

we get

(16.4) $$i_{xx} = CLi_{tt} + (CR + GL)i_t + GRi$$

This is called the *transmission-line equation* for the current i and is a second-order *hyperbolic equation* (unless C or L are zero, in which case it becomes a parabolic equation).

The voltage also has its own equation, which is

(16.5) $$v_{xx} = CLv_{tt} + (CR + GL)v_t + GRv$$

Note, too, if $G = R = 0$, we have the equations

(16.6) $$\begin{aligned} v_{tt} &= \alpha^2 v_{xx} \\ i_{tt} &= \alpha^2 i_{xx} \end{aligned} \qquad \alpha^2 = 1/CL$$

PROBLEMS

1. Derive the transmission-line equation (16.5) for v from the system of two first-order equations (16.3).
2. From your knowledge of the various terms of the wave equation, what would you expect the solution of the following problem to look like for various values of time?

   PDE $\quad u_{tt} = u_{xx} - u_t \qquad 0 < x < 1 \qquad 0 < t < \infty$

   BCs $\quad \begin{cases} u(0,t) = 0 \\ u(1,t) = 0 \end{cases} \quad 0 < t < \infty$

   ICs $\quad \begin{cases} u(x,0) = \sin(\pi x) \\ u_t(x,0) = 0 \end{cases} \quad 0 \le x \le 1$

3. What would the solution of the following problem look like for various values of time?

   PDE $\quad u_{tt} = u_{xx} \qquad 0 < x < 1$

   BCs $\quad \begin{cases} u(0,t) = 0 \\ u(1,t) = \sin t \end{cases} \quad 0 < t < \infty$

$$\text{ICs} \quad \begin{cases} u(x,0) = 0 \\ u_t(x,0) = 0 \end{cases} \quad 0 \leqslant x \leqslant 1$$

What is the physical interpretation of this problem?

4. How many solutions of $u_{tt} = u_{xx}$ can you find by looking for solutions of the form

$$u(x,t) = e^{ax + bt}$$

Is the sum of two solutions a solution?

OTHER READING

Elementary Partial Differential Equations by P. Berg and J. McGregor. Holden-Day, 1966. An excellent reference text.

The D'Alembert Solution of the Wave Equation

PURPOSE OF LESSON: To find the solution of the initial-value problem

$$\text{PDE} \qquad u_{tt} = c^2 u_{xx} \qquad -\infty < x < \infty \qquad 0 < t < \infty$$

$$\text{ICs} \qquad \begin{cases} u(x,0) = f(x) \\ u_t(x,0) = g(x) \end{cases} \qquad -\infty < x < \infty$$

This problem (which has no boundaries) describes the motion of an infinite string with given initial conditions and was solved in about 1750 by the French mathematician D'Alembert. This so-called D'Alembert solution

$$u(x,t) = \frac{1}{2}\left[f(x - ct) + f(x + ct) \right] + \frac{1}{2c} \int_{x-ct}^{x+ct} g(\xi)\, d\xi$$

is easy to compute once we have the two initial conditions. In addition, this solution has interesting interpretations in terms of moving wave motion.

If the reader recalls the parabolic case, we started solving diffusion problems when the space variable was bounded (by separation of variables) and then went on to solve the unbounded case (where $-\infty < x < \infty$) by the Fourier transform. In the hyperbolic case (wave problem), we will do the opposite. We start by solving the one-dimensional wave equation in free space. That is, the pure initial-value problem

(17.1)

$$\text{PDE} \qquad u_{tt} = c^2 u_{xx} \qquad -\infty < x < \infty \qquad 0 < t < \infty$$

$$\text{ICs} \qquad \begin{cases} u(x,0) = f(x) \\ u_t(x,0) = g(x) \end{cases} \qquad -\infty < x < \infty$$

We could solve this problem by using the Fourier transform (transforming x) or the Laplace transform (transforming t), but we will introduce yet a new technique (canonical coordinates), which will introduce the reader to several new and exciting ideas; we now solve problem (17.1). This technique is basically the same as the moving-coordinate method of Lesson 15.

D'Alembert's Solution to the One-Dimensional Wave Equation

We solve problem (17.1) by breaking it into steps.

STEP 1 [Replacing (x,t) by new *canonical coordinates* (ξ, η)]
To solve (17.1), it turns out that if we replace the two independent variables x and t by the two new space-time coordinates (ξ, η)

$$\xi = x + ct$$
$$\eta = x - ct$$

the PDE

$$u_{tt} = c^2 u_{xx}$$

is transformed into

$$u_{\xi\eta} = 0$$

This is really very easy to see, since a simple application of the chain rule gives

(17.2)
$$
\begin{aligned}
u_x &= u_\xi + u_\eta \\
u_t &= c(u_\xi - u_\eta) \\
u_{xx} &= u_{\xi\xi} + 2u_{\xi\eta} + u_{\eta\eta} \\
u_{tt} &= c^2(u_{\xi\xi} - 2u_{\xi\eta} + u_{\eta\eta})
\end{aligned}
$$

and, hence, substituting the expressions for u_{tt} and u_{xx} into the wave equation gives

$$u_{\xi\eta} = 0$$

This completes step 1.

STEP 2 (Solving the transformed equations)
The idea now is that this new equation can be solved easily by two straightforward integrations (first with respect to ξ and then with respect to η). Integration with respect to ξ gives

$$u_\eta(\xi,\eta) = \phi(\eta) \quad \text{(an arbitrary function of } \eta)$$

and, secondly, integration with respect to η gives

$$u(\xi,\eta) = \Phi(\eta) + \psi(\xi)$$

where $\Phi(\eta)$ is the antiderivative of $\phi(\eta)$, and $\psi(\xi)$ is any function of ξ. In other words, we can say that the general solution (all solutions) of

$$u_{\xi\eta} = 0$$

is

(17.3) $$u(\xi,\eta) = \phi(\eta) + \psi(\xi)$$

where $\phi(\eta)$, $\psi(\xi)$ are arbitrary functions of η and ξ, respectively. For example, the reader can check that the functions

$$u(\xi,\eta) = \sin \eta + \xi^2$$
$$u(\xi,\eta) = 1/\eta + \tan \xi$$
$$u(\xi,\eta) = \eta^2 + e^\xi$$

are all solutions of $u_{\xi\eta} = 0$. This completes step 2.

STEP 3. (transforming back to the original coordinates x and t)
To find the *general solution* (all solutions) to $u_{tt} = c^2 u_{xx}$, we substitute

$$\xi = x + ct$$
$$\eta = x - ct$$

into

$$u(\xi,\eta) = \phi(\eta) + \psi(\xi)$$

to get

(17.4) $$u(x,t) = \phi(x - ct) + \psi(x + ct)$$

This is the general solution of the wave equation, and it is interesting in that it physically represents the sum of *any* two *moving waves*, each moving in opposite directions with velocity c. For example, the functions

$$u(x,t) = \sin (x - ct) \quad \text{(one right-moving wave)}$$
$$u(x,t) = (x + ct)^2 \quad \text{(one left-moving wave)}$$
$$u(x,t) = \sin (x - ct) + (x + ct)^2 \quad \text{(two oppositely moving waves)}$$

would be three typical solutions.

Figure 17.1 illustrates the simple (right) moving wave

$$u(x,t) = e^{-(x-ct)^2}$$

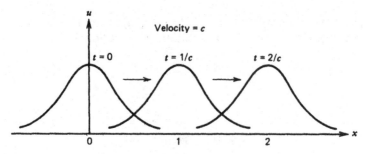

FIGURE 17.1 Right-moving wave $u(x, t) = e^{-(x - ct)^2}$

STEP 4 (Substituting the general solution into the ICs)

If the reader remembers ODE theory, the general strategy for solving initial-value problems was to find the general solution and then substitute this equation into the ICs in order to find the arbitrary constants. In our current problem, we have an analogous situation; in order to solve our initial-value problem, we first found the general solution

$$u(x,t) = \phi(x - ct) + \psi(x + ct)$$

to the PDE (which contains *two arbitrary functions*) and now substitute this expression into the two ICs

$$u(x,0) = f(x)$$
$$u_t(x,0) = g(x)$$

to find the arbitrary functions ϕ, ψ. Doing this, we get

(17.5)
$$\phi(x) + \psi(x) = f(x)$$
$$-c\phi'(x) + c\psi'(x) = g(x)$$

We now integrate the second equation of (17.5) to get a new expression in $\phi(x)$ and $\psi(x)$ [and then algebraically solve for $\phi(x)$ and $\psi(x)$ from the two equations].

Carrying out this integration on the second equation of (17.5) by integrating from x_0 to x, we get

(17.6)
$$-c\phi(x) + c\psi(x) = \int_{x_0}^{x} g(\xi)\, d\xi + K$$

and so if we algebraically solve for $\phi(x)$ and $\psi(x)$ from the first equation of (17.5) and (17.6), we have

(17.7)
$$\phi(x) = \frac{1}{2} f(x) - \frac{1}{2c} \int_{x_0}^{x} g(\xi) \, d\xi$$

$$\psi(x) = \frac{1}{2} f(x) + \frac{1}{2c} \int_{x_0}^{x} g(\xi) \, d\xi$$

and, hence, the solution to our problem (17.1) is

(17.8) $$u(x,t) = \frac{1}{2} [f(x - ct) + f(x + ct)] + \frac{1}{2c} \int_{x-ct}^{x+ct} g(\xi) \, d\xi$$

This is what we were aiming for, and it is called the **D'Alembert solution** to (17.1). The reader can play around with the limits on the integral to verify for himself or herself that they come out to $x - ct$ and $x + ct$; this completes the problem.

Before we complete the lesson, however, we present a few examples to show how the D'Alembert solution is applied to specific problems.

Examples of the D'Alembert Solution

1. Motion of an Initial Sine Wave

Consider the initial conditions

$$u(x,0) = \sin x$$
$$u_t(x,0) = 0$$

The initial sine wave would have the solution

$$u(x,t) = \frac{1}{2} [\sin (x - ct) + \sin (x + ct)]$$

This can be interpreted as dividing the initial shape $u(x,0) = \sin x$ into two equal parts

$$\frac{\sin x}{2} \quad \text{and} \quad \frac{\sin x}{2}$$

and then adding the two resultant waves as one moves to the left and the other to the right (each with velocity c). The reader might try to imagine the resulting wave form.

2. Motion of a Simple Square Wave

In this case, if we start with the initial conditions

$$u(x,0) = \begin{cases} 1 & -1 < x < 1 \\ 0 & \text{everywhere else} \end{cases}$$

$$u_t(x,0) = 0$$

and then split the initial wave into two half waves and let each move in an opposite direction, we arrive at the wave motion shown in Figure 17.2.

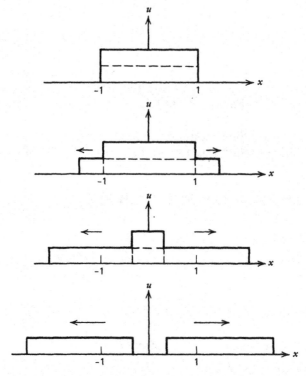

FIGURE 17.2 Initial wave decomposed into two traveling waves.

3. Initial Velocity Given

Suppose now the initial position of the string is at *equilibrium* and we impose an *initial velocity* (as in a piano string) of sin x

$$u(x,0) = 0$$
$$u_t(x,0) = \sin x$$

Here, the solution would be

$$u(x,t) = \frac{1}{2c} \int_{x-ct}^{x+ct} \sin \xi \, d\xi$$

$$= \frac{1}{2c} [\cos (x + ct) - \cos (x - ct)]$$

which represents the sum of two moving cosine waves. The reader should ask himself or herself if this solution is reasonable.

This completes Lesson 17; in Lesson 18, we show how the D'Alembert solution can be used to draw useful interpretations in the xt-plane.

NOTES

1. Note that a second-order PDE has two arbitrary *functions* in its general solution, whereas the general solution of a second-order ODE has two arbitrary *constants*. In other words, there are more solutions to a PDE than to an ODE.
2. The general technique of changing coordinate systems in a PDE in order to find a simpler equation is common in PDE theory. The new coordinates (ξ, η) in this problem are known as *canonical coordinates*, and we will discuss them further in later lessons, especially when we study hyperbolic problems.
3. The strategy of finding the *general solution* to a PDE and then substituting it into the boundary and initial conditions is *not* a common technique in solving PDEs. The solution discussed in this lesson is the only one that utilizes this strategy. Usually, we cannot find the general solution to the PDE, and even if we could, it is generally too complicated to substitute it into the side conditions.

PROBLEMS

1. Verify that the general D'Alembert solution (17.8) satisfies the initial-value problem (17.1).
2. Substitute (17.7) into the general solution

$$u(x,t) = \phi(x - ct) + \psi(x + ct)$$

to get the D'Alembert solution.
3. What is the solution to the initial-value problem

$$\text{PDE} \quad u_{tt} = u_{xx} \quad -\infty < x < \infty \quad 0 < t < \infty$$

$$\text{ICs} \quad \begin{cases} u(x,0) = e^{-x^2} \\ u_t(x,0) = 0 \end{cases} \quad -\infty < x < \infty$$

What does the solution look like for various values of time?

4. What is the solution of the initial-value problem:

$$\text{PDE} \qquad u_{tt} = u_{xx} \qquad -\infty < x < \infty \qquad 0 < t < \infty$$

$$\text{ICs} \qquad \begin{cases} u(x,0) = 0 \\ u_t(x,0) = xe^{-x^2} \end{cases} \qquad -\infty < x < \infty$$

Graph the solution $u(x,t)$ for various values of time.

5. Algebraically solve for $\phi(x)$ and $\psi(x)$ from the first equation in (17.5) and (17.6) to arrive at (17.7).

OTHER READING

Advanced Mathematics for Engineers by C. Wylie. McGraw-Hill, 1958. A readable presentation of some of the elementary ideas on PDEs.

More on the D'Alembert Solution

PURPOSE OF LESSON: To illustrate how the D'Alembert solution can be used to find the wave motion of a *semi-infinite*-string problem

$$\text{PDE} \qquad u_{tt} = c^2 u_{xx} \qquad 0 < x < \infty \qquad 0 < t < \infty$$

$$\text{BC} \qquad u(0,t) = 0 \qquad 0 < t < \infty$$

$$\text{ICs} \qquad \begin{cases} u(x,0) = f(x) \\ u_t(x,0) = g(x) \end{cases} \qquad 0 \le x < \infty$$

In addition, the D'Alembert solution is interpreted in the *xt*-plane.

In the previous lesson, we found that the expression

$$(18.1) \qquad u(x,t) = \frac{1}{2}[f(x - ct) + f(x + ct)] + \frac{1}{2c}\int_{x-ct}^{x+ct} g(\xi)\, d\xi$$

describes the displacement u of an infinite string in terms of its initial displacement $u(x,0) = f(x)$ and velocity $u_t(x,0) = g(x)$. This lesson will show the reader some interesting interpretations of this equation in the *xt*-plane (space-time plane) and how the equation can be modified to find the solution of the vibrating *semi-infinite* string.

We start with our interpretation of (18.1) in the *xt*-plane.

The Space-Time Interpretation of D'Alembert's Solution

We proved in the last lesson the solution of the pure initial-value problem

$$\text{PDE} \qquad u_{tt} = c^2 u_{xx} \qquad -\infty < x < \infty \qquad 0 < t < \infty$$

$$(18.2)$$

$$\text{ICs} \qquad \begin{cases} u(x,0) = f(x) \\ u_t(x,0) = g(x) \end{cases} \qquad -\infty < x < \infty$$

is given by

$$u(x,t) = \frac{1}{2}\left[f(x - ct) + f(x + ct)\right] + \frac{1}{2c}\int_{x-ct}^{x+ct} g(\xi)\,d\xi$$

We now present an interpretation of this solution in the xt-plane looking at two specific cases.

CASE 1 (Initial position given; initial velocity zero)
Suppose the string has initial conditions

$$u(x,0) = f(x)$$
$$u_t(x,0) = 0$$

Here, the D'Alembert solution is

$$u(x,t) = \frac{1}{2}\left[f(x - ct) + f(x + ct)\right]$$

and the solution u at a point (x_0, t_0) can be interpreted via Figure 18.1 as being the average of the initial displacement $f(x)$ at the points $(x_0 - ct_0, 0)$ and $(x_0 + ct_0, 0)$ found by backtracking along the lines

$$x - ct = x_0 - ct_0$$
$$x + ct = x_0 + ct_0$$ (characteristic curves)

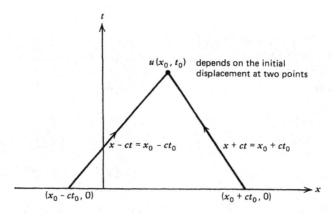

FIGURE 18.1 Interpretation of $u(x, t) = \dfrac{1}{2}[f(x - ct) + f(x + ct)]$ in the xt-plane.

For example, using this interpretation, the initial-value problem

$$\text{PDE} \qquad u_{tt} = c^2 u_{xx} \qquad -\infty < x < \infty \qquad 0 < t < \infty$$

(18.3)

$$\text{ICs} \quad \begin{cases} u(x,0) = \begin{cases} 1 & -1 < x < 1 \\ 0 & \text{everywhere else} \end{cases} \\ u_t(x,0) = 0 \end{cases}$$

would give us the solution in the xt-plane shown in Figure 18.2.

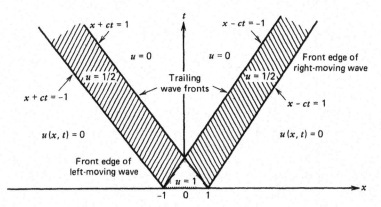

FIGURE 18.2   Solution of the initial-value problem (18.1) in the xt-plane.

Figure 18.2 is the xt-plane version of the solution graphed in Figure 17.2 in the previous lesson.

We now interpret the D'Alembert solution when the initial position is zero, but the velocity is arbitrary.

CASE 2   (Initial displacement zero; velocity arbitrary)
Consider now the IC

$$u(x,0) = 0$$
$$u_t(x,0) = g(x)$$

Here, the solution is

$$u(x,t) = \frac{1}{2c} \int_{x-ct}^{x+ct} g(\xi) \, d\xi$$

and, hence, the solution u at (x_0, t_0) can be interpreted as integrating the initial velocity between $x_0 - ct_0$ and $x_0 + ct_0$ on the initial line $t = 0$ (Figure 18.3).

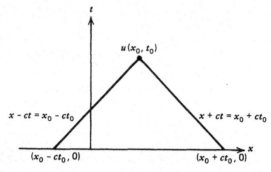

FIGURE 18.3 Interpretation of initial velocity in the *xt*-plane.

Again, using this interpretation, the solution to the initial-value problem

(18.4)

$$\text{PDE} \qquad u_{tt} = c^2 u_{xx} \qquad -\infty < x < \infty \qquad 0 < t < \infty$$

$$\text{ICs} \qquad \begin{cases} u(x,0) = 0 \\ u_t(x,0) = \begin{cases} 1 & -1 < x < 1 \\ 0 & \text{everywhere else} \end{cases} \end{cases}$$

has a solution in the *xt*-plane illustrated in Figure 18.4.

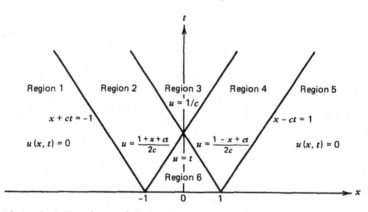

FIGURE 18.4 Solution to problem (18.4) in the *xt*-plane.

Problem 18.4 corresponds to imposing an initial *impulse* (velocity = 1) on the string for $-1 < x < 1$ and watching the resulting wave motion (as in the piano string). To find the displacement, we compute the D'Alembert solution

$$u(x,t) = \frac{1}{2c} \int_{x-ct}^{x+ct} g(\xi) \, d\xi$$

$$= \frac{1}{2c} \int_{x-ct}^{x+ct} 0 \, d\xi \qquad (x,t)\varepsilon \text{ Region 1}$$

$$= \frac{1}{2c} \int_{-1}^{x+ct} d\xi \qquad (x,t)\varepsilon \text{ Region 2}$$

(18.5)
$$= \frac{1}{2c} \int_{-1}^{1} d\xi \qquad (x,t)\varepsilon \text{ Region 3}$$

$$= \frac{1}{2c} \int_{x-ct}^{1} d\xi \qquad (x,t)\varepsilon \text{ Region 4}$$

$$= \frac{1}{2c} \int_{x-ct}^{x+ct} 0 \, d\xi \qquad (x,t)\varepsilon \text{ Region 5}$$

$$= \frac{1}{2c} \int_{x-ct}^{x+ct} d\xi \qquad (x,t)\varepsilon \text{ Region 6}$$

This solution is graphed at various values of time in Figure 18.5.

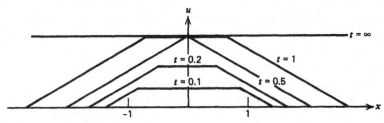

FIGURE 18.5   Solution of problem (18.4) for various values of time.

This completes our interpretation of the D'Alembert solution in the xt-plane. In the remainder of the lesson, we will solve the initial-boundary-value problem for the semi-infinite string

$$\text{PDE} \qquad u_{tt} = c^2 u_{xx} \qquad 0 < x < \infty \qquad 0 < t < \infty$$

(18.6)
$$\text{BC} \qquad u(0,t) = 0 \qquad 0 < t < \infty$$

$$\text{ICs} \qquad \begin{cases} u(x,0) = f(x) \\ u_t(x,0) = g(x) \end{cases} \qquad 0 < x < \infty$$

by *modifying* the D'Alembert formula.

Solution of the Semi-Infinite String via the D'Alembert Formula

The object now is to find the wave motion of the vibrating string whose left end is fixed at *zero* and has given initial conditions. To find the solution of (18.6), we proceed in a manner similar to that used with the infinite string, which is to find the general solution to the PDE

$$u(x,t) = \phi(x - ct) + \psi(x + ct)$$

If we now substitute this general solution into the initial conditions as we did in Lesson 17, we arrive at (same equations)

(18.7)

$$\phi(x - ct) = \frac{1}{2} f(x - ct) - \frac{1}{2c} \int_{x_0}^{x-ct} g(\xi)\, d\xi$$

$$\psi(x + ct) = \frac{1}{2} f(x + ct) + \frac{1}{2c} \int_{x_0}^{x+ct} g(\xi)\, d\xi$$

We now have a problem we didn't encounter when dealing with the *infinite* string. Since we are looking for the solution $u(x,t)$ everywhere in the first quadrant ($x > 0$, $t > 0$) of the xt-plane, it is obvious that we must find

$$\phi(x - ct) \quad \text{for all } -\infty < x - ct < \infty$$
$$\psi(x + ct) \quad \text{for all } 0 < x + ct < \infty$$

Unfortunately, the first equation of (18.7) only gives us $\phi(x - ct)$ for $x - ct \geq 0$, since our initial data $f(x)$ and $g(x)$ are only known for *positive* arguments.

As long as $x - ct \geq 0$, we have no problem, since we can substitute (18.7) into the general solution $u(x,t) = \phi(x - ct) + \psi(x + ct)$ to get

$$u(x,t) = \phi(x - ct) + \psi(x + ct)$$
$$= \frac{1}{2} [f(x - ct) + f(x + ct)] + \frac{1}{2c} \int_{x-ct}^{x+ct} g(\xi)\, d\xi$$

The question is, what to do when $x < ct$? This is where the BC $u(0,t) = 0$ comes into use. When $x < ct$, we use the BC of the problem to find $\phi(x - ct)$. Substituting the general solution $u(x,t) = \phi(x - ct) + \psi(x + ct)$ into the BC $u(0,t) = 0$ gives

$$\phi(-ct) = -\psi(ct)$$

and, hence, by functional substitution

$$\phi(x - ct) = -\frac{1}{2} f(ct - x) - \frac{1}{2c} \int_{x_0}^{ct-x} g(\xi)\, d\xi + K$$

Substituting this value of ϕ into the general solution

$$u(x,t) = \phi(x - ct) + \psi(x + ct)$$

gives

$$u(x,t) = \frac{1}{2}[f(x + ct) - f(ct - x)] + \frac{1}{2c}\int_{ct-x}^{x+ct} g(\xi)\, d\xi \qquad 0 < x < ct$$

and, hence, combining the solutions for $x < ct$ and $x > ct$, we have our desired result

(18.8) $$u(x,t) = \begin{cases} \frac{1}{2}[f(x - ct) + f(x + ct)] + \frac{1}{2c}\int_{x-ct}^{x+ct} g(\xi)\, d\xi & x \geq ct \\[4mm] \frac{1}{2}[f(x + ct) - f(ct - x)] + \frac{1}{2c}\int_{ct-x}^{x+ct} g(\xi)\, d\xi & x < ct \end{cases}$$

See Figure 18.6.

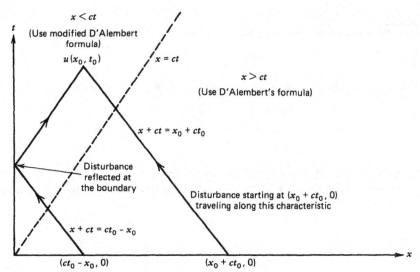

FIGURE 18.6   Interpretation of the semi-infinite string in the xt-plane.

This completes our lesson; we will examine the interpretation of equation (18.8) in the notes.

NOTES

1. Equation (18.8) is what we would expect from the semi-infinite string with BC $u(0,t) = 0$. For $x \geq ct$, the solution is the same as the D'Alembert solution for the infinite wave, while for $x < ct$, the solution $u(x,t)$ is modified as a result of the wave *reflecting* from the boundary. (The sign of the wave is changed when it's reflected.)

2. Solution (18.8) would not be the same if the BC $u(0,t) = 0$ were changed. Solutions can *also* be found with other BCs, such as

$$\text{or} \quad \begin{array}{l} u(0,t) = f(t) \\ u_x(0,t) = 0 \end{array}$$

The reader can consult the reference in the reading list for additional information.

3. The straight lines

$$x + ct = \text{constant}$$
$$x - ct = \text{constant}$$

are known as **characteristics**, and it is along these lines that disturbances are propagated. Characteristics are generally associated with hyperbolic equations.

PROBLEMS

1. Solve the semi-infinite string problem

$$\text{PDE} \quad u_{tt} = u_{xx} \quad 0 < x < \infty \quad 0 < t < \infty$$

$$\text{BC} \quad u(0,t) = 0 \quad 0 < t < \infty$$

$$\text{ICs} \quad \begin{cases} u(x,0) = xe^{-x^2} \\ u_t(x,0) = 0 \end{cases} \quad 0 < x < \infty$$

Draw the solution for various values of time.

2. The solution of the semi-infinite string problem in problem 1 can also be found by

(a) extending the IC to the whole real axis $-\infty < x < \infty$ via

$$u(x,0) = -xe^{-x^2} \quad -\infty < x < 0$$
$$u_t(x,0) = 0 \quad -\infty < x < 0$$

(b) averaging the two left- and right-moving waves as we did in the previous lesson

(c) looking at the solution for $x \geq 0$

Use this idea to graph the solution (for various values of t) for the IC shown in the following diagram.

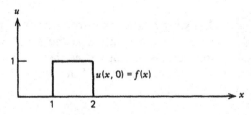

3. Solve the semi-infinite string problem

$$\text{PDE} \qquad u_{tt} = c^2 u_{xx} \qquad 0 < x < \infty \qquad 0 < t < \infty$$

$$\text{BC} \qquad u_x(0,t) = 0 \qquad 0 < t < \infty$$

$$\text{ICs} \qquad \begin{cases} u(x,0) = f(x) \\ u_t(x,0) = 0 \end{cases} \qquad 0 \leq x < \infty$$

in a manner analogous to the way the semi-infinite string problem was solved in the lesson; what is the interpretation of this problem?

4. Suppose the vibration of a string is described by $u_{tt} = u_{xx}$ and has an initial displacement as given by the following diagram.

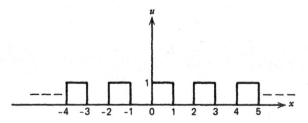

Assuming the initial velocity $u_t(x,0) = 0$, describe the solution of this problem in the xt-plane. Note that in this problem, the IC $u(x,0)$ is discontinuous.

OTHER READING

Techniques in Partial Differential Equations by C. Chester. McGraw-Hill, 1971. Chapter 2 discusses many variations of D'Alembert's equation, including the semi-infinite problem discussed in this lesson.

Boundary Conditions Associated with the Wave Equation

PURPOSE OF LESSON: To illustrate how the wave equation (with a bounded space variable)

$$u_{tt} = c^2 u_{xx} \qquad 0 < x < L \qquad 0 < t < \infty$$

is generally associated with one of *three* general kinds of BCs

1. *Controlled end points* (first kind)

$$u(0,t) = g_1(t)$$
$$u(L,t) = g_2(t)$$

2. *Force specified on the boundaries* (second kind)

$$u_x(0,t) = g_1(t)$$
$$u_x(L,t) = g_2(t)$$

3. *Elastic attachment* (third kind)

$$u_x(0,t) - \gamma_1 u(0,t) = g_1(t)$$
$$u_x(L,t) - \gamma_2 u(L,t) = g_2(t)$$

(or a mixture of these) and to illustrate the nature of solutions associated with these problems.

So far, the only kind of wave motion we have discussed is the one-dimensional transverse vibrations of a string. The reader should realize that this is only the tip of the iceberg as far as wave motion is concerned. A few other types of important vibrations are:
1. Sound waves (longitudinal waves)
2. Electromagnetic waves of light and electricity
3. Vibrations in solids (longitudinal, transverse, and torsional)
4. Probability waves in quantum mechanics
5. Water waves (transverse waves)

6. Vibrating string (transverse waves)

The purpose of this lesson is to discuss some of the various types of BCs that are associated with physical problems of this kind. Here, we will stick to *one-dimensional problems* where the BCs (linear ones) are generally grouped into one of three kinds:

1. *Controlled end points* (first kind)

$$u(0,t) = g_1(t)$$
$$u(L,t) = g_2(t)$$

2. *Force given on the boundaries* (second kind)

$$u_x(0,t) = g_1(t)$$
$$u_x(L,t) = g_2(t)$$

3. *Elastic attachment on the boundaries* (third kind)

$$u_x(0,t) - \gamma_1 u(0,t) = g_1(t)$$
$$u_x(L,t) - \gamma_2 u(L,t) = g_2(t)$$

We start by discussing BCs of the first kind.

1. Controlled End Points

We are now involved with problems like

$$\text{PDE} \qquad u_{tt} = c^2 u_{xx} \qquad 0 < x < 1 \qquad 0 < t < \infty$$

(19.1) $\qquad$ BCs $\qquad \begin{cases} u(0,t) = g_1(t) \\ u(1,t) = g_2(t) \end{cases} \qquad 0 < t < \infty$

$$\text{ICs} \qquad \begin{cases} u(x,0) = f(x) \\ u_t(x,0) = g(x) \end{cases} \qquad 0 \le x \le 1$$

where we *control* the end points so that they move in a given manner (Figure 19.1).

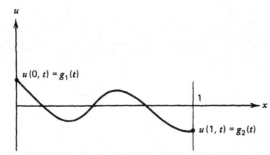

FIGURE 19.1 Controlling the ends of a vibrating string.

A typical problem of this kind would involve suddenly twisting (at $t = 1$) the right end of a fastened rod so many degrees and observing the resulting torsional vibration (Figure 19.2).

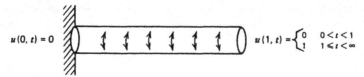

FIGURE 19.2 Torsional vibrations of a rod.

In the area of mathematical control theory, an important problem involves *determining* the boundary function $g_2(t)$, so that a vibrating string can be shaken to zero in minimum time.

2. Force Given on the Boundaries

Inasmuch as the *vertical forces* on the string at the left and right ends are given by $Tu_x(0,t)$ and $Tu_x(L,t)$, respectively, by allowing the ends of the string to slide vertically on frictionless sleeves, the boundary conditions become

$$u_x(0,t) = 0$$
$$u_x(L,t) = 0$$

See Figure 19.3.

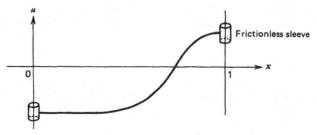

FIGURE 19.3 Free BC on the string.

Boundary conditions similar to these are presented in the following two examples:

(a) Free end of a longitudinally vibrating spring
Consider a vibrating spring with the bottom end unfastened (Figure 19.4).

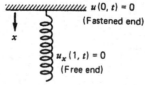

FIGURE 19.4 Free end of a vibrating spring

(b) Forced end of a vibrating spring

If a force of $v(t)$ dynes is applied at the end $x = 1$ (a positive force is measured downward), then the BC would be

$$u_x(1,t) = \frac{1}{k}v(t) \qquad (k \text{ is Young's modulus})$$

In the case of a forced BC, the ends of the string (or spring) are not *required* to maintain a given position, but the force that's applied tends to move the boundaries in the given direction. Physical problems like these come about in physics when an electric field (a force) is applied to vibrating electrons.

3. Elastic Attachment on the Boundaries

Consider finally a violin string whose ends are attached to an elastic arrangement like the one shown in Figure 19.5.

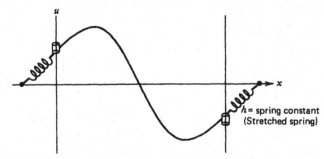

FIGURE 19.5   Diagram illustrating elastic attachment.

Here, the spring attachments at each end give rise to vertical forces proportional to the displacements

$$\text{Displacement at the left end} = u(0,t)$$
$$\text{Displacement at the right end} = u(L,t)$$

Setting the vertical tensions of the spring at the two ends

$$\text{Upward tension at the left end} = Tu_x(0,t)$$
$$\text{Upward tension at the right end} = -Tu_x(L,t)$$

$(T = \text{string tension})$

equal to these displacements (multiplied by the spring constant h) gives us our desired BCs

(19.2)

$$u_x(0,t) = \frac{h}{T}u(0,t)$$

$$u_x(L,t) = -\frac{h}{T}u(L,t)$$

Note that $u(0,t)$ positive means that $u_x(0,t)$ is positive, while if $u(L,t)$ is positive, then $u_x(L,t)$ is negative. We can rewrite these two homogeneous BCs as

(19.3)

$$u_x(0,t) - \frac{h}{T} u(0,t) = 0$$

$$u_x(L,t) + \frac{h}{T} u(L,t) = 0$$

If the two spring attachments are displaced according to the functions $\theta_1(t)$ and $\theta_2(t)$, we would have the *nonhomogeneous* BCs

(19.4)

$$u_x(0,t) = \frac{h}{T} [u(0,t) - \theta_1(t)]$$

$$u_x(L,t) = -\frac{h}{T} [u(L,t) - \theta_2(t)]$$

See Figure 19.6.

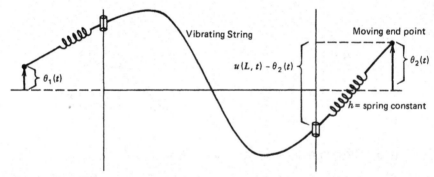

FIGURE 19.6   Diagram illustrating nonhomogeneous elastic BCs.

This completes our discussion of the most common types of BCs associated with hyperbolic problems. In the next few lessons, we will solve problems having BCs similar to these.

NOTES

1. Another BC not discussed in this lesson occurs when the vibrating string experiences a force at the ends proportional to the string velocity (and in the opposite direction). Here, we have the BC (at the left end)

$$Tu_x(0,t) = -\beta u_t(0,t)$$

2. A nonlinear elastic attachment at the left end of the string would be

$$Tu_x(0,t) = \phi[u(0,t)]$$

where $\phi(u)$ is an arbitrary function of u; for example,

$$Tu_x(0,t) = -hu^3(0,t)$$

says that the restoring force at the left end of the string is proportional to the cube of the displacement and not to u (as it was in the linear case with Hooke's law).

3. If a mass m is attached to the lower end of a longitudinally vibrating string, the BC would be

$$mu_{tt}(L,t) = -ku_x(L,t) + mg$$

PROBLEMS

1. From your intuition of the various kinds of BCs, draw a rough sketch of the solution to

$$\text{PDE} \qquad u_{tt} = c^2 u_{xx} \qquad 0 < x < 1 \qquad 0 < t < \infty$$

$$\text{BCs} \qquad \begin{cases} u(0,t) = 0 \\ u(1,t) = \sin t \end{cases} \qquad 0 < t < \infty$$

$$\text{ICs} \qquad \begin{cases} u(x,0) = 0 \\ u_t(x,0) = 0 \end{cases} \qquad 0 \leq x \leq 1$$

for various values of time.

2. Draw a rough sketch of the solution to

$$\text{PDE} \qquad u_{tt} = u_{xx} \qquad 0 < x < 1 \qquad 0 < t < \infty$$

$$\text{BCs} \qquad \begin{cases} u(0,t) = 0 \\ u_x(1,t) = 0 \end{cases} \qquad 0 < t < \infty$$

$$\text{ICs} \qquad \begin{cases} u(x,0) = \sin(\pi x/2) \\ u_t(x,0) = 0 \end{cases} \qquad 0 \leq x \leq 1$$

for various values of time. Can you guess the solution to this problem?

3. What is the general nature of the BC

$$u_x(0,t) = \frac{h}{T}[u(0,t) - \theta_1(t)]$$

when

(a) $h \to \infty$
(b) $h \to 0$

Does this agree with your intuition?

4. Draw a rough sketch of the solution to

$$\text{PDE} \qquad u_{tt} = u_{xx} \qquad 0 < x < 1 \qquad 0 < t < \infty$$

$$\text{BCs} \quad \begin{cases} u(0,t) = 0 \\ u_x(1,t) = -u(1,t) \end{cases} \qquad 0 < t < \infty$$

$$\text{ICs} \quad \begin{cases} u(x,0) = x \\ u_t(x,0) = 0 \end{cases} \qquad 0 \leqslant x \leqslant 1$$

OTHER READING

Analysis and Solution of Partial Differential Equations by R. L. Street. Brooks/Cole, 1973. An excellent text with an extensive chapter (Chapter 2) on initial and boundary conditions.

The Finite Vibrating String (Standing Waves)

PURPOSE OF LESSON: To show how transverse vibrations of a finite string described by the IBVP

$$\text{PDE} \qquad u_{tt} = \alpha^2 u_{xx} \qquad 0 < x < L \qquad 0 < t < \infty$$

$$\text{BCs} \qquad \begin{cases} u(0,t) = 0 \\ u(L,t) = 0 \end{cases} \qquad 0 < t < \infty$$

$$\text{ICs} \qquad \begin{cases} u(x,0) = f(x) \\ u_t(x,0) = g(x) \end{cases} \qquad 0 \leqslant x \leqslant L$$

can be found by the standard technique of separation of variables and to show how the solution $u(x,t)$ can be interpreted as the infinite sum

$$u(x,t) = \sum_{n=1}^{\infty} X_n(x)T_n(t)$$

of simple vibrations where the shape $X_n(x)$ of these fundamental vibrations are solutions (eigenfunctions) of a certain Sturm-Liouville boundary-value problem.

So far, we have studied the wave equation $u_{tt} = c^2 u_{xx}$ for the unbounded domain $-\infty < x < \infty$ and have found (D'Alembert's solution) solutions to be certain *traveling* waves (moving in opposite directions). When we study the same wave equation in a *bounded region* of space $0 < x < L$, we find that the waves no longer *appear* to be moving due to their repeated interaction with the boundaries and, in fact, often appear to be what are known as *standing waves*. For instance, consider what happens when a guitar string (fixed at both ends $x = 0,L$) described by the simple hyperbolic IBVP

$$\text{PDE} \qquad u_{tt} = \alpha^2 u_{xx} \qquad 0 < x < L \qquad 0 < t < \infty$$

$$\text{BCs} \qquad \begin{cases} u(0,t) = 0 \\ u(L,t) = 0 \end{cases} \qquad 0 < t < \infty$$

$$\text{ICs} \quad \begin{cases} u(x,0) = f(x) \\ u_t(x,0) = g(x) \end{cases} \quad 0 \leqslant x \leqslant L$$

is set in motion. What happens is that the traveling-wave solution to the PDE and IC keeps reflecting from the boundaries in such a way that the wave motion does not appear to be moving, but, in fact, appears to be vibrating in one position; for example, a few typical standing waves are shown in Figure 20.1.

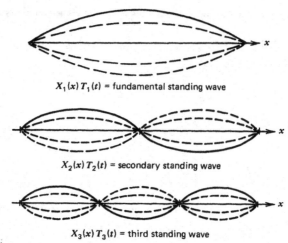

$X_1(x)\,T_1(t)$ = fundamental standing wave

$X_2(x)\,T_2(t)$ = secondary standing wave

$X_3(x)\,T_3(t)$ = third standing wave

FIGURE 20.1 Three typical standing waves $X(x)\,T(t)$.

If we *knew* the shapes $X_n(x)$ of these standing waves and how each one of them vibrated $T_n(t)$, then all we would have to do to find the solution of the vibrating guitar string is sum the simple vibrations $X_n(x)T_n(t)$

$$u(x,t) = \sum_{n=1}^{\infty} c_n X_n(x) T_n(t)$$

in such a way (find the coefficients c_n) that the sum agrees with the ICs $u(x,0) = f(x)$, $u_t(x,0) = g(x)$ when $t = 0$.

We will now solve the guitar-string problem by the method of separation of variables.

Separation-of-Variables Solution to the Finite Vibrating String

To solve the IBVP

$$\text{PDE} \quad u_{tt} = \alpha^2 u_{xx} \quad 0 < x < L \quad 0 < t < \infty$$

(20.1)
$$\text{BCs} \quad \begin{cases} u(0,t) = 0 \\ u(L,t) = 0 \end{cases} \quad 0 < t < \infty$$

$$\text{ICs} \quad \begin{cases} u(x,0) = f(x) \\ u_t(x,0) = g(x) \end{cases} \quad 0 \leq x \leq L$$

we start by seeking standing-wave solutions to the PDE; that is, solutions of the form

$$u(x,t) = X(x)T(t)$$

Substituting this expression into the wave equation and separating variables gives us the two ODEs

$$T'' - \alpha^2 \lambda T = 0$$
$$X'' - \lambda X = 0$$

where the constant λ can now be any number $-\infty < \lambda < \infty$.

Investigating the solutions of these two ODEs for all different values of λ yields the diagram in Figure 20.2.

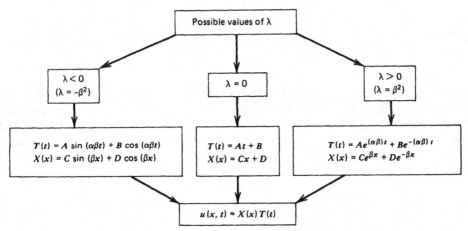

FIGURE 20.2 Standing-wave solutions for different values of λ.

The idea now is to *prune away* all those standing waves that either are unbounded as $t \to \infty$ or else yield only the zero solution when substituted into the BCs $u(0,t) = u(L,t) = 0$. It will be left as an exercise for the reader to verify that only negative values of λ give feasible (nonzero and bounded) solutions. Hence, our goal is to find the constants A, B, C, and D and the negative separation constant λ so that the expression

(20.2) $\quad u(x,t) = [C \sin(\beta x) + D \cos(\beta x)][A \sin(\alpha \beta t) + B \cos(\alpha \beta t)]$

satisfies the BCs. This will give us the collection of fundamental vibrations of the string, and the final goal will then be to sum them (the sum will still satisfy the PDE and the BC), so that the sum agrees with the IC when $t = 0$.

Plugging (20.2) into $u(0,t) = u(L,t) = 0$ gives

$$
\begin{aligned}
u(0,t) &= X(0)T(t) = D[A \sin (\alpha\beta t) + B \cos (\alpha\beta t)] = 0 \Rightarrow D = 0 \\
(20.3) \quad u(L,t) &= X(L)T(t) \\
&= C \sin (\beta L)[A \sin (\alpha\beta t) \\
&\quad + B \cos (\alpha\beta t)] = 0 \Rightarrow \sin (\beta L) = 0
\end{aligned}
$$

In other words, the separation constant β (we can forget about λ and find β) must satisfy $\sin (\beta L) = 0$ or

$$
\beta_n = \frac{n\pi}{L} \qquad n = 0, 1, 2, \ldots
$$

Note that if we choose $C = 0$ in the second equation of (20.3), we would get $X(x)T(t) = 0$. Hence, we have now found a *sequence* of simple vibrations (which we subscript with n)

(20.4) $\quad u_n(x,t) = X_n(x)T_n(t) = \sin (n\pi x/L)[a_n \sin (n\pi\alpha t/L) + b_n \cos (n\pi\alpha t/L)]$
$$
n = 1, 2, 3, \ldots
$$

or

$$
u_n(x,t) = R_n \sin (n\pi x/L) \cos [n\pi\alpha(t - \delta_n)/L]
$$

(where the constants a_n, b_n, R_n, and δ_n are arbitrary) all of which satisfy the wave equation and the BCs. The reader should be able to see that this sequence of functions constitutes a family of standing waves (which have the property that each point on the wave vibrates with the same frequency) whose shapes look like Figure 20.3.

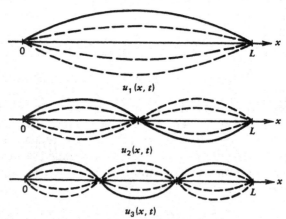

$u_1 (x, t)$

$u_2(x, t)$

$u_3(x, t)$

FIGURE 20.3   Standing waves $u_n(x,t) = X_n(x)T_n(t)$.

Since any sum of these vibrations is also a solution to the PDE and BCs (since the PDE and BCs are linear and homogeneous), we add them together in such a way that the resulting sum *also* agrees with the ICs. This will then be the solution to our problem. Substituting the sum

$$u(x,t) = \sum_{n=1}^{\infty} \sin (n\pi x/L)[a_n \sin (n\pi\alpha t/L) + b_n \cos (n\pi\alpha t/L)]$$

into the ICs

$$u(x,0) \doteq f(x)$$
$$u_t(x,0) = g(x)$$

gives the two equations

$$\sum_{n=1}^{\infty} b_n \sin (n\pi x/L) = f(x)$$

$$\sum_{n=1}^{\infty} a_n(n\pi\alpha/L) \sin (n\pi x/L) = g(x)$$

and using the orthogonality condition

$$\int_0^L \sin (m\pi x/L) \sin (n\pi x/L) \, dx = \begin{cases} 0 & m \neq n \\ L/2 & m = n \end{cases}$$

we can find the coefficients a_n and b_n

(20.5)
$$a_n = \frac{2}{n\pi\alpha} \int_0^L g(x) \sin (n\pi x/L) \, dx$$

$$b_n = \frac{2}{L} \int_0^L f(x) \sin (n\pi x/L) \, dx$$

So we're done, the solution is

(20.6)
$$u(x,t) = \sum_{n=1}^{\infty} \sin (n\pi x/L)[a_n \sin (n\pi\alpha t/L) + b_n \cos (n\pi\alpha t/L)]$$

where the coefficients a_n and b_n are given by (20.5). This completes the problem, but before we close, we will make a few useful observations.

NOTES

1. If the initial *velocity* of the string is zero, then the solution (20.6) takes the form

$$u(x,t) = \sum_{n=1}^{\infty} b_n \sin (n\pi x/L) \cos (n\pi\alpha t/L)$$

and has the following interpretation. Suppose we break the initial string position

$$u(x,0) = f(x)$$

into simple sine components

$$f(x) = \sum_{n=1}^{\infty} b_n \sin (n\pi x/L)$$

and let each sine term vibrate on its own according to

$$u_n(x,t) = b_n \sin (n\pi x/L) \cos (n\pi\alpha t/L)$$

(this is a fundamental vibration). If we now add each individual vibration of this type, we will get the solution to our problem. For example, suppose our initial string position $f(x)$ is

$$f(x) = \sin (\pi x/L) + 0.5 \sin (3\pi x/L) + 0.25 \sin (5\pi x/L)$$

The overall response to this IC would then be the sum of the responses to each term; that is,

$$u(x,t) = \sin (\pi x/L) \cos (\pi\alpha t/L) + 0.5 \sin (3\pi x/L) \cos (3\pi\alpha t/L)$$
$$+ 0.25 \sin (5\pi x/L) \cos (5\pi\alpha t/L)$$

2. The n-th term in the solution (20.6)

$$\sin (n\pi x/L) [a_n \sin (n\pi\alpha t/L) + b_n \cos (n\pi\alpha t/L)]$$

is called the **n-th mode of vibration** or the **n-th harmonic**. By using a trigonometric identity, we can write this harmonic as

$$R_n \sin (n\pi x/L) \cos [n\pi\alpha(t - \delta_n)/L]$$

where R_n and δ_n are the new arbitrary constants (amplitude and phase angle). This new form of the n-th mode is more useful for analyzing the vibrations. Note that the frequency ω_n (rad/sec) of the n-th mode is

$$\omega_n = \frac{n\pi\alpha}{L} = \frac{n\pi}{L}\sqrt{\frac{T}{\rho}}$$

(T,ρ are tension and density of the string, respectively)

Note, too, that this frequency is *n times* the fundamental frequency ($n = 1$). The property that all sound frequencies are multiples of a basic one is not shared by all types of vibrations. This has something to do with the pleasing sound of a violin or guitar string in contrast to a drumhead, where the higher-order frequencies are not multiple frequencies of the fundamental one.

PROBLEMS

1. Find the solution to the vibrating-string problem (20.1) if the ICs are given by

$$u(x,0) = \sin(\pi x/L) + 0.5\sin(3\pi x/L)$$
$$u_t(x,0) = 0$$

   Graph this solution for various values of time. Is the solution periodic in time? What is the period?

2. What is the solution of the vibrating-string problem (20.1) if the ICs are

$$u(x,0) = 0$$
$$u_t(x,0) = \sin(3\pi x/L)$$

   What does the graph of the solution look like for various values of time?

3. Show that for $\lambda \geqslant 0$ in Figure 20.2, the solutions $X(x)T(t)$ are either unbounded or zero.

4. What is the solution of the vibrating-string problem if the ICs are

$$u(x,0) = \sin(3\pi x/L)$$
$$u_t(x,0) = (3\pi\alpha/L)\sin(3\pi x/L)$$

5. A guitar string of length $L = 1$ is pulled upward at the middle so that it reaches height h. Assuming the position of the string is initially

$$u(x,0) = \begin{cases} 2hx & 0 \leqslant x \leqslant 0.5 \\ 2h(1-x) & 0.5 \leqslant x \leqslant 1 \end{cases}$$

Initial position of the string

what is the subsequent motion of the string if it is suddenly released?

6. Solve the damped vibrating-string problem

$$\text{PDE} \qquad u_{tt} = \alpha^2 u_{xx} - \beta u_t, \qquad 0 < x < 1 \qquad 0 < t < \infty$$

$$\text{BCs} \qquad \begin{cases} u(0,t) = 0 \\ u(1,t) = 0 \end{cases} \qquad 0 < t < \infty$$

$$\text{ICs} \qquad \begin{cases} u(x,0) = f(x) \\ u_t(x,0) = 0 \end{cases} \qquad 0 \le x \le 1$$

Does the solution seem reasonable? Does it satisfy the above PDE, BCs, and ICs?

7. How would you solve the *nonhomogeneous* PDE with given boundary and initial conditions

$$\text{PDE} \qquad u_{tt} = \alpha^2 u_{xx} + Kx \qquad 0 < x < 1 \qquad 0 < t < \infty$$

$$\text{BCs} \qquad \begin{cases} u(0,t) = 0 \\ u(1,t) = 0 \end{cases} \qquad 0 < t < \infty$$

$$\text{ICs} \qquad \begin{cases} u(x,0) = f(x) \\ u_t(x,0) = 0 \end{cases} \qquad 0 \le x \le 1$$

OTHER READING

Advanced Engineering Mathematics by C. Wylie. McGraw-Hill, 1970. A very readable text that contains many interesting examples; see in particular Chapter 7.

The Vibrating Beam (Fourth-Order PDE)

PURPOSE OF LESSON: To illustrate how higher-order PDEs come about in the study of vibrating-beam problems and to solve the problem of a vibrating beam with simply supported ends by separation of variables. It is also pointed out how the vibrations of the beam *compare* with the vibrations of the violin string.

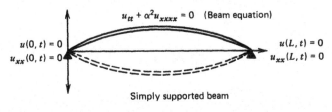

$$u_{tt} + \alpha^2 u_{xxxx} = 0 \quad \text{(Beam equation)}$$

$u(0, t) = 0$

$u_{xx}(0, t) = 0$

$u(L, t) = 0$

$u_{xx}(L, t) = 0$

Simply supported beam

The major difference between the transverse vibrations of a violin string and the transverse vibrations of a thin beam is that the beam offers resistance to bending. Without going into the mechanics of thin beams, we can show that this resistance is responsible for changing the wave equation to the fourth-order beam equation

(21.1) $$u_{tt} = -\alpha^2 u_{xxxx}$$

where

$\alpha^2 = K/\rho$

K = rigidity constant (the larger K, the more rigid the beam and the faster the vibrations)

ρ = linear density of the beam (mass/unit length).

The derivation of this equation can be found in reference 1 of Other Reading. Since this is the first time the reader has seen an application of PDEs higher than second order in this text, it will be useful to solve a typical vibrating-beam problem. Later, we will talk about other types of beam problems.

The Simply Supported Beam

Consider the small vibrations of a thin beam whose ends are simply fastened to two foundations. By "simply fastened," we mean that the ends of the beam are held stationary, but the slopes at the end points can move (the beam is held by a pin-type arrangement, Figure 21.1).

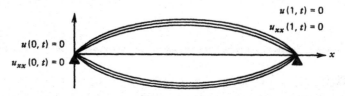

FIGURE 21.1 A simply supported beam.

It seems clear that the BCs at the ends of the beam should be

$$u(0,t) = 0$$
$$u(1,t) = 0$$

but what *isn't* so obvious is that the two BCs

$$u_{xx}(0,t) = 0$$
$$u_{xx}(1,t) = 0$$

also hold at the two ends. Using the theory of thin beams (see reference 1 of Other Reading), we can show that the *bending moment* of the beam is represented by u_{xx} and a simply fastened beam should have zero moments at the end points. Hence, the vibrating beam in Figure 21.1 can be described by the IBVP (α is set equal to one for simplicity)

$$\text{PDE} \qquad u_{tt} = -u_{xxxx} \qquad 0 < x < 1 \qquad 0 < t < \infty$$

(21.2)   BCs $\begin{cases} u(0,t) = 0 \\ u_{xx}(0,t) = 0 \\ u(1,t) = 0 \\ u_{xx}(1,t) = 0 \end{cases}$ $0 < t < \infty$

ICs $\begin{cases} u(x,0) = f(x) \\ u_t(x,0) = g(x) \end{cases}$ $0 \le x \le 1$

To solve this problem, we use the separation of variables method and look for arbitrary periodic solutions; that is, vibrations of the form

(21.3) $$u(x,t) = X(x)[A \sin (\omega t) + B \cos (\omega t)]$$

Note that by choosing the solution in the form (21.3), we are essentially saying that the separation constant in the separation-of-variables method has been chosen to be negative.

We now substitute equation (21.3) into the beam equation to get the ODE in $X(x)$.

$$X^{iv} - \omega^2 X = 0$$

which has the general solution

$$X(x) = C \cos \sqrt{\omega}x + D \sin \sqrt{\omega}x + E \cosh \sqrt{\omega}x + F \sinh \sqrt{\omega}x$$

To find the constants C, D, E, and F, we substitute this expression into the BCs, giving

$$\left.\begin{array}{l} u(0,t)=0 \Rightarrow X(0)T(t)=0 \Rightarrow X(0)=0 \quad \Rightarrow C+E=0 \\ u_{xx}(0,t)=0 \Rightarrow X''(0)T(t)=0 \Rightarrow X''(0)=0 \quad \Rightarrow -C+E=0 \end{array}\right\} \Rightarrow C=E=0$$

$$u(1,t) = 0 \Rightarrow D \sin \sqrt{\omega} + F \sinh \sqrt{\omega} = 0$$
$$u_{xx}(1,t) = 0 \Rightarrow -D \sin \sqrt{\omega} + F \sinh \sqrt{\omega} = 0$$

From these last two equations, we arrive at the expressions

$$F \sinh \sqrt{\omega} = 0$$
$$D \sin \sqrt{\omega} = 0$$

from which we can conclude

$$F = 0$$
$$\sin \sqrt{\omega} = 0 \Rightarrow \omega = (n\pi)^2 \quad n = 1, 2, \ldots$$

In other words, the *natural frequencies* of the simply supported beam are

$$\omega_n = (n\pi)^2$$

and the *fundamental solutions* u_n (solutions of the PDE and BCs) are

$$u_n(x,t) = X_n(x)T_n(t) = [a_n \sin (n\pi)^2 t + b_n \cos (n\pi)^2 t] \sin (n\pi x)$$

Now, since the PDE and BCs are linear and homogeneous, we can conclude that the sum

(21.4)
$$u(x,t) = \sum_{n=1}^{\infty} [a_n \sin (n\pi)^2 t + b_n \cos (n\pi)^2 t] \sin (n\pi x)$$

also satisfies the PDE and BCs. Hence, all that remains to do is choose the constants a_n and b_n in such a way that the ICs are satisfied. Substituting equation (21.4) into the ICs gives us

(21.5)
$$u(x,0) = f(x) = \sum_{n=1}^{\infty} b_n \sin (n\pi x)$$

$$u_t(x,0) = g(x) = \sum_{n=1}^{\infty} (n\pi)^2 a_n \sin (n\pi x)$$

and using the fact that the family $\{\sin (n\pi x)\}$ is orthogonal on the interval $[0,1]$ gives us

(21.6)
$$a_n = \frac{2}{(n\pi)^2} \int_0^1 g(x) \sin (n\pi x) \, dx$$

$$b_n = 2 \int_0^1 f(x) \sin (n\pi x) \, dx$$

Hence, the solution is given by (21.4), and a_n and b_n are given by (21.6).

In order for the reader to understand this problem, we present a simple example.

Sample Vibrating Beam

Consider the simply supported beam shown in Figure 21.2 with ICs

$$u(x,0) = \sin (\pi x) + 0.5 \sin (3\pi x)$$
$$u_t(x,0) = 0$$

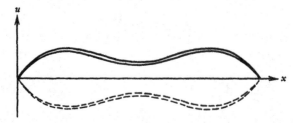

FIGURE 21.2 Simple vibrations of a simply supported beam.

We could find the solution by substituting the values of $f(x)$ and $g(x)$ into equation (21.6), but it seems easier to look at equations (21.5) and simply make the observation that

$$
\begin{aligned}
a_n &= 0 \quad \text{for all } n = 1, 2, \ldots \\
b_1 &= 1 \\
b_2 &= 0 \\
b_3 &= 0.5 \\
b_n &= 0 \quad n = 4, 5, \ldots
\end{aligned}
$$

Hence, the solution is

$$u(x,t) = \cos{(\pi^2 t)} \sin{(\pi x)} + 0.5 \cos{(9\pi^2 t)} \sin{(3\pi x)}$$

It is interesting to see how this solution compares with the vibrating string with the same ICs. If we look back to Lesson 20, we find that the solution to the vibrating-string problem is given by

$$u(x,t) = \cos{(\pi t)} \sin{(\pi x)} + 0.5 \cos{(3\pi t)} \sin{(3\pi x)}$$

In other words, the *vibrating beam* vibrates at higher frequencies than does the *vibrating string*. It would be interesting for the reader to imagine just how each of these vibrations looks. Note, however, that both higher frequencies are integer multiples of the fundamental frequencies.

NOTES

1. Beams are generally fastened in one of three ways
   (a) Free (unfastened)
   (b) Simply fastened
   (c) Rigidly fastened
   Some sketches are given in Figure 21.3 along with their BCs.
2. Another important vibrating-beam problem is the *cantilever-beam problem* shown in Figure 21.3. The solution to this vibrating beam is not the usual sum of products of sines and cosines, but due to the nonstandard BCs,

$$
\begin{aligned}
u(0,t) &= 0 \\
u_x(0,t) &= 0 \\
u_{xx}(1,t) &= 0 \\
u_{xxx}(1,t) &= 0
\end{aligned}
$$

we arrive at the more complicated solution

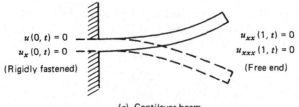

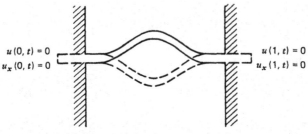

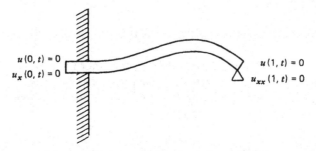

(c) Beam rigidly fastened at left; simply fastened at right.

FIGURE 21.3a–21.3c Typical beam problems.

$$u(x,t) = \sum_{n=1}^{\infty} X_n(x) [a_n \sin(\omega_n t) + b_n \cos(\omega_n t)]$$

where the eigenfunctions (basic shapes of vibrations) are given by linear combinations of sines, cosines, hyperbolic sines, and hyperbolic cosines. The solution to this problem can be found in reference 3 of Other Reading.

PROBLEMS

1. Solve the *cantilever-beam* problem

$$\text{PDE} \qquad u_{tt} + u_{xxxx} = 0 \qquad 0 < x < 1 \qquad 0 < t < \infty$$

$$\text{BCs} \quad \begin{cases} u(0,t) = 0 \\ u_x(0,t) = 0 \\ u_{xx}(1,t) = 0 \\ u_{xxx}(1,t) = 0 \end{cases} \quad 0 < t < \infty$$

$$\text{ICs} \quad \begin{cases} u(x,0) = f(x) \\ u_t(x,0) = g(x) \end{cases} \quad 0 \leq x \leq 1$$

HINT   Although the eigenfunctions $X_n(x)$ in this problem are not the usual sine functions, we can still use the Sturm-Liouville theory to say that the eigenfunctions are orthogonal on [0,1].

2.   What is the solution to the simply supported beam (at both ends) with ICs

$$\begin{aligned} u(x,0) &= \sin(\pi x) \\ u_t(x,0) &= \sin(\pi x) \end{aligned} \quad 0 \leq x \leq 1$$

3.   What is the solution to the simply supported beam problem with ICs

$$\begin{aligned} u(x,0) &= 1 - x^2 \\ u_t(x,0) &= 0 \end{aligned} \quad 0 \leq x \leq 1$$

4.   Let the left end ($x = 0$) of a beam be rigidly fastened to a wall and let the right end ($x = 1$) be simply fastened according to the BCs shown in Figure 21.3. Solve the beam problem with these BCs and tell how to find the *natural frequencies of vibration* of this beam. Knowing the natural frequencies of the beam is important, since various kinds of inputs of the same frequency can give rise to resonance.

OTHER READING

1.   *Analysis and Solution of Partial Differential Equations* by R. L. Street. Brooks-Cole, 1973. Chapter 5 contains a derivation of the vibrating-beam problem.

2.   *Mathematical Methods in Physics and Engineering* by J. W. Dettman. McGraw-Hill, 1962; Dover, 1988. This text contains a large section on the Sturm-Liouville problem.

3.   *Advanced Mathematics for Engineers* by C. R. Wylie. McGraw-Hill, 1961. This book contains the solution of the cantilever-beam problem.

Dimensionless Problems

PURPOSE OF LESSON: To show how boundary-value problems, initial-value problems, and other types of physical models can be written in dimensionless form. In this form, we replace the original variables of the problem by new dimensionless ones (they have no units).

By writing a problem in dimensionless form, specific equations from physical, chemistry, biology, and economics that originally look different become one and the same. For this reason, the mathematical study of PDEs generally doesn't concern itself with the physical parameters in the equations. It is up to the chemist, physicist, or biologist to transform his or her equation into those in the textbook.

The basic idea behind dimensional analysis is that by introducing new (dimensionless) variables in a problem, the problem becomes *purely mathematical* and contains none of the physical constants that originally characterized it. In this way, many different equations in physics, biology, engineering, and chemistry that contain special nuances via physical parameters are all transformed into the same simple form (Figure 22.1).

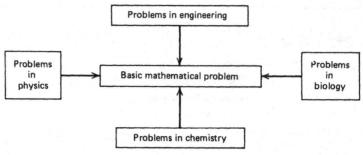

FIGURE 22.1 Several problems converted to one basic nondimensional form.

To see how this process works, let's consider a simple example.

Converting a Diffusion Problem to Dimensionless Form

Suppose we start with the initial-boundary-value problem where the temperature is initially $u(x,0) = \sin(\pi x/L)$, but the boundaries are then instantly raised to

T_1 and T_2. In other words, we have

$$\text{PDE} \quad u_t = \alpha^2 u_{xx} \quad 0 < x < L \quad 0 < t < \infty$$

(22.1) $\quad$ BCs $\quad \begin{cases} u(0,t) = T_1 \\ u(L,t) = T_2 \end{cases} \quad 0 < t < \infty$

$$\text{IC} \quad u(x,0) = \sin(\pi x/L) \quad 0 \leqslant x \leqslant L$$

See Figure 22.2.

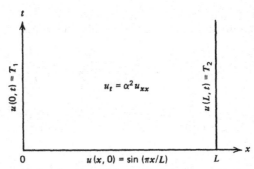

FIGURE 22.2 Domain for the heat-flow problem.

Our goal is to change problem (22.1) to a new equivalent formulation that has the properties
1. No physical parameters (like α) in the new equation
2. The initial and boundary conditions are simpler

To do this, we will introduce three new dimensionless variables U, ξ, and τ that take the place of u, x, and t, respectively

$$u \longrightarrow U \qquad \text{(dimensionless temperature)}$$
$$x \longrightarrow \xi \qquad \text{(dimensionless length)}$$
$$t \longrightarrow \tau \qquad \text{(dimensionless time)}$$

We carry out these three transformations one at a time for simplicity.

Transforming the Dependent Variable $u \to U$

We define $U(x,t)$ by

$$U(x,t) = \frac{u(x,t) - T_1}{T_2 - T_1}$$

It's clear that this new temperature $U(x,t)$ has no units, since we are dividing °C by °C. It's also clear why we chose $U(x,t)$. You can see that the new BCs for $U(x,t)$ at $x = 0$ and L will be $U(0,t) = 0$ and $U(L,t) = 1$. In fact, let's examine the new transformed problem for $U(x,t)$. With a little effort, we can see that the original problem (22.1) has now been transformed into

$$\begin{array}{llll} \text{PDE} & U_t = \alpha^2 U_{xx} & 0 < x < L & 0 < t < \infty \end{array}$$

(22.2) $\quad$ BCs $\quad \begin{cases} U(0,t) = 0 \\ U(L,t) = 1 \end{cases} \quad 0 < t < \infty$

$$\text{IC} \quad U(x,0) = \frac{\sin(\pi x/L) - T_1}{T_2 - T_1} \quad 0 \leqslant x \leqslant L$$

If we wished, we could stop here and solve for $U(x,t)$ and then solve for $u(x,t)$ from the formula

$$u(x,t) = T_1 + (T_2 - T_1)U(x,t)$$

Let's continue, however, and transform the *independent* variables x and t. Next, we transform the space variable x.

Transforming the Space Variable $x \rightarrow \xi$

It seems obvious how we should pick the dimensionless-space variable ξ. Since $0 \leqslant x \leqslant L$, we pick

$$\xi = x/L$$

By computing the derivatives

$$U_x = U_\xi \xi_x = \frac{1}{L} U_\xi$$

$$U_{xx} = \frac{1}{L^2} U_{\xi\xi}$$

it is clear that the *next* problem (in U, ξ, and t) is

$$\begin{array}{llll} \text{PDE} & U_t = (\alpha/L)^2 U_{\xi\xi} & 0 < \xi < 1 & 0 < t < \infty \end{array}$$

(22.3) $\quad$ BCs $\quad \begin{cases} U(0,t) = 0 \\ U(1,t) = 1 \end{cases} \quad 0 < t < \infty$

$$\text{IC} \quad U(\xi,0) = \frac{\sin(\pi\xi) - T_1}{T_2 - T_1} \quad 0 \leqslant \xi \leqslant 1$$

We are now two-thirds of the way toward our goal. The final step is to introduce a *dimensionless time* τ, so that the constant $[\alpha/L]^2$ disappears from the differential equation.

Transforming the Time Variable $t \rightarrow \tau$

How to introduce a new dimensionless time isn't quite so clear as choosing the first two variables. However, since our goal is to eliminate the constant $[\alpha/L]^2$ from the PDE, we proceed as follows:

1. Try a transformation of the form $\tau = ct$, where c is an unknown constant
2. Compute $u_t = u_\tau \tau_t = cu_\tau$
3. Substitute this derivative into the PDE to obtain

$$cu_\tau = [\alpha/L]^2 u_{\xi\xi}$$

and, hence, pick $c = [\alpha/L]^2$. This gives us our new time

$$\tau = [\alpha/L]^2 t$$

Using this transformation on our previous problem (22.3), we have the completely dimensionless problem (U, ξ, and τ)

$$\text{PDE} \qquad U_\tau = U_{\xi\xi} \qquad 0 < \xi < 1 \qquad 0 < \tau < \infty$$

(22.4) $\qquad$ BCs $\qquad \begin{cases} U(0,\tau) = 0 \\ U(1,\tau) = 1 \end{cases} \quad 0 < \tau < \infty$

$$\text{IC} \qquad U(\xi,0) = \phi(\xi) \qquad 0 \leqslant \xi \leqslant 1$$

where $\phi(\xi) = \dfrac{\sin(\pi\xi) - T_1}{T_2 - T_1}$

This new dimensionless problem has the following properties:

1. No parameters in the PDE
2. Simple BCs
3. IC hasn't essentially been changed (still a known function)
4. Problem is simpler and more compact than the original one

The solution to this problem can be found once and for all, so if a scientist transformed his or her original problem (22.1) to this dimensionless one (22.4) and found the answer $U(\xi,\tau)$ in a textbook or research journal, he or she could find the solution $u(x,t)$ to the original problem (22.1) by merely computing

$$u(x,t) = T_1 + (T_2 - T_1)U(x/L, \alpha^2 t/L^2)$$

This completes our discussion of transforming problems into dimensionless form. There aren't any set rules on how the new variables are defined; we more or less have to use physical intuition and try various possibilities.

We finish this lesson with a simple example of how to transform into dimensionless form, solve the new problem, and transform back to the original laboratory coordinates.

Example of Transforming a Hyperbolic Problem to Dimensionless Form

Consider the vibrating string

$$\text{PDE} \qquad u_{tt} = \alpha^2 u_{xx} \qquad 0 < x < L \qquad 0 < t < \infty$$

$$(22.5) \qquad \text{BCs} \qquad \begin{cases} u(0,t) = 0 \\ u(L,t) = 0 \end{cases} \qquad 0 < t < \infty$$

$$\text{ICs} \qquad \begin{cases} u(x,0) = \sin(\pi x/L) + 0.5 \sin(3\pi x/L) \\ u_t(x,0) = 0 \end{cases} \qquad 0 \leq x \leq L$$

By transforming the independent variables (no need to transform u) into to new ones

$$\xi = x/L \qquad \text{and} \qquad \tau = [\alpha/L]t$$

we get the new problem

$$\text{PDE} \qquad u_{\tau\tau} = u_{\xi\xi} \qquad 0 < \xi < 1 \qquad 0 < \tau < \infty$$

$$(22.6) \qquad \text{BCs} \qquad \begin{cases} u(0,\tau) = 0 \\ u(1,\tau) = 0 \end{cases} \qquad 0 < \tau < \infty$$

$$\text{ICs} \qquad \begin{cases} u(\xi,0) = \sin(\pi\xi) + 0.5 \sin(3\pi\xi) \\ u_\tau(\xi,0) = 0 \end{cases} \qquad 0 \leq \xi \leq 1$$

which has the solution

$$u(\xi,\tau) = \cos(\pi\tau) \sin(\pi\xi) + 0.5 \cos(3\pi\tau) \sin(3\pi\xi)$$

If we now transform back to coordinates x and t, we have the solution to our original problem (22.5)

$$u(x,t) = \cos(\pi\alpha t/L) \sin(\pi x/L) + 0.5 \cos(3\pi\alpha t/L) \sin(3\pi x/L)$$

NOTES

1. Dimensional analysis is especially important in numerical analysis, since most computer programs are written in a general form and don't solve problems with a great many physical parameters. Anyone using these programs must transform the problem into the form accepted by the program, solve the transformed problem, and then transform the numerical results back to his or her own coordinates.

2. Dimensional analysis allows mathematicians to work with PDEs without bothering with a lot of parameters and constants that are not relevant to the mathematical analysis.
3. It's not always necessary to transform *all* the variables into dimensionless form; sometimes only one or two have to be transformed.

PROBLEMS

1. Transform the vibrating string problem (22.5) into dimensionless form (22.6) by means of the transformations

$$\xi = x/L \qquad \tau = [\alpha/L]\, t$$

2. Find the dimensionless formulation for the problem

$$\text{PDE} \qquad u_t = \alpha^2 u_{xx} \qquad 0 < x < L$$

$$\text{BCs} \qquad \begin{cases} u(0,t) = T_1 \\ u(L,t) = 0 \end{cases} \qquad 0 < t < \infty$$

$$\text{IC} \qquad u(x,0) = T_2 \qquad 0 \leq x \leq L$$

3. Transform problem (22.1) into (22.2) by means of the change of variable

$$U(x,t) = \frac{u(x,t) - T_1}{T_2 - T_1}$$

4. Can you think of a *physical reason* why the new time variable $\tau = \alpha t$ would eliminate the parameter α^2 in the wave equation

$$u_{tt} = \alpha^2 u_{xx}$$

Remember what α means in terms of the velocity of the wave; intuition plays a major role in finding the most desirable new coordinates.
5. How could you pick a new space variable ξ so that v is eliminated in the equation

$$u_t + v u_x = 0$$

OTHER READING

Dimensional Analysis and Theory of Models by H. L. Langhaar. John Wiley & Sons, 1951. A well-written book that contains many more aspects of dimensional analysis than does this lesson.

Classification of PDEs (Canonical Form of the Hyperbolic Equation)

PURPOSE OF LESSON: To show how the second-order linear PDE in two independent variables

$$Au_{xx} + Bu_{xy} + Cu_{yy} + Du_x + Eu_y + Fu = G$$

(*A, B, C, D, E, F*, and *G* are functions of *x* and *y* and could be constants) can be categorized as either

1. Hyperbolic (if $B^2 - 4AC > 0$)
2. Parabolic (if $B^2 - 4AC = 0$)
3. Elliptic (if $B^2 - 4AC < 0$)

and to show how new coordinates $\xi = \xi(x,y)$ and $\eta = \eta(x,y)$ are introduced (in place of *x* and *y*) that simplify the equation. When this PDE is written in terms of the new coordinates ξ and η, it takes on one of three canonical forms (depending on whether $B^2 - 4AC$ is positive, zero, or negative, respectively)

1. $\begin{cases} u_{\xi\xi} - u_{\eta\eta} = \Psi(\xi, \eta, u, u_\xi, u_\eta) \\ u_{\xi\eta} = \Phi(\xi, \eta, u, u_\xi, u_\eta) \end{cases}$ $\begin{pmatrix} \text{two canonical forms for} \\ \text{the hyperbolic equation} \end{pmatrix}$

2. $u_{\eta\eta} = \Phi(\xi, \eta, u, u_\xi, u_\eta)$ $\begin{pmatrix} \text{the canonical form for} \\ \text{the parabolic equation} \end{pmatrix}$

3. $u_{\xi\xi} + u_{\eta\eta} = \Phi(\xi, \eta, u, u_\xi, u_\eta)$ $\begin{pmatrix} \text{the canonical form for} \\ \text{the elliptic equation} \end{pmatrix}$

where Φ is and Ψ are functions of the first derivatives u_ξ and u_η, the dependent variable *u*, and the new independent variables ξ and η. The exact functions Φ and Ψ depend, of course, on the original equation.

The reader may think offhand that a chapter dealing with the classification of PDEs should occur at the beginning of the book; this is probably true, and many books do begin by discussing this subject. However, it is true, too, that most students do not get very excited about studying something they know nothing about, and for that reason, we have waited until now to introduce the topic of classifying PDEs.

The purpose here is to *classify* the PDE

$$(23.1) \qquad Au_{xx} + Bu_{xy} + Cu_{yy} + Du_x + Eu_y + Fu = G$$

(where A, B, C, D, E, F, and G are, in general, functions of x and y) as
  1. Hyperbolic at a point (x_0, y_0) if $B^2(x_0, y_0) - 4A(x_0, y_0)C(x_0, y_0) > 0$.
  2. Parabolic at a point (x_0, y_0) if $B^2(x_0, y_0) - 4A(x_0, y_0)C(x_0, y_0) = 0$.
  3. Elliptic at a point (x_0, y_0) if $B^2(x_0, y_0) - 4A(x_0, y_0)C(x_0, y_0) < 0$.

and depending on which is true, to transform the equation into a corresponding canonical (simple) form. In order for the reader to understand the classification scheme, we first give four examples of hyperbolic, parabolic, and elliptic equations.

Examples of Hyperbolic, Parabolic, and Elliptic Equations*

  1. The *heat equation* $u_t = u_{xx}$ is a second-order linear equation of the form (23.1) with coefficients

$$A = 1 \quad B = 0 \quad C = 0 \quad D = 0$$
$$E = -1 \quad F = 0 \quad G = 0$$

so $B^2 - 4AC = 0$ for all x and t; hence, the equation is *parabolic* for all x and t. Note that we have called the time variable y in the general equation. In fact, we would have gotten the same result if we called the x in the heat equation the y in the general equation and then called the time variable t the x in the general equation.

  2. The *wave equation* $u_{tt} = u_{xx}$ is also of the form (23.1) with coefficients

$$A = 1 \quad B = 0 \quad C = -1 \quad D = E = F = G = 0$$

Hence, $B^2 - 4AC = 4$ for all x and t, and the equation is *hyperbolic* for all x and t.

---

* It should be pointed out that the examples are prototypes of the more general parabolic, hyperbolic, and elliptic equations and that the behavior of these examples characterizes much of the general situation.

3. The *Laplace's equation* $u_{xx} + u_{yy} = 0$ is elliptic for all x and y, since $B^2 - 4AC = -4 < 0$.

4. The linear equation $xu_{xx} + u_{yy} = \sin x$ with *variable coefficients* is also of the form (23.1), but now $B^2 - 4AC = -4x$, and, hence, the equation is

> Elliptic for $x > 0$
>
> Parabolic for $x = 0$
>
> Hyperbolic for $x < 0$

This example brings out the fact that equations with *variable coefficients* can change form in different regions of the domain.

The reader should note, too, that whether equation (23.1) is hyperbolic, parabolic, or elliptic depends *only* on the coefficients of the *second derivatives*; it has nothing to do with the *first-derivative terms*, the *term in u*, or *the non-homogeneous term*.

We now come to the major portion of this lesson: rewriting hyperbolic equations in their canonical form. It turns out that if an equation is hyperbolic (in a given region of space), then it is possible to introduce new coordinates ξ and η (characteristic coordinates) in place of the original x and y, so that the equation takes on the simple form

$$(23.2) \qquad u_{\xi\eta} = \Phi(\xi, \eta, u, u_\xi, u_\eta)$$

This equation contains *only* one second derivative $u_{\xi\eta}$, while the function $\Phi(\xi, \eta, u, u_\xi, u_\eta)$ is some function of the new independent variables ξ and η, the dependent variable u, and the *first derivatives* u_ξ, u_η. The exact form of the function Φ depends, of course, on the original equations, and finding it, along with the new coordinates ξ and η, is the object of this lesson.

The Canonical Form of the Hyperbolic Equation

We start with the general PDE

$$(23.3) \qquad Au_{xx} + Bu_{xy} + Cu_{yy} + Du_x + Eu_y + Fu = G$$

where $B^2 - 4AC > 0$ in our domain of interest. The object here is to introduce new coordinates

$$\xi = \xi(x, y)$$
$$\eta = \eta(x, y)$$

so that the general PDE contains only one second derivative $u_{\xi\eta}$ (it turns out that if we tried to transform a *hyperbolic equation* into the *elliptic or parabolic canonical forms*, the technique wouldn't work).

First of all, we compute the partial derivatives

$$u_x = u_\xi \xi_x + u_\eta \eta_x$$
$$u_y = u_\xi \xi_y + u_\eta \eta_y$$

(23.4) $$u_{xx} = u_{\xi\xi}\xi_x^2 + 2u_{\xi\eta}\xi_x\eta_x + u_{\eta\eta}\eta_x^2 + u_\xi\xi_{xx} + u_\eta\eta_{xx}$$
$$u_{xy} = u_{\xi\xi}\xi_x\xi_y + u_{\xi\eta}(\xi_x\eta_y + \xi_y\eta_x) + u_{\eta\eta}\eta_x\eta_y + u_\xi\xi_{xy} + u_\eta\eta_{xy}$$
$$u_{yy} = u_{\xi\xi}\xi_y^2 + 2u_{\xi\eta}\xi_y\eta_y + u_{\eta\eta}\eta_y^2 + u_\xi\xi_{yy} + u_\eta\eta_{yy}$$

Substituting these values into the original equation (23.3), we have

(23.5) $$\overline{A}u_{\xi\xi} + \overline{B}u_{\xi\eta} + \overline{C}u_{\eta\eta} + \overline{D}u_\xi + \overline{E}u_\eta + \overline{F}u = \overline{G}$$

where

(23.6)
$$\overline{A} = A\xi_x^2 + B\xi_x\xi_y + C\xi_y^2$$
$$\overline{B} = 2A\xi_x\eta_x + B(\xi_x\eta_y + \xi_y\eta_x) + 2C\xi_y\eta_y$$
$$\overline{C} = A\eta_x^2 + B\eta_x\eta_y + C\eta_y^2$$
$$\overline{D} = A\xi_{xx} + B\xi_{xy} + C\xi_{yy} + D\xi_x + E\xi_y$$
$$\overline{E} = A\eta_{xx} + B\eta_{xy} + C\eta_{yy} + D\eta_x + E\eta_y$$
$$\overline{F} = F$$
$$\overline{G} = G$$

These calculations, although mathematically straightforward, are quite cumbersome. The reader will get a chance to carry them out in the problem set.

The next step in our process is to set the coefficients $\overline{A}$ and $\overline{C}$ in equation (23.5) equal to zero and solve for the transformation $\xi = \xi(x,y)$, $\eta = \eta(x,y)$. This will give us the coordinates that reduce the original PDE to canonical form; so, setting

$$\overline{A} = A\xi_x^2 + B\xi_x\xi_y + C\xi_y^2 = 0$$
$$\overline{C} = A\eta_x^2 + B\eta_x\eta_y + C\eta_y^2 = 0$$

we then rewrite these two equations in the form

$$A[\xi_x/\xi_y]^2 + B[\xi_x/\xi_y] + C = 0$$
$$A[\eta_x/\eta_y]^2 + B[\eta_x/\eta_y] + C = 0$$

Solving these equations for $[\xi_x/\xi_y]$ and $[\eta_x/\eta_y]$, we find

(23.7)
$$[\xi_x/\xi_y] = \frac{-B + \sqrt{B^2 - 4AC}}{2A}$$

(characteristic equations)

$$[\eta_x/\eta_y] = \frac{-B - \sqrt{B^2 - 4AC}}{2A}$$

Note that $[\xi_x/\xi_y]$ and $[\eta_x/\eta_y]$ *each* have *two* solutions to their quadratic equations, but we only have to find one solution for each in order for $\overline{A}$ and $\overline{C}$ to be zero. The only restriction is that we don't pick the *same* roots, or else we will end up with the two coordinates the same.

We have now reduced the problem to finding the two functions $\xi(x,y)$ and $\eta(x,y)$ so that their *ratio* $[\xi_x/\xi_y]$ and $[\eta_x/\eta_y]$ satisfy equation (23.7). Finding *functions* satisfying these conditions is really quite easy once we look for a few moments at Figure 23.1.

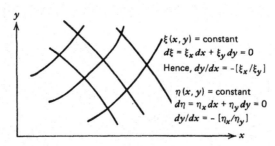

$\xi(x,y)$ = constant
$d\xi = \xi_x\,dx + \xi_y\,dy = 0$
Hence, $dy/dx = -[\xi_x/\xi_y]$

$\eta(x,y)$ = constant
$d\eta = \eta_x\,dx + \eta_y\,dy = 0$
$dy/dx = -[\eta_x/\eta_y]$

FIGURE 23.1   Characteristic curves $\xi(x,y) = c$ and $\eta(x,y) = c$.

To understand how we find ξ and η from this figure, consider the simple equation

$$u_{xx} - 4u_{yy} + u_x = 0 \qquad B^2 - 4AC = 16 > 0$$

whose *characteristic equations* are

$$\frac{dy}{dx} = -[\xi_x/\xi_y] = \frac{B - \sqrt{B^2 - 4AC}}{2A} = -2$$

$$\frac{dy}{dx} = -[\eta_x/\eta_y] = \frac{B + \sqrt{B^2 - 4AC}}{2A} = 2$$

To find ξ and η, we first solve for y (integrating), getting

$$y = -2x + c_1$$
$$y = 2x + c_2$$

and, hence, to find ξ and η, we solve for the constants c_1 and c_2, leaving them on the right-hand side of the equations while moving everything else to the left. The *functions* of x and y on the left are ξ and η; that is,

$$\xi = y + 2x = c_1$$
$$\eta = y - 2x = c_2$$

It is clear that the functions ξ and η satisfy the above characteristic equations. These particular new coordinates are drawn in Figure 23.2. This completes the discussion on how to find the new coordinates; the last step is to find the new equation.

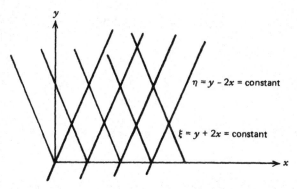

FIGURE 23.2 The new characteristic coordinates for $u_{xx} - 4u_{yy} + u_x = 0$.

The last part is very easy: All we must do to find the canonical equation is take the new coordinates $\xi(x,y)$ and $\eta(x,y)$ and substitute them into the equation

$$\overline{A}u_{\xi\xi} + \overline{B}u_{\xi\eta} + \overline{C}u_{\eta\eta} + \overline{D}u_{\xi} + \overline{E}u_{\eta} + \overline{F}u = \overline{G}$$

where $\overline{A}, \overline{B}, \overline{C}, \overline{D}, \overline{E}, \overline{F},$ and $\overline{G}$ are given by (23.6).

Before we complete this lesson, let's apply the general procedure to see how it works in a specific example.

Rewriting the Hyperbolic Equation $y^2u_{xx} - x^2u_{yy} = 0$ in Canonical Form

Suppose we start with the equation

$$y^2u_{xx} - x^2u_{yy} = 0 \qquad x > 0 \qquad y > 0$$

which is a hyperbolic equation in the first quadrant. We consider the problem of finding new coordinates that will change the original equation to canonical form for x and y in the first quadrant.

STEP 1 Solve the two characteristic equations

$$\frac{dy}{dx} = \frac{B + \sqrt{B^2 - 4AC}}{2A} = -\frac{x}{y}$$

(remember, this step is equivalent to setting $\overline{A} = \overline{C} = 0$)

$$\frac{dy}{dx} = \frac{B - \sqrt{B^2 - 4AC}}{2A} = \frac{x}{y}$$

Integrating these two equations by the ODE technique of separating variables gives the implicit relationship (we can't actually solve for y *explicitly* in terms of x here)

$$y^2 - x^2 = \text{constant}$$
$$y^2 + x^2 = \text{constant}$$

and, hence, the new coordinates ξ and η are

$$\xi = y^2 - x^2$$
$$\eta = y^2 + x^2$$

These two new coordinates are drawn in Figure 23.3.

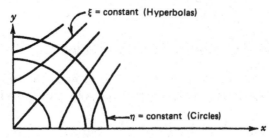

FIGURE 23.3　New characteristic coordinates.

This gives us the *new coordinates*; in order to find the *new equation*, we compute

$\overline{A} = 0$　　this must be true; we set it equal to zero and solved for ξ and η
$\overline{B} = 2A\xi_x\eta_x + B(\xi_x\eta_y + \xi_y\eta_x) + 2C\xi_y\eta_y = -16x^2y^2$
$\overline{C} = 0$　　same here; we set it equal to zero
$\overline{D} = A\xi_{xx} + B\xi_{xy} + C\xi_{yy} + D\xi_x + E\xi_y = -2(x^2 + y^2)$
$\overline{E} = A\eta_{xx} + B\eta_{xy} + C\eta_{yy} + D\eta_x + E\eta_y = 2(y^2 - x^2)$
$\overline{F} = F = 0$
$\overline{G} = G = 0$

and substitute them into the equation

$$\overline{A}u_{\xi\xi} + \overline{B}u_{\xi\eta} + \overline{C}u_{\eta\eta} + \overline{D}u_\xi + \overline{E}u_\eta + \overline{F}u = \overline{G}$$

to get

$$u_{\xi\eta} = \frac{-(x^2 + y^2)u_\xi + (y^2 - x^2)u_\xi}{8x^2y^2}$$

STEP 2　Finally, solving for x and y in terms of ξ and η, we get

(23.8)

$$u_{\xi\eta} = \frac{\eta u_\xi - \xi u_\eta}{2(\xi^2 - \eta^2)}$$

This is the end of the line; we now know how to find the new coordinates and the new equation.

NOTES

1. The general hyperbolic equation actually has *two* canonical forms; the other one can be found by making yet another transformation

$$\alpha = \alpha(\xi,\eta) = \xi + \eta$$
$$\beta = \beta(\xi,\eta) = \xi - \eta$$

and rewriting the *first canonical form* in terms of α and β. Doing this for equation (23.8) gives

$$u_\xi = u_\alpha \alpha_\xi + u_\beta \beta_\xi = u_\alpha + u_\beta$$
$$u_\eta = u_\alpha \alpha_\eta + u_\beta \beta_\eta = u_\alpha - u_\beta$$
$$u_{\xi\eta} = u_{\alpha\alpha}\alpha_\eta + u_{\alpha\beta}\beta_\eta + u_{\beta\alpha}\alpha_\eta + u_{\beta\beta}\beta_\eta = u_{\alpha\alpha} - u_{\beta\beta}$$

and hence,

(23.9)

$$u_{\alpha\alpha} - u_{\beta\beta} = \frac{-\beta u_\alpha - \alpha u_\beta}{2\alpha\beta}$$

If we wish, we could solve for the α and β coordinates in terms of the original x and y; in other words,

$$\alpha = \xi + \eta = (y^2 - x^2) + (y^2 + x^2) = 2y^2$$
$$\beta = \xi - \eta = (y^2 - x^2) - (y^2 + x^2) = -2x^2$$

2. One question the reader may ask is why someone would be interested in classifying and transforming a PDE into canonical form.
    (a) The three major classifications of linear PDEs as hyperbolic, parabolic, and elliptic equations essentially classify physical problems into three basic physical types: wave propagation, diffusion, and steady-state problems. The mathematical solutions of these three types of equations are quite different.
    (b) Much of the theoretical work on the properties of solutions to hyperbolic problems assumes the equation has been written in canonical form. In other words, it's the equation

$$u_{\xi\xi} - u_{\eta\eta} = \Psi(\xi, \eta, u, u_\xi, u_\eta)$$

that is studied. If we have an equation and want to study properties of the solution, we must convert it to canonical form and use existing results.

(c) Many computer programs have been written to find the numerical solution of the *canonical hyperbolic equation*. The function $\Phi(\xi, \eta, u, u_\xi, u_\eta)$ is fed into the computer in the form of a subroutine so it is necessary to convert the PDE to canonical form before starting. After finding the solution in terms of the new coordinates, we can always convert back to the original coordinates.

PROBLEMS

1. State whether the folowing PDEs are hyperbolic, parabolic, or elliptic:
   (a) $u_{xx} - u_{xy} = 0$
   (b) $u_{tt} = u_{xx} + u_x + hu$
   (c) $u_{xx} + 3u_{yy} = \sin x$
   (d) $u_{xx} + u_{yy} = f(x,y)$
   (e) $u_{rr} + \dfrac{1}{r}u_r + \dfrac{1}{r^2}u_{\theta\theta} = f(r,\theta)$

2. Verify equations (23.4), (23.5), and (23.6) in the lesson.
3. Verify that the equation

$$3u_{xx} + 7u_{xy} + 2u_{yy} = 0$$

is hyperbolic for all x and y and find the new *characteristic coordinates*.
4. Continue with problem 3 by finding the new canonical equation

$$u_{\xi\eta} = \Phi(\xi, \eta, u, u_\xi, u_\eta)$$

5. Continue with problem 4 by finding the *alternative* canonical form

$$u_{\alpha\alpha} - u_{\beta\beta} = \Psi(\alpha, \beta, u, u_\alpha, u_\beta)$$

6. Find the new characteristic coordinates for

$$u_{xx} + 4u_{xy} = 0$$

Solve the transformed equation in the new coordinate system and then transform back to the original coordinates to find the solution to the original problem.

The Wave Equation in Two and Three Dimensions (Free Space)

PURPOSE OF LESSON: To solve the initial-value problem

$$\text{PDE} \qquad u_{tt} = c^2[u_{xx} + u_{yy} + u_{zz}] \qquad \begin{cases} -\infty < x < \infty \\ -\infty < y < \infty \\ -\infty < z < \infty \end{cases}$$

$$\text{ICs} \qquad \begin{cases} u(x,y,z,0) = \phi(x,y,z) \\ u_t(x,y,z,0) = \psi(x,y,z) \end{cases}$$

in three dimensions and show how this solution satisfies *Huygen's principle*. The method of descent is then used to solve the corresponding problem in two dimensions

$$\text{PDE} \qquad u_{tt} = c^2[u_{xx} + u_{yy}] \qquad \begin{cases} -\infty < x < \infty \\ -\infty < y < \infty \end{cases}$$

$$\text{ICs} \qquad \begin{cases} u(x,y,0) = \phi(x,y) \\ u_t(x,y,0) = \psi(x,y) \end{cases}$$

It is then shown that the two-dimensional solution does *not* satisfy Huygen's principle. Finally, the method of descent is used once more to show that the one-dimensional version of this problem has the D'Alembert solution (which we have seen before).

Earlier, we discussed the infinite vibrating string with ICs and showed how it gave rise to the D'Alembert solution. The reader should realize that *another* application of the one-dimensional wave equation would be in describing plane waves in three dimensions. For instance, sound waves that are reasonably far from their source are essentially *longitudinally vibrating plane waves* and, hence, are described by this equation. The general situation is

1. One-dimensional waves are called *plane* waves
2. Two-dimensional waves are called *cylindrical* waves
3. Three-dimensional waves are called *spherical* waves

In other words, the one-dimensional wave equation might describe either plane waves in higher dimensions or else a one-dimensional vibrating string. The problem of this lesson is to generalize the D'Alembert solution to two and three dimensions.

Waves in Three Dimensions

We start by considering spherical waves in three dimensions that have given ICs; that is, we would like to solve the *initial-value* problem

$$\text{PDE} \qquad u_{tt} = c^2(u_{xx} + u_{yy} + u_{zz}) \qquad \begin{cases} -\infty < x < \infty \\ -\infty < y < \infty \\ -\infty < z < \infty \end{cases}$$

(24.1)

$$\text{ICs} \qquad \begin{cases} u(x,y,z,0) = \phi(x,y,z) \\ u_t(x,y,z,0) = \psi(x,y,z) \end{cases}$$

To solve this problem, we first solve the *simpler* one (set $\phi = 0$)

$$\text{PDE} \qquad u_{tt} = c^2 \nabla^2 u$$

(24.2)

$$\text{ICs} \qquad \begin{cases} u(x,y,z,0) = 0 \\ u_t(x,y,z,0) = \psi(x,y,z) \end{cases}$$

where ∇^2 is the differential operator

$$\nabla^2 \equiv \frac{\partial^2}{\partial x^2} + \frac{\partial^2}{\partial y^2} + \frac{\partial^2}{\partial z^2}$$

This problem can be solved by the Fourier transform and has the solution

(24.3)
$$u(x,y,z,t) = t\bar{\psi}$$

where $\bar{\psi}$ is the *average* of the initial disturbance ψ over the *sphere* of radius ct centered at (x,y,z); that is,

$$\bar{\psi} = \frac{1}{4\pi c^2 t^2} \int_0^\pi \int_0^{2\pi} \psi(x + ct \sin\phi \cos\theta, \ y + ct \sin\phi \sin\theta,$$

$$z + ct \cos\phi) \ (ct)^2 \sin\phi \ d\theta \ d\phi$$

The arguments of ψ range over the surface of the sphere as θ and ϕ range from $[0,2\pi]$ and $[0,\pi]$, respectively (Figure 24.1).

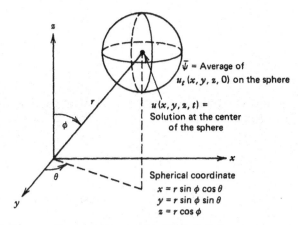

$\overline{\psi}$ = Average of
$u_t(x, y, z, 0)$ on the sphere

$u(x, y, z, t) =$
Solution at the center
of the sphere

Spherical coordinate
$x = r \sin \phi \cos \theta$
$y = r \sin \phi \sin \theta$
$z = r \cos \phi$

FIGURE 24.1 Solution as the average of initial disturbances on a sphere.

The interpretation of this solution is that the *initial* disturbance ψ radiates outward spherically (velocity c) at each point, so that after so many seconds, the point (x, y, z) will be *influenced* by those initial disturbances on a sphere (of radius ct) around that point (Figure 24.2).

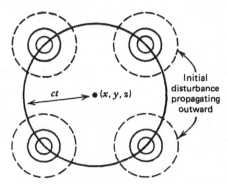

ct

$\bullet (x, y, z)$

Initial
disturbance
propagating
outward

FIGURE 24.2 Initial disturbance ψ propagating outward from each point.

The actual value of the solution (24.3) would most likely have to be computed numerically on a computer for most initial disturbances. It might be interesting for the reader to try to evaluate this solution for a few simple functions ψ.

Now, to finish the problem. What about the other half; that is,

(24.4)   PDE $u_{tt} = c^2 \nabla^2 u$ $(x, y, z) \epsilon R^3$

ICs $\begin{cases} u = \phi \\ u_t = 0 \end{cases}$

This is easy: A famous theorem developed by *Stokes* says all we have to do to solve this problem is change the ICs to $u = 0$, $u_t = \phi$, *and then* differentiate this solution with respect to time. In other words, we solve

$$(24.5)$$

$$\text{PDE} \qquad u_{tt} = c^2 \nabla^2 u$$

$$\text{ICs} \quad \begin{cases} u = 0 \\ u_t = \phi \end{cases}$$

to get $u = t\bar{\phi}$ and then differentiate with respect to time. This gives us the solution to problem (24.4)

$$u = \frac{\partial}{\partial t} [t\bar{\phi}]$$

We can see how this works for the one-dimensional wave equation where the solution (D'Alembert's solution) of (24.5) is

$$u(x,t) = \frac{1}{2c} \int_{x-ct}^{x+ct} \phi(s) \, ds$$

Therefore, if we differentiate this equation (Leibnitz rule, problem 7), we get

$$u_t(x,t) = \frac{1}{2} [\phi(x + ct) + \phi(x - ct)]$$

which is the solution of (24.4).

Knowing this, we now have the solution to our general three-dimensional problem

$$\text{PDE} \qquad u_{tt} = c^2 \nabla^2 u \qquad (x,y,z)\epsilon R^3$$

$$\text{ICs} \quad \begin{cases} u = \phi \\ u_t = \psi \end{cases}$$

It's just

$$u(x,y,z,t) = t\bar{\psi} + \frac{\partial}{\partial t} [t\bar{\phi}]$$

where $\bar{\phi}$ and $\bar{\psi}$ are the averages of the functions ϕ and ψ over the *sphere* of radius ct centered at (x,y,z).

This is known as *Poisson's formula* for the free-wave equation in three dimensions, and it is the three-dimensional generalization of the D'Alembert formula. The most important aspect of the Poisson formula is the fact that the two integrals in $\bar{\phi}$ and $\bar{\psi}$ are integrated over the *surface* of a sphere, which enables us to make the following important interpretation of the solution. When time is $t = t_1$, the solution u at (x, y, z) depends only on the initial disturbances ϕ and ψ on a sphere of radius ct_1 around (x, y, z) (Figure 24.3).

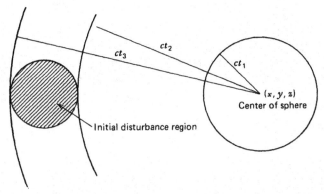

FIGURE 24.3   Diagram showing how initial disturbances affect a point (x, y, z).

Suppose now the initial disturbances ϕ and ψ are *zero* except for a small sphere (see Figure 24.3). As time increases, the *radius* of the sphere around (x, y, z) increases with velocity c, and so after t_2 seconds, it will finally *intersect* the initial disturbance region, and, hence, $u(x, y, z, t)$ becomes nonzero. For $t_2 < t < t_3$, the solution at (x, y, z) will be nonzero, since the sphere intersects the disturbance region, but when $t = t_3$, the solution at (x, y, z) abruptly becomes zero again. In other words, the wave disturbance originating from the initial-disturbance region has a *sharp trailing edge*. This general principle is known as **Huygen's principle** for three dimensions, and it is the reason why sound waves in three dimensions stimulate our ears but die off instanteously when the wave has passed. It turns out that waves *always* have sharp leading edges, but the *trailing* edges are sharp *only* in dimensions 3, 5, 7, ... We already know from the D'Alembert solution that the initial disturbance

$$u(x, 0) = \phi(x)$$
$$u_t(x, 0) = \psi(x)$$

in *one dimension* does *not* have a sharp trailing edge [since the D'Alembert solution *integrates* ψ from $(x - ct)$ to $(x + ct)$].

We will now show that the Huygen's principle does *not* apply to cylindrical waves. This situation occurs when a water wave originates from a point where the trailing edge of the wave is not sharp but gradually damps to zero.

Two-Dimensional Wave Equation

To solve the two-dimensional problem

$$
\begin{array}{ll}
\text{PDE} & u_{tt} = c^2(u_{xx} + u_{yy}) \qquad \left\{ \begin{array}{l} -\infty < x < \infty \\ -\infty < y < \infty \end{array} \right.
\end{array}
$$

(24.6)

$$
\text{ICs} \qquad \left\{ \begin{array}{l} u(x,y,0) = \phi(x,y) \\ u_t(x,y,0) = \psi(x,y) \end{array} \right.
$$

we merely let the initial disturbances ϕ and ψ in the *three-dimensional* problem depend on only the two variables x and y. Doing this, the *three-dimensional formula*

$$
u = \overline{t\psi} + \frac{\partial}{\partial t}[\overline{t\phi}]
$$

for u will describe *cylindrical waves* and, hence, give us the solution for the *two-dimensional problem*; this technique is called the **method of descent**. Carrying out the computations (which are by no means trivial), we get

$$
u(x,y,t) = \frac{1}{2\pi c}\left\{ \int_0^{2\pi}\int_0^{ct} \frac{\psi(x^1,y^1)}{\sqrt{(ct)^2 - r^2}}\, r\, dr\, d\theta \right.
$$

$$
\begin{array}{l} x^1 = x + r\cos\theta \\ y^1 = y + r\sin\theta \end{array} \qquad \left. + \frac{\partial}{\partial t}\left[\frac{1}{2\pi c}\int_0^{2\pi}\int_0^{ct} \frac{\phi(x^1,y^1)}{\sqrt{(ct)^2 - r^2}}\, r\, dr\, d\theta \right] \right\}
$$

This is the solution for the free-wave equation in *two dimensions*, and although we would probably have to evaluate it numerically, it has an interesting interpretation in terms of Huygen's principle. Note that in this solution, the two integrals of the initial conditions ϕ and ψ are integrated over the *interior* of a circle (the key word is interior) with center at (x,y,z) and radius ct. In other words, if we analyze what this means in a manner similar to the three-dimensional case, we see that initial disturbances give rise to sharp leading waves, but not to *sharp trailing waves*. Thus, Huygen's principle doesn't hold in two dimensions.

Finally, if we assume the initial conditions ϕ and ψ depend only on *one* variable, this gives rise to *plane waves* and, hence, the preceding equation descends one more dimension to the well-known D'Alembert solution

$$
u(x,t) = \frac{1}{2}[\phi(x + ct) + \phi(x - ct)] + \frac{1}{2c}\int_{x-ct}^{x+ct} \psi(s)\, ds
$$

Again, carrying out the actual computation in the method of descent is nontrivial. Note in the D'Alembert solution, the initial position ϕ gives rise to sharp trailing edges, but the initial velocity does not. In other words, one dimension is a little unusual in that the initial position satisfies Huygen's principle, but the initial velocity does not. We generally say here that Huygen's principle does *not* hold in one dimension.

188 Hyperbolic-Type Problems

NOTES

The method of descent wasn't illustrated in detail in this lesson, since we didn't show how the lower-dimensional integral could be constructed from the higher-dimensional one. The actual calculus can be found in reference 2 of Other Reading. The general idea is that the solutions of problems in higher-dimensional spaces can be used to find the solution to problems in lower-dimensional ones by assuming certain boundary and initial conditions are *independent* of certain variables. The reader should realize that this isn't the only problem to which the method of descent applies.

PROBLEMS

1. Show that in one dimension, we can find the solution of

    $$\text{PDE} \qquad u_{tt} = c^2 u_{xx} \qquad -\infty < x < \infty$$

    $$\text{ICs} \quad \begin{cases} u(x,0) = \phi(x) \\ u_t(x,0) = 0 \end{cases}$$

    by differentiating with respect to t the solution of

    $$\text{PDE} \qquad u_{tt} = c^2 u_{xx}$$

    $$\text{ICs} \quad \begin{cases} u(x,0) = 0 \\ u_t(x,0) = \phi(x) \end{cases}$$

2. Apply the results of problem 1 to find the solution of

    $$\text{PDE} \qquad u_{tt} = u_{xx} \qquad -\infty < x < \infty$$

    $$\text{ICs} \quad \begin{cases} u(x,0) = x \\ u_t(x,0) = 0 \end{cases}$$

3. Illustrate by picture and words the spherical wave solution of the three-dimensional problem

    $$\text{PDE} \qquad u_{tt} = c^2 \nabla^2 u \qquad (x,y,z)\varepsilon R^3$$

    $$\text{ICs} \quad \begin{cases} u = 0 \\ u_t = \end{cases} \begin{cases} 1 & x^2 + y^2 + z^2 \leqslant 1 \\ 0 & \text{elsewhere} \end{cases}$$

4. (To be worked with problem 3.) What is the two-dimensional solution of the analagous cylindrical-wave problem

$$\text{PDE} \qquad u_{tt} = c^2 \nabla^2 u_{xx} \qquad (x,y)\varepsilon R^2$$

$$\text{ICs} \qquad \begin{cases} u(x,0) = 0 \\ u_t(x,0) = \end{cases} \begin{cases} 1 & x^2 + y^2 \leq 1 \\ 0 & \text{elsewhere} \end{cases}$$

5. (To be worked with problem 3.) What is the one-dimensional solution of the analogous plane wave problem

$$\text{PDE} \qquad u_{tt} = c^2 u_{xx} \qquad -\infty < x < \infty$$

$$\text{ICs} \qquad \begin{cases} u(x,0) = 0 \\ u_t(x,0) = \end{cases} \begin{cases} 1 & |x| \leq 1 \\ 0 & \text{elsewhere} \end{cases}$$

6. What is the physical interpretation of why Huygen's principle does *not* hold in two dimensions?
7. Use *Leibniz's* rule

$$\frac{d}{dt} \int_{f(t)}^{g(t)} F(\xi,t)\, d\xi = \int_{f(t)}^{g(t)} \frac{\partial F}{\partial t} (\xi,t)\, d\xi + g'(t)F[g(t),t] - f'(t)F[f(t),t]$$

to differentiate the integral

$$\frac{1}{2c} \int_{x-ct}^{x+ct} \phi(s)\, ds$$

with respect to t.

OTHER READING

1. *Partial Differential Equations* by P. Garbedian. John Wiley & Sons, 1964. An excellent account of Huygen's principle; the book is considerably more advanced than this lesson.

2. *Equations of Mathematical Physics* by A. N. Tikhonov and A. A. Samarskii. Macmillan, 1963; Dover, 1990. An excellent reference for problems in mathematical physics.

LESSON **25**

The Finite Fourier Transforms (Sine and Cosine Transforms)

PURPOSE OF LESSON: To introduce two new integral transforms (finite sine and cosine transforms)

$$S_n = S[f] = \frac{2}{L} \int_0^L f(x) \sin(n\pi x/L)\, dx \qquad \text{(finite sine transform)}$$

$$C_n = C[f] = \frac{2}{L} \int_0^L f(x) \cos(n\pi x/L)\, dx \qquad \text{(finite cosine transform)}$$

$$f(x) = \sum_{n=1}^{\infty} S_n \sin(n\pi x/L) \qquad \text{(inverse sine transform)}$$

$$f(x) = \frac{C_o}{2} + \sum_{n=1}^{\infty} C_n \cos(n\pi x/L) \qquad \text{(inverse cosine transform)}$$

and to show how to solve boundary-value problems (particularly nonhomogeneous ones) by means of these transforms. Earlier, we learned about the regular Fourier and Laplace transforms and how problems are solved by transforming partial differential equations into ODEs. The usual Fourier transform requires the variable being transformed to range from $-\infty$ to ∞; hence, it is used to solve problems in free space (no boundaries). In this lesson, we show how to solve boundary value problems (with boundaries) by transforming the bounded variables (first time we've done this).

First, let's forget the motivation; let's just define the transforms, their inverses, and use them. We will talk about why they work later. In short, however, transform methods can be thought of as resolving the functions of the problem into their various frequencies—solving an entire spectrum of problems for each frequency and then adding up the results.

We first start with a function $f(x)$ defined on an interval $[0, L]$. The finite sine and cosine transforms of this function are defined by

Finite sine transform	$S[f] = S_n = \dfrac{2}{L} \displaystyle\int_0^L f(x) \sin (n\pi x/L)\, dx$ $\qquad\qquad\qquad\qquad n = 1, 2, \ldots$
Finite cosine transform	$C[f] = C_n = \dfrac{2}{L} \displaystyle\int_0^L f(x) \cos (n\pi x/L)\, dx$ $\qquad\qquad\qquad\qquad n = 0, 1, \ldots$

The reader will note that these transforms do nothing more than transform a function into the Fourier sine and cosine coefficients. The inverse transforms of these transforms are the Fourier sine and cosine series

Inverse sine transform	$f(x) = \displaystyle\sum_{n=1}^{\infty} S_n \sin (n\pi x/L)$
Inverse cosine transform	$f(x) = \dfrac{C_0}{2} + \displaystyle\sum_{n=1}^{\infty} C_n \cos (n\pi x/L)$

Note that the summation in the inverse cosine starts at $n = 0$, while the inverse sine starts at $n = 1$.

Examples of the Sine Transform

$$f(x) = 1 \quad 0 \leqslant x \leqslant 1 \quad S_n = S[1] = 2\int_0^1 \sin (n\pi x)\, dx = \begin{cases} 0 & n \text{ even} \\ 4/n\pi & n \text{ odd} \end{cases}$$

See the graphs of $f(x)$ and its transform in Figure 25.1.

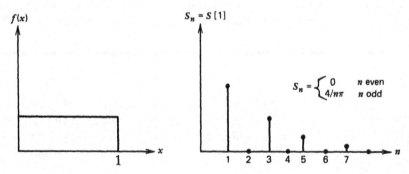

FIGURE 25.1   Graph of $f(x) = 1$ and its transform.

The inverse transform would be

$$f(x) = 1 = \frac{4}{\pi} \sum_{n=1}^{\infty} \left[\frac{1}{2n-1} \right] \sin(n\pi x)$$

Do you know what the graph of this function would be *outside* the interval [0,1]? Think about it. Note, too, that the sine transform of $f(x)$ is a function defined only at the positive integers (that is, it is just a sequence of numbers). *In other words*, the finite sine and cosine transforms transform *functions* into *sequences*.

Properties of the Transforms

Before solving problems, we must derive some of the useful properties of these transforms.

If $u(x,t)$ is a function of *two* variables, then (note we're transforming the x-variable)

$$S[u] = S_n(t) = \frac{2}{L} \int_0^L u(x,t) \sin(n\pi x/L) \, dx$$

$$C[u] = C_n(t) = \frac{2}{L} \int_0^L u(x,t) \cos(n\pi x/L) \, dx$$

(Note that we transformed the *x-variable* and now have a sequence of functions in time alone.)

What about derivatives? Here are a few of the more useful laws:

$$S[u_t] = \frac{dS[u]}{dt} \qquad S[u_{tt}] = \frac{d^2 S[u]}{dt^2}$$

$$S[u_{xx}] = -[n\pi/L]^2 S[u] + \frac{2n\pi}{L^2} [u(0,t) + (-1)^{n+1} u(L,t)]$$

$$C[u_{xx}] = -[n\pi/L]^2 C[u] - \frac{2}{L} [u_x(0,t) + (-1)^{n+1} u_x(L,t)]$$

Solving Problems via Finite Transforms

Solution of a Nonhomogeneous BVP via the Finite Sine Transform

Consider the *nonhomogeneous* wave equation

PDE $\qquad u_{tt} = u_{xx} + \sin(\pi x) \qquad 0 < x < 1 \qquad 0 < t < \infty$

BCs $\qquad \begin{cases} u(0,t) = 0 \\ u(1,t) = 0 \end{cases} \quad 0 < t < \infty$

ICs $\quad \begin{cases} u(x,0) = 1 \\ u_t(x,0) = 0 \end{cases} \quad 0 \leq x \leq 1$

To solve this problem, carry out the following steps:

STEP 1. (Determine the transform)
Since the x-variable ranges from 0 to 1, we use a finite transform. Also, you will see why, in this case, we use the sine transform. We *could* solve this problem with the Laplace transform by transforming t (it would involve about the same level of difficulty as the finite sine transform).

STEP 2. (Carry out the transformation)
Here, transforming the PDE, we get [we will call $S_n(t) = S[u]$ for convenience]

$$S[u_{tt}] = S[u_{xx}] + S[\sin (\pi x)]$$

Using the identities for the sine transform, we have

$$\frac{d^2 S_n(t)}{dt^2} = -(n\pi)^2 S_n(t) + 2n\pi[u(0,t) + (-1)^{n+1}u(1,t)] + D_n(t)$$
$$= -(n\pi)^2 S_n(t) + D_n(t)$$

where

$$D_n(t) = S[\sin (\pi x)] = \begin{cases} 1 & n = 1 \\ 0 & n = 2, 3, \ldots \end{cases}$$
$$\text{(these are the coefficients in the Fourier sine series)}$$

If we now transform the initial conditions of the boundary-value problem, we will arrive at the initial conditions for our ordinary differential equation

$$S[u(x,0)] = S_n(0) = \begin{cases} 4/n\pi & n = 1, 3, \ldots \\ 0 & n = 2, 4, \ldots \end{cases}$$

$$S[u_t(x,0)] = \frac{dS_n(0)}{dt} = 0$$

So, solving our new initial-value problem(s)

$$\text{ODE} \qquad \frac{d^2 S_n}{dt^2} + (n\pi)^2 S_n = \begin{cases} 1 & n = 1 \\ 0 & n = 2, 3, \ldots \end{cases}$$

$$
\text{ICs} \quad \begin{cases} S_n(0) = \begin{cases} 4/n\pi & n = 1, 3, \dots \\ 0 & n = 2, 4, \dots \end{cases} \\ \dfrac{dS_n(0)}{dt} = 0 \quad n = 1, 2, 3, \dots \end{cases}
$$

we have

$$
S_1(t) = A \cos(\pi t) + (1/\pi)^2
$$

where

$$
A = \frac{4}{\pi} - \frac{1}{\pi^2} = 1.17
$$

$$
S_n(t) = \begin{cases} 0 & n = 2, 4, \dots \\ \dfrac{4}{n\pi} \cos(n\pi t) & n = 3, 5, 7, \dots \end{cases}
$$

Hence, the solution $u(x,t)$ of the problem is

$$
u(x,t) = [A \cos(\pi t) + (1/\pi)^2] \sin(\pi x) + \frac{4}{\pi} \sum_{n=1}^{\infty} \frac{1}{2n+1}
$$
$$
\cos[(2n+1)\pi t] \sin[(2n+1)\pi x]
$$

NOTES

1. In order to apply the finite sine or cosine transform, the BCs at $x = 0, L$ must both be of the form

$$
\begin{aligned} u(0,t) &= f(t) \\ u(L,t) &= g(t) \end{aligned} \qquad \text{(use sine transform)}
$$
$$
\begin{aligned} u_x(0,t) &= f(t) \\ u_x(L,t) &= g(t) \end{aligned} \qquad \text{(use cosine transform)}
$$

In other words, the BCs $u(0,t) = f(t)$ and $u_x(L,t) = g(t)$ wouldn't work. Also, BCs like $u_x(0,t) + hu(0,t) = 0$ don't apply. There are *other* transforms for BCs like these; refer to the generalized sine and cosine transforms in reference 2 of Other Reading.
2. In order to apply the finite sine and cosine transforms, the equation shouldn't contain first-order derivatives in x (since the sine transform of the first derivative involves the cosine transform, and vice versa).

3. The finite sine- and cosine-transform method essentially resolves all functions in the original problem (like u_{tt}, u_{xx}, the ICs, BCs) into a Fourier sine or cosine series, solves a sequence of problems (ODE) for the Fourier coefficients, and then adds up the results.

PROBLEMS

1. Solve the diffusion problem with insulated boundaries; that is,

   $$\text{PDE} \qquad u_t = u_{xx} \qquad 0 < x < 1 \qquad 0 < t < \infty$$

   $$\text{BCs} \qquad \begin{cases} u_x(0,t) = 0 \\ u_x(1,t) = 0 \end{cases} \qquad 0 < t < \infty$$

   $$\text{IC} \qquad u(x,0) = 1 + \cos(\pi x) + 0.5\cos(3\pi x) \qquad 0 \leqslant x \leqslant 1$$

2. Solve the *general* problem

   $$\text{PDE} \qquad u_t = \alpha^2 u_{xx} + bu + f(x,t) \qquad 0 < x < 1 \qquad 0 < t < \infty$$

   $$\text{BCs} \qquad \begin{cases} u(0,t) = 0 \\ u(1,t) = 0 \end{cases} \qquad 0 < t < \infty$$

   $$\text{IC} \qquad u(x,0) = 0 \qquad 0 \leqslant x \leqslant 1$$

3. Derive the basic laws for $S[u_{xx}]$ and $C[u_{xx}]$ that were given on page 198. Can you see why it would be hard to solve differential equations that contained first derivatives u_x in the equation?

4. What is the finite sine transform of $f(x) = \sin(\pi x) + 0.5\sin(3\pi x)$? Graph the sine transform. Pick $L = 1$ in the transform.

5. What is the cosine transform of the function $f(x) = x$, $0 \leqslant x \leqslant 1$? What would the graph of the inverse transform look like for *all* values of x. [You know it reproduces $f(x) = x$ for $0 \leqslant x \leqslant 1$, but what about outside the interval $[0, 1]$?]

6. Solve the problem

   $$\text{PDE} \qquad u_t = u_{xx} + \sin(3\pi x) \qquad 0 < x < 1 \qquad 0 < t < \infty$$

   $$\text{BCs} \qquad \begin{cases} u(0,t) = 0 \\ u(1,t) = 1 \end{cases} \qquad 0 < t < \infty$$

   $$\text{IC} \qquad u(x,0) = \sin(\pi x) \qquad 0 \leqslant x \leqslant 1$$

OTHER READING

1. *Partial Differential Equations* by Tyn Myint-U. Elsevier, 1973. Several examples are worked by means of the finite sine and cosine transforms. This book has good problems and is clearly written.

2. *Operational Mathematics* by R. Churchill. McGraw-Hill, 1972. An excellent book covering the topic of integral transforms and their applications; more advanced than this book, but useful to anyone seriously interested in the technique.

LESSON **26**

Superposition (The Backbone of Linear Systems)

> **PURPOSE OF LESSON:** To introduce the idea of superposition and show how it can simplify problems by breaking them into subproblems, enabling us to solve the subproblems one at a time and then add the results to obtain the solution of the original problem. It is also shown that the two basic methods for solving linear equations, *separation of variables* and *integral transforms*, use the principle of superposition.

For an engineer who wishes to find the response u to a linear system from an *input* f, a common approach is

1. Break f into elementary parts, $f = \Sigma f_k$.
2. Find the system response u_k to f_k.
3. Add (superimpose) the simple responses u_k to get $u = \Sigma u_k$.

It turns out if the system is *linear*, then the sum u is the response we get if the function f were imputted directly; this is the **principle of superposition** (Figure 26.1).

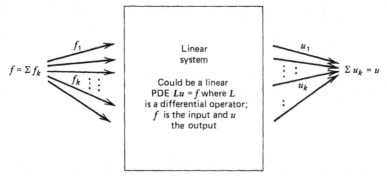

FIGURE 26.1   Basic idea of superposition.

We can use this basic idea to solve initial-boundary-value problems by breaking the problem into subproblems, solving each subpart individually, and then adding the results of each part (of course, the differential equation and the boundary conditions must be linear).

Superposition Used to Break an IBVP into Two Simpler Problems

Suppose we have the linear problem (call it P)

(P)

$$\text{PDE} \qquad u_t = u_{xx} + \sin(\pi x) \qquad 0 < x < 1 \qquad 0 < t < \infty$$

$$\text{BCs} \qquad \begin{cases} u(0,t) = 0 \\ u(1,t) = 0 \end{cases} \qquad 0 < t < \infty$$

$$\text{IC} \qquad u(x,0) = \sin(2\pi x) \qquad 0 \leq x \leq 1$$

Here, we have a *nonhomogeneous* heat equation, so separation of variables is not a viable method of attack. We could, of course, use the finite sine transform on the variable x or the Laplace transform on t, but still another idea would be to consider two subproblems

(P₁)

$$\text{PDE} \qquad u_t = u_{xx} + \sin(\pi x)$$

$$\text{BCs} \qquad \begin{cases} u(0,t) = 0 \\ u(1,t) = 0 \end{cases}$$

$$\text{IC} \qquad u(x,0) = 0$$

and

(P₂)

$$\text{PDE} \qquad u_t = u_{xx}$$

$$\text{BCs} \qquad \begin{cases} u(0,t) = 0 \\ u(1,t) = 0 \end{cases}$$

$$\text{IC} \qquad u(x,0) = \sin(2\pi x)$$

These two problems can be solved individually with a little effort, and it should be clear here that the *sum* of the solutions to P_1 and P_2 is the solution to the original problem P; that is,

$$u(x,t) = \underbrace{\frac{1}{\pi^2}(1 - e^{-\pi^2 t})\sin(\pi x)}_{\text{Solution to } P_1} + \underbrace{e^{-(2\pi)^2 t}\sin(2\pi x)}_{\text{Solution to } P_2}$$

In general, we should be able to show that the solution to

$$u_t = u_{xx} + f(x,t)$$
$$u(0,t) = 0$$
$$u(1,t) = 0$$
$$u(x,0) = \phi(x)$$

is the *sum* of the solutions to

$$u_t = u_{xx} + f(x,t)$$
$$u(0,t) = 0$$
$$u(1,t) = 0$$
$$u(x,0) = 0$$

and

$$u_t = u_{xx}$$
$$u(0,t) = 0$$
$$u(1,t) = 0$$
$$u(x,0) = \phi(x)$$

Separation of Variables and Integral Transforms as Superpositions

We may not think of superposition when using separation of variables or integral transforms, but, in fact, we are using the idea of an *infinite* superposition. In separation of variables, we generally break down the initial conditions into an infinite number of simple parts and find the response to each part. We then sum these individual responses to find the solution to the problem.

On the other hand, integral transforms *also* use superposition, for instance, let's show how the *finite sine transform* uses this principle. Consider the *non-homogeneous* heat equation

PDE $\qquad u_t = u_{xx} + f(x,t) \qquad 0 < x < 1 \qquad 0 < t < \infty$

BCs $\qquad \begin{cases} u(0,t) = 0 \\ u(1,t) = 0 \end{cases} \quad 0 < t < \infty$

IC $\qquad u(x,0) = 0 \qquad 0 \leqslant x \leqslant 1$

and consider its solution by use of the finite sine transform. What we're really doing is resolving the *input* $f(x,t)$ into components, finding the response U_n due to each component, and adding these responses. Mathematically, it may not be quite so obvious, but watch carefully. We start by expanding the PDE

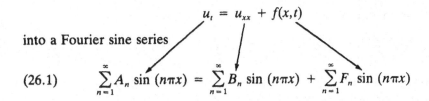

into a Fourier sine series

(26.1) $\displaystyle\sum_{n=1}^{\infty} A_n \sin(n\pi x) = \sum_{n=1}^{\infty} B_n \sin(n\pi x) + \sum_{n=1}^{\infty} F_n \sin(n\pi x)$

where

$$A_n(t) = 2 \int_0^1 u_t(x,t) \sin(n\pi x) \, dx$$

$$B_n(t) = 2 \int_0^1 u_{xx}(x,t) \sin(n\pi x) \, dx$$

$$F_n(t) = 2 \int_0^1 f(x,t) \sin(n\pi x) \, dx$$

Note that the coefficients A_n, B_n, and F_n are actually *functions* of t, since we started with functions of x and t. Note, too, that we have resolved the input $f(x,t)$ into simple components $F_n(t)$. What we would like to find is the response $U_n(t)$ to each $F_n(t)$. Then we would add the $U_n(t)$ to get the solution $u(x,t)$.

To find the simple responses $U_n(t)$, we must take our resolved PDE (26.1) and perform a little calculus on the coefficients $A_n(t)$ and $B_n(t)$, so that the integrands contain u instead of u_t and u_{xx}. Integration by parts gives us

$$A_n(t) = 2 \int_0^1 u_t \sin(n\pi x) \, dx = \frac{d}{dt}\left[2 \int_0^1 u(x,t) \sin(n\pi x) \, dx \right] = \frac{dU_n(t)}{dt}$$

$$B_n(t) = 2 \int_0^1 u_{xx} \sin(n\pi x) \, dx = -(n\pi)^2 U_n(t) + 2n\pi \left[u(0,t) + (-1)^{n+1} u(1,t) \right]$$

where $U_n(t)$ is the sine transform of $u(x,t)$. Substituting our BCs

$$u(0,t) = 0$$
$$u(1,t) = 0$$

into this last equation, we have

$$B_n(t) = -(n\pi)^2 U_n(t)$$

and, hence, the resolved PDE (26.1) becomes

$$\sum_{n=1}^{\infty} [U'_n + (n\pi)^2 U_n - F_n(t)] \sin(n\pi x) = 0$$

Since this is an identity in x, the coefficients must be zero; that is,

$$U'_n + (n\pi)^2 U_n = F_n(t)$$

Hence, we have our input-output relationship between F_n and U_n. Before we can solve for $U_n(t)$, however, we must go to the IC

$$u(x,0) = 0$$

If we expand $u(x,0)$ as a sine series and set it equal to zero, we get

$$\sum_{n=1}^{\infty} U_n(0) \sin{(n\pi x)} = 0$$

Hence, our initial conditions are

$$U_n(0) = 0 \qquad n = 1, 2, \ldots$$

We have now *resolved* our original IBVP into the simple input-output problems

$$\begin{array}{ll} \text{ODEs} & U'_n(t) + (n\pi)^2\, U_n(t) = F_n(t) \\ \text{ICs} & U_n(0) = 0 \qquad n = 1, 2, \ldots \end{array}$$

We can solve each of these problems by using an integrating factor (or Laplace transform, if we like); in any case, we get

$$U_n(t) = e^{-(n\pi)^2 t} \int_0^t e^{(n\pi)^2 \tau} F_n(\tau)\, d\tau$$

We have *now* found the responses $U_n(t)$ to the simple inputs $F_n(t)$. The final step, as the reader knows, is to sum these simple responses

$$u(x,t) = \sum_{n=1}^{\infty} U_n(t) \sin{(n\pi x)}$$

to obtain the solution to the original problem. The reader should note that each $U_n(t)$ is weighted by $\sin{(n\pi x)}$. These, of course, are the same weights used when $f(x,t)$ was decomposed into $F_n(t)$.

NOTES

In the finite sine transform, the resolutions were infinite series, whereas in most other integral transforms, the resolutions are integrals (continuous resolutions).

It would be interesting if the reader carried out a similar superposition interpretation for the problem

$$\text{PDE} \quad u_t = u_{xx} \qquad -\infty < x < \infty$$
$$\text{IC} \quad u(x,0) = \phi(x) \qquad -\infty < x < \infty$$

for the exponential Fourier transform (on x). This is essentially what was done in lesson 12 when we solved this problem with Green's function.

PROBLEMS

1. Show that if u_1 and u_2 are the solutions to problems P_1 and P_2 in the lesson, then $u_1 + u_2$ satisfies problem P.
2. Solve the IBVP

$$\text{PDE} \quad u_{tt} = u_{xx} + \sin(3\pi x) \quad 0 < x < 1 \quad 0 < t < \infty$$

$$\text{BCs} \quad \begin{cases} u(0,t) = 0 \\ u(1,t) = 0 \end{cases} \quad 0 < t < \infty$$

$$\text{ICs} \quad \begin{cases} u(x,0) = \sin(\pi x) \\ u_t(x,0) = 0 \end{cases} \quad 0 \le x \le 1$$

by superposition (each subproblem can be solved in any manner you like).
3. Suppose u_1 and u_2 are solutions of the following equations; for which equations is $u_1 + u_2$ a solution?
   (a) $u_t = u_{xx}$
   (b) $u_t = u_{xx} + e^t$
   (c) $u_t = e^{-t} u_{xx}$
   (d) $u_t = u_{xx} + u^2$
   What conclusions can you reach from your answers?
4. Find four initial-boundary-value problems whose solutions sum to the solution of the following problem:

$$u_t = u_{xx} + f(x,t) \qquad 0 < x < 1 \qquad 0 < t < \infty$$

$$\begin{aligned} u(0,t) &= g_1(t) \\ u(1,t) &= g_2(t) \end{aligned} \quad 0 < t < \infty$$

$$u(x,0) = \phi(x) \qquad 0 \le x \le 1$$

5. Solve the IVP

$$\text{ODE} \quad U'_n(t) + (n\pi)^2 U_n(t) = F_n(t)$$
$$\text{IC} \quad U_n(0) = 0$$

Can you verify the solution?

6. Suppose u_1 and u_2 both satisfy the *linear homogeneous* BCs

$$u_x(0,t) + h_1 u(0,t) = 0$$
$$u_x(1,t) + h_2 u(1,t) = 0$$

Does $u_1 + u_2$ also satisfy the BCs?

OTHER READING

A First Course in Partial Differential Equations by H. F. Weinberger. Ginn and Company, 1965. This text contains a good section on linearity and superposition; it also has a large section on complex variables. See Chapter 2 in particular.

First-Order Equations (Method of Characteristics)

PURPOSE OF LESSON: To introduce the notion of first-order partial differential equations (so far, we've studied only second order) and introduce an important technique for solving initial-value problems (method of characteristics). The problem we will solve is the initial-value problem

$$\text{PDE} \quad a(x,t)u_x + b(x,t)u_t + c(x,t)u = 0 \quad \begin{array}{l} -\infty < x < \infty \\ 0 < t < \infty \end{array}$$

$$\text{IC} \quad u(x,0) = \phi(x) \quad -\infty < x < \infty$$

Note that this is the first time we've solved problems with variable coefficients. It turns out that if we change coordinates from (x,t) to appropriate new coordinates (s,τ) (characteristic coordinates), then our differential equation becomes an *ordinary differential equation*. Hence, we solve the ODE to find $u(s,\tau)$, and, then, the last step is to plug in the values of s and τ in terms of x and t to get $u(x,t)$.

If the reader recalls, when we solved the diffusion equation

$$u_t = \alpha^2 u_{xx} - vu_x \quad -\infty < x < \infty \quad 0 < t < \infty$$

the constant α^2 stood for the amount of diffusion, while v stood for the velocity of the medium. Hence, if $\alpha = 0$ (no diffusion), it is clear (since we only have convection) that the solution will travel along the x-axis with velocity v. In other words, if the initial solution is $u(x,0) = \phi(x)$, then the solution to

$$u_t = -vu_x$$

would be $u(x,t) = \phi(x - vt)$.

So (one interpretation) we can think of the first-order equation

$$a(x,t)u_x + b(x,t)u_t = 0$$

as the *concentration* along a stream where the velocity is given by

$$v = -a(x,t)/b(x,t)$$

Of course, if $a(x,t)$ and $b(x,t)$ are constants, then we have traveling wave solutions; on the other hand, if $a(x,t)$, $b(x,t)$ change in x and t, then the stream velocity varies as we go along the stream and with time (the reader can see that the initial curve can get very distorted).

So much for the convection analog; let's get back to our basic problem

PDE $a(x,t)u_x + b(x,t)u_t + c(x,t)u = 0$ $-\infty < x < \infty$ $0 < t < \infty$
IC $u(x,0) = \phi(x)$ $-\infty < x < \infty$

The solution to this (linear first-order homogeneous) equation is based on a physical fact, namely, that an *initial disturbance* at some point x propagates along a *line* (curve) in the tx-plane (called a *characteristic*). See Figure 27.1.

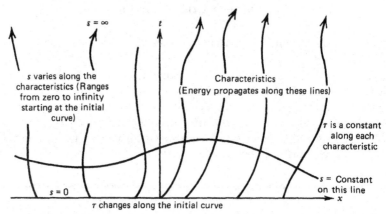

FIGURE 27.1   The initial solution at x affects the solution only along a line in the xt-plane.

This phenomenon contrasts with many other equations (such as the heat equation $u_t = u_{xx}$) where an initial disturbance at a point affects the solution everywhere else later in time. Also, if the reader recalls, the initial-position disturbance of the violin string at a point x affected the solution along *two* lines in the tx-plane (corresponding to two moving waves). See Figure 27.2.

Solving this IVP, we get

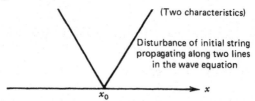

FIGURE 27.2   Initial disturbance $u(x, 0)$ at a point giving rise to two waves.

With this in mind, the idea is to introduce *two new coordinates s* and τ (to replace x and t) that have the properties

s will change along the *characteristic* curves.

τ will change along the *initial* curve (most likely the line $t = 0$).

First, consider the new coordinate s. By choosing s having the above property, the PDE

$$a(x,t)u_x + b(x,t)u_t + c(x,t)u = 0$$

is transformed into the ODE

$$\frac{du}{ds} + c(x,t)u = 0$$

Of course, the question becomes how to find these characteristics? The answer is simple; we pick the *characteristic curves* $\{[x(s),t(s)]: 0 < s < \infty\}$ so that

$$\frac{dx}{ds} = a(x,t)$$

$$\frac{dt}{ds} = b(x,t)$$

By doing this, it is clear that

$$\frac{du}{ds} = u_x \frac{dx}{ds} + u_t \frac{dt}{ds} = a(x,t)u_x + b(x,t)u_t$$

In other words, along the curves $\{[x(s),t(s)]: 0 < s < \infty\}$, the PDE becomes an ODE. We can make these ideas clear with an example.

Example
Suppose we want to solve the following IVP with constant coefficients:

PDE $\quad u_x + u_t + 2u = 0 \quad\quad -\infty < x < \infty \quad\quad 0 < t < \infty$

IC $\quad u(x,0) = \sin x \quad\quad -\infty < x < \infty$

STEP 1 Find the characteristics (along which the initial data propagate)

$$\frac{dx}{ds} = 1 \quad\quad \frac{dt}{ds} = 1 \quad\quad 0 < s < \infty$$

Solving these equations gives (remember s is now the independent variable)

$$x(s) = s + c_1 \qquad t(s) = s + c_2$$

To evaluate the constants c_1 and c_2, let $s = 0$, and referring to Figure 27.1, let

$$x(0) = \tau$$
$$t(0) = 0$$

Doing this gives values of $c_1 = \tau$ and $c_2 = 0$. Hence, our characteristic curves are

$$x = s + \tau \qquad t = s$$

To see what these curves look like in the tx-plane, we can eliminate s from these equations to get

$$x - t = \tau \qquad -\infty < \tau < \infty$$

(for each value of τ we get a straight line; for example, when $\tau = 0$, we get $t = x$). See Figure 27.3

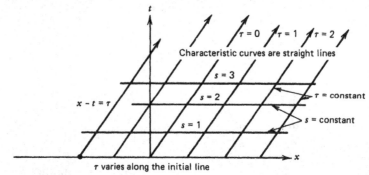

FIGURE 27.3   New coordinate system (s,τ).

STEP 2   Using this new coordinate system, we change the PDE to the ODE

$$\text{ODE} \qquad \frac{du}{ds} + 2u = 0 \qquad 0 < s < \infty$$

$$\text{IC} \qquad u(0) = \sin \tau$$

Solving this IVP, we get

$$u(s,\tau) = \sin \tau \, e^{-2s}$$

(Note that we have written u as a function of both s and τ).

So this is the solution; the only problem is that it is in the *wrong coordinate system*. It would be better if we found u as a function of x and t. This is done in Step 3.

STEP 3   We now solve the transformation

$$x = s + \tau \qquad t = s$$

for s and τ in terms of x and t. We have

$$s = t \qquad \tau = x - t$$

Hence, our answer is

$$u(x,t) = \sin (x - t)e^{-2t} \qquad \text{(check it)}$$

You can see that this means that the initial wave $u(x,0) = \sin x$ is moving to the right (undistorted in this case) and damping to zero (Figure 27.4).

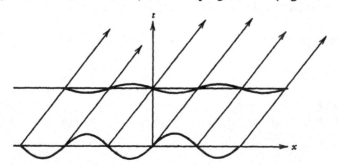

FIGURE 27.4   Solution moving with constant velocity to the right and damping to zero.

We now summarize what we have been saying.

General Strategy for Solving the First-Order Equation

Suppose we are given the IVP

PDE $a(x,t)u_x + b(x,t)u_t + c(x,t)u = 0 \qquad -\infty < x < \infty \qquad 0 < t < \infty$

IC $u(x,0) = f(x) \qquad -\infty < x < \infty$

STEP 1   Solve the two ODEs (characteristic equations)

$$\frac{dx}{ds} = a(x,t) \qquad \frac{dt}{ds} = b(x,t)$$

Find the constants of integration by setting $x(0) = \tau$ and $t(0) = 0$. We now have the transformation from (x,t) into (s,τ)

$$x = x(s,\tau)$$
$$t = t(s,\tau)$$

STEP 2   Solve the ODE

$$\text{ODE} \quad \frac{du}{ds} + c[x(s,\tau),\, t(s,\tau)]u = 0 \qquad 0 < s < \infty$$

$$\text{IC} \quad u(0) = f(\tau)$$

Note that we have substituted the values of x and t in terms of s and τ into the coefficient $c(x,t)$.

STEP 3   After solving the ODE (with the ICs), we get a solution $u(s,\tau)$. We must now solve for s and τ in terms of x and t (from the transformation we found in Step 1) and substitute these values into $u(s,\tau)$.

We now apply these steps in the following example (note variable coefficients)

Another Example

$$\text{PDE} \quad xu_x + u_t + tu = 0 \qquad -\infty < x < \infty \qquad 0 < t < \infty$$
$$\text{IC} \quad u(x,0) = F(x) \qquad \text{(an \textit{arbitrary} initial wave)}$$

STEP 1

$$\frac{dx}{ds} = x \text{ has solution} \qquad x(s) = c_1 e^s$$

$$\frac{dt}{ds} = 1 \text{ has solution} \qquad t(s) = s + c_2$$

Letting $x(0) = \tau$ and $t(0) = 0$ gives $c_1 = \tau$ and $c_2 = 0$. Hence, the transformation into new coordinates is

$$x = \tau e^s$$
$$t = s$$

STEP 2   Solve

$$\frac{du}{ds} + su = 0 \qquad 0 < s < \infty$$

$$u(0) = F(\tau)$$

The solution is

$$u(s,\tau) = F(\tau)e^{-s^2/2}$$

STEP 3 Solving for s and τ gives $s = t$, $\tau = xe^{-t}$. Hence, the solution

$$u(x,t) = F(xe^{-t})e^{-t^2/2}$$

In other words, if the IC were $u(x,0) = \sin x$, then the solution would be

$$u(x,t) = \sin(xe^{-t})e^{-t^2/2}$$

NOTE

The reader may wonder why second-order PDEs were discussed first in this book and first-order equations second? From a mathematical point of view, we could have easily switched them around. However, since the techniques for solving second-order equations don't rely heavily on first-order ideas and since second-order equations are more important, we decided to study second-order equations first.

In the study of ordinary differential equations, the situation is different. The method of solution, the general theory, and so forth, follow naturally from first- to second-order equations, so text books always cover the first-order case first.

PROBLEMS

1. Solve the simple concentration problem

$$\begin{array}{lll} \text{PDE} & u_x + u_t = 0 & -\infty < x < \infty \quad 0 < t < \infty \\ \text{IC} & u(x,0) = \cos x & -\infty < x < \infty \end{array}$$

   What does the solution look like? Does it satisfy the PDE and IC?
2. Solve the problem

$$\begin{array}{lll} \text{PDE} & xu_x + tu_t + 2u = 0 & -\infty < x < \infty \quad 1 < t < \infty \\ \text{IC} & u(x,1) = \sin x & -\infty < x < \infty \end{array}$$

   (note that time starts at $t = 1$ here). What do the characteristics look like? Plot the solution for different values of time. And, of course, check the answer.
3. Solve the problem in *higher dimensions* (surface waves)

$$\text{PDE} \qquad au_x + bu_y + cu_t + du = 0 \qquad -\infty < x < \infty$$
$$-\infty < y < \infty$$
$$0 < t < \infty$$

$$\text{IC} \qquad u(x,y,0) = e^{-(x^2+y^2)} \qquad -\infty < x < \infty$$
$$-\infty < y < \infty$$

where a, b, c, and d are given constants (check the answer).
4. Solve

$$u_x + u_t + tu = 0 \qquad -\infty < x < \infty \qquad 0 < t < \infty$$
$$u(x,0) = F(x) \qquad -\infty < x < \infty$$

Check the answer.
5. It is possible to specify the solution u on a curve other than the usual initial line $t = 0$. In fact, the differential equation doesn't have to involve the time variable at all (maybe u depends only on *space* variables). Try to solve the more general problem

$$\text{PDE} \qquad u_x + 2u_y + 2u = 0 \qquad -\infty < x < \infty$$
$$-\infty < y < \infty$$

Initial
curve $\qquad u(x,y) = F(x,y)$ on the curve C: $y = x$

[the function $F(x,y)$ is assumed given].

OTHER READING

Techniques in Partial Differential Equations by C. R. Chester. McGraw-Hill, 1971. A clear, concise description of first-order PDEs; an application to gas dynamics is given in Chapter 13.

Nonlinear First-Order Equations (Conservation Equations)

PURPOSE OF LESSON: To introduce the idea of first-order nonlinear equations and to show how certain ones (so-called conservation equations) can be used to describe physical phenomena. One particular conservation equation

$$u_t + g(u)u_x = 0$$

along with initial conditions

$$u(x,0) = \phi(x)$$

is used to describe the flow of automobiles along a freeway. This example points out that PDEs can be used to describe physical phenomena other than the usual ones in physics, biology, and engineering. It turns out that this particular nonlinear equation can give rise to discontinuous solutions (shock waves) that propagate along the road.

One of the most useful partial differential equations in mathematics is the **conservation equation**

$$\boxed{u_t + f_x = 0}$$

This equation says that the *increase* of a physical quantity u_t is equal to the *change* in flux $-f_x$ of that quantity across the boundary (flux is measured left to right). In *fluid dynamics*, $u(x,t)$ could stand for the density of a fluid at x, while $f(x,t)$ could be the flux (amount of liquid passing a point x at time t). In order not to get involved in the many details of fluid dynamics (such as assumptions of nonviscosity, incompressibility, and so forth), this lesson shows how the conservation equation can be used to predict traffic flow (flow of cars rather than the flow of water molecules). We first derive the conservation equation.

Derivation of the Conservation Equation

Suppose we have a stretch of highway on which cars are moving from left to right, and we assume there are no exit or entrance ramps. Suppose, too, we let

$u(x,t)$ = density of cars at x (cars per unit length at x)
$f(x,t)$ = flux of cars at x (cars per minute passing the point x)

Then, it is fairly obvious that for a road segment $[a,b]$, the *change* in the number of cars (with respect to time) is given by both of the following expressions:

$$\text{Change in the number of cars in } [a,b] = \frac{d}{dt} \int_a^b u(x,t) \, dx$$

and

$$\text{Change in the number of cars in } [a,b] = f(a,t) - f(b,t) = -\int_a^b \frac{\partial f}{\partial x}(x,t) \, dx$$

(The last equation is a result of the fundamental theorem of calculus.) Setting these two integrals equal to each other, we have

$$\int_a^b \frac{\partial u}{\partial t}(x,t) \, dx = -\int_a^b \frac{\partial f}{\partial x}(x,t) \, dx$$

Finally, since the interval $[a,b]$ was arbitrary, we have the integrands themselves equal; hence, the one-dimensional conservation equation

$$u_t + f_x = 0$$

If we carried out a similar analysis in two of three dimensions, we would have

$u_t + f_x + f_y = 0$ (Conservation equation in two dimensions)
$u_t + f_x + f_y + f_z = 0$ (Conservation equation in three dimensions)

Conservation Equation Applied to the Traffic Problem

There is nothing very complicated about the equation $u_t + f_x = 0$; it's just that the equation has two unknowns! So, the question is, how do we use this equation and what are we looking for?

In traffic control, the amount of cars passing a given point (flux) is generally found experimentally as a function of the car density u. In other words, experiments can be performed to find $f(u)$. It seems obvious that as the density u increases, the flux f increases (to a point, anyway). A typical model for traffic flow would be the equation

$$f(u) = Au(1 - u)$$

See Figure 28.1.

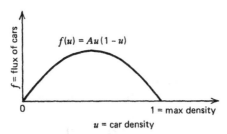

FIGURE 28.1 Typical flux versus density curve.

Other flow rates could be
1. $f(u) = ku$ (linear flow rate)
2. $f(u) = u^2$ (Quadratic flow rate)
For a complete description of flow rates in traffic control, see reference 4 of the recommended readings.

We are now ready to see how the conservation equation $u_t + f_x = 0$ can be applied to the traffic problem. To do this, we rewrite the change in flux f_x as $f_x = (df/du)u_x$ (chain rule) and substitute it in the conservation equation. Doing this, we arrive at

$$u_t + \frac{df}{du} u_x = u_t + g(u)u_x = 0$$

For example, if we find that the flux depends on the density using the equation $f(u) = u^2$, then the conservation equation becomes

$$u_t + 2uu_x = 0$$

Hence, if the initial density of the cars was $u(x,0) = \phi(x)$, then in order to find the density at any time t, we would solve the IVP

PDE $u_t + 2uu_x = 0$ $-\infty < x < \infty$ $0 < t < \infty$
IC $u(x,0) = \phi(x)$ $-\infty < x < \infty$

With these ideas in mind, we now solve a nonlinear initial-value problem.

The Nonlinear Initial-Value Problem

$$\text{PDE} \quad u_t + g(u)u_x = 0 \quad -\infty < x < \infty \quad 0 < t < \infty$$
$$\text{IC} \quad u(x,0) = \phi(x) \quad -\infty < x < \infty$$

If the reader recalls the previous lesson, where we studied the convection equation

$$u_t + vu_x = 0$$

the function $u(x,t)$ represented the concentration in a stream that was moving with velocity v. We can see an analogy with the nonlinear equation

$$u_t + g(u)u_x = 0$$

where we can imagine a particle of water starting at the point x_0 and moving upstream or downstream with velocity $g(u)$. Hence, after t seconds, the position x of the particle will be

$$x = x_0 + g(u)t \quad \text{(Characteristic equation)}$$

Remembering that the concentration $u(x,t)$ will *not change* along each characteristic, *if* we know the initial concentration $u(x_0,0)$, then the characteristic equation for x can be written

$$x = x_0 + g[u(x_0,0)]t \quad \text{[Characteristic curve starting from } (x_0,0)]$$

For example, if we wanted to solve the initial-value problem (IVP)

$$\text{PDE} \quad u_t + 3uu_x = 0$$

$$\text{IC} \quad u(x,0) = \begin{cases} 0 & x < 0 \\ 1 & x \geq 0 \end{cases}$$

then to find the characteristic that starts from an initial point, say $(2,0)$, we would write

$$x = 2 + g[u(2,0)]t$$
$$= 2 + g[1]t$$
$$= 2 + 3t$$

Here, we can claim that the solution $u(x,t)$ of our initial-value problem has the value one along the line $x = 2 + 3t$. That is $u(2 + 3t, t) = 1$ for all $t > 0$. Both the general characteristics and this specific example can be seen in Figure 28.2.

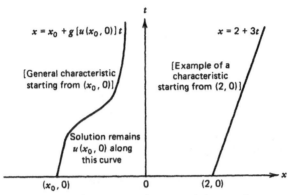

FIGURE 28.2 Characteristic curves of $u_t + g(u)u_x = 0$.

It is now clear that by knowing these characteristic curves starting from each point (and knowing the solution doesn't change along these curves), we can piece together the solution $u(x,t)$ for all time t. We won't actually find the explicit equation for the solution $u(x,t)$ in terms of x and t but will use our knowledge of the characteristics of the equation to solve some interesting problems.

Traffic-Flow Problem

We will now solve a traffic-flow problem where the flux is given by $f(u) = u^2$ and the initial density of cars is given. In other words, we have

$$\text{PDE} \quad u_t + 2uu_x = 0 \quad -\infty < x < \infty \quad 0 < t < \infty$$

$$\text{IC} \quad u(x,0) = \begin{cases} 1 & x \leqslant 0 \\ 1 - x & 0 < x < 1 \\ 0 & 1 \leqslant x \end{cases} \quad -\infty < x < \infty$$

See Figure 28.3.

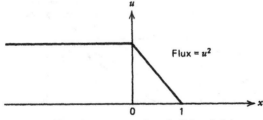

FIGURE 28.3 Initial density of cars moving left to right.

We begin by finding the characteristics starting from each initial point $(x_0, 0)$. For $x_0 < 0$, the characteristics are

$$x = x_0 + g[u(x_0, 0)]t$$
$$= x_0 + g[1]t$$
$$= x_0 + 2t$$

Solving for t, we get the lines

$$t = \frac{1}{2}(x - x_0)$$

These straight lines can be seen in Figure 28.4. Now consider those initial points $0 < x_0 < 1$. Here, the characteristic curves are

$$x = x_0 + g[u(x_0, 0)]t$$
$$= x_0 + g[(1 - x_0)]t$$
$$= x_0 + 2(1 - x_0)t$$

Again, solving for t, we have

$$t = \frac{x - x_0}{2(1 - x_0)}$$

Finally, for $1 \leqslant x_0 < \infty$, we have the characteristics

$$x = x_0 + g[u(x_0, 0)]t$$
$$= x_0 + g[0]t$$
$$= x_0$$

which represents *vertical lines* starting at x_0. The characteristics of this problem can be seen in Figure 28.4.

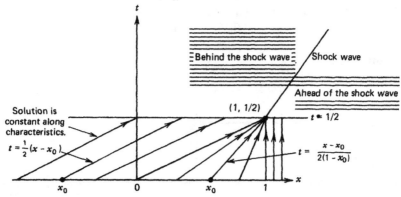

FIGURE 28.4 Characteristic of the equation $u_t + 2uu_x = 0$.

218 **Hyperbolic-Type Problems**

It is now clear how the traffic behaves for $0 < t < 1/2$. Looking at Figure 28.5, we can draw a few values of the solution for various values of time.

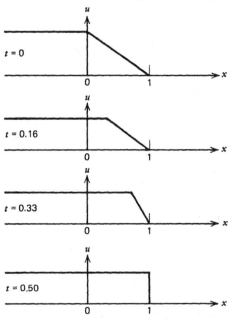

FIGURE 28.5   Density of traffic at different values of time.

Note that the characteristics run together starting at $t = 1/2$; hence, to find the solution past $t = 1/2$, we must use another type of analysis. When characteristics run together, we have the phenomenon of *shock waves* (discontinuous solutions), and what we must ask is, how fast does the leading edge of the shock wave propagate along the road? A more detailed discussion of shock waves and further references can be found in reference 2 of the recommended reading for this lesson. Although not obvious, it turns out that the *speed* of the discontinuity is given by

$$S = \frac{f(u_R) - f(u_L)}{u_R - u_L}$$

where u_R and u_L are the values of the solution at the right and left of the wave front, and $f(u_R)$ and $f(u_L)$ are the values of the flux at these points. So, in our example, the speed of the wave front would be [remembering that $f(u) = u^2$]

$$S = \frac{0 - 1}{0 - 1} = 1$$

This means that for $t > 1/2$, the wave front will move from left to right with a velocity of one.

Nonlinear First-Order Equations (Conservation Equations)   219

The complete picture of the solution to

$$\text{PDE} \qquad u_t + 2uu_x = 0 \qquad -\infty < x < \infty$$

$$\text{IC} \qquad u(x,0) = \begin{cases} 1 & x \le 0 \\ 1 - x & 0 < x < 1 \\ 0 & 1 < x \end{cases}$$

can be illustrated by Figure 28.5

NOTES

1. The shock wave in our example occurred because the flux grows very large as a function of the density u. If, on the other hand, the flux were given by $f(u) = u$, then the PDE describing the flow would be $u_t + u_x = 0$, and the solution would be an undistorted moving wave (the initial wave would just move to the right). If we think for a moment about what $f(u) = u$ means, then it is obvious the solution will move in this manner.
2. We can actually verify by direct substitution that

$$u = \phi[x - g(u)t]$$

is an *implicit solution* of the nonlinear problem

$$u_t + g(u)u_x = 0 \qquad -\infty < x < \infty \qquad 0 < t < \infty$$
$$u(x,0) = \phi(x) \qquad -\infty < x < \infty$$

For example, the implicit solution of the initial-value problem

$$\text{PDE} \qquad u_t + uu_x = 0$$
$$\text{IC} \qquad u(x,0) = x$$

is

$$\begin{aligned} u &= \phi[x - g(u)t] \\ &= x - g(u)t \\ &= x - ut \end{aligned}$$

In this particular example (although we can't do this in many examples), we can solve for $u(x,t)$ *explicitly* in terms of x and t. Doing this, we get

$$u(x,t) = \frac{x}{1 + t}$$

Check it.

PROBLEMS

1. Solve the IVP

$$\text{PDE} \quad u_t + uu_x = 0 \quad -\infty < x < \infty \quad 0 < t < \infty$$

$$\text{IC} \quad u(x,0) = \begin{cases} 0 & x < 0 \\ x & 0 \leqslant x \end{cases}$$

Draw the solution for different values of time. What is your interpretation of this solution? What is the relationship between the flux and density in this problem?

2. Solve

$$\text{PDE} \quad u_t + u^2 u_x = 0 \quad -\infty < x < \infty \quad 0 < t < \infty$$

$$\text{IC} \quad u(x,0) = \begin{cases} 0 & x < 0 \\ x & 0 \leqslant x \end{cases}$$

What is the flux-density relationship? Would you expect the solution to behave as it does? Compare the solutions of problem 1 and problem 2.

3. Suppose a nonviscous liquid is traveling through a pipe and suppose the liquid leaks through the walls of the pipe according to the law $F(u) = ku$ (g/cm sec). The conservation equation (nonconservation equation now) becomes

$$u_t + f_x = -F(u)$$

What is the solution of this equation if $f(u) = u$ and $u(x,0) = \phi(x)$ (a general initial density)? What is the interpretation of your solution?

4. What is the solution to this nonconservation equation if the loss to the outside is changed to $F(x,t) = 1/x$? Does your solution check? Does it make sense physically?

5. Verify that $u = \phi[x - g(u)t]$ is an implicit solution of the nonlinear problem

$$\text{PDE} \quad u_t + g(u)u_x = 0$$
$$\text{IC} \quad u(x,0) = \phi(x)$$

OTHER READING

1. *An Introduction to Fluid Dynamics* by G. K. Batchelor. Univesity Press, 1970. An excellent, comprehensive book on fluid dynamics.

2. *Formation and Decay of Shock Waves* by P. Lax. *Mathematical Monthly*, March

1972. A readable account of how shock waves propagate in space, written by a leading mathematician.

3. *Introduction to Partial Differential Equations with Applications* by E. C. Zachmanoglou and D. W. Thoe. Williams and Wilkins, 1976; Dover, 1986. Chapter 3 presents a nice description of first-order nonlinear equations with examples.

4. *Mathematical Theory of Traffic Control* by F. A. Haight. Academic Press, 1963. A mathematical description of traffic control; one of the few books of this kind.

Systems of PDEs

PURPOSE OF LESSON: To point out that many physical systems (particularly in fluid dynamics) cannot be described by a single PDE but, in fact, are modeled by a system of interlocking equations. In these equations, several unknown functions like pressure $= p(x,y,z,t)$, density $= \rho(x,y,z,t)$ and temperature $= T(x,y,z,t)$ (and their partial derivatives) are described by physical laws, and we seek to find all of these functions simultaneously.

It will also be shown how the *linear* system of equations

$$u_t + Au_x = 0$$

can be solved by decomposing it into a new system of independent equations

$$v_t + \Lambda v_x = 0$$

(This way, each equation can be solved separately.)

In many areas of science, several unknown quantities (and their derivatives) are related by more than one equation. For example, in fluid dynamics, the four equations

(29.1)

$$u_t + uu_x + vu_y + \frac{1}{\rho}p_x = 0 \qquad \text{(Conservation of momentum in } x\text{-direction)}$$

$$v_t + uv_x + vv_y + \frac{1}{\rho}p_y = 0 \qquad \text{(Conservation of momentum in } y\text{-direction)}$$

$$\rho_t + (\rho u)_x + (\rho v)_y = 0 \qquad \text{(Conservation of mass)}$$

$$\left(\frac{p}{\rho^\gamma}\right)_t + u\left(\frac{p}{\rho^\gamma}\right)_x + v\left(\frac{p}{\rho^\gamma}\right)_y = 0 \qquad \text{(Conservation of energy)}$$

where

$$p(x,y,t) = \text{Pressure of a fluid}$$
$$u(x,y,t) = x\text{-velocity of a fluid}$$
$$v(x,y,t) = y\text{-velocity of a fluid}$$
$$\rho(x,y,t) = \text{Density of a fluid}$$

constitute a *nonlinear* system of PDEs known as **Euler's equations of motion for a fluid**. The problem we face here is to find the unknown functions p, u, v, and ρ *simultaneously* that satisfy the four equations (along with initial and boundary conditions).

There are other reasons for studying systems of equations. If the reader recalls ODE theory, a second-order ordinary equation can be written as a system of two first-order equations. Although things are not quite so simple in PDE theory, it is often possible to write a single higher-order PDE as a system of first-order equations. For example, the *telegraph equation*

$$u_{tt} = c^2 u_{xx} + a u_t + b u$$

can be rewritten in equivalent form as three first-order PDEs by introducing the three variables u_1, u_2, u_3

$$u_1 = u$$
$$u_2 = u_x$$
$$u_3 = u_t$$

It is a simple matter to see that the telegraph equation is equivalent to the three equations

$$\frac{\partial u_1}{\partial x} = u_2$$

(29.2)

$$\frac{\partial u_1}{\partial t} = u_3$$

$$\frac{\partial u_3}{\partial t} = c^2 \frac{\partial u_2}{\partial x} + a u_3 + b u_1$$

In order to solve systems like equations (29.2), it is necessary to be familiar with the *eigenvalues* and *eigenvectors* of a *matrix*. For that reason, we present a short review of linear algebra.

Linear Algebra Review

General Case

Definition: If A is an n by n matrix, then the n eigenvalues of A are the n roots of the polynomial equation

$$\det (A - \lambda I) = 0$$

where $\det (A - \lambda I)$ is the determinant of the matrix $(A - \lambda I)$. Some eigenvalues may be numerically the same.

Special Case

Example:

Let $\quad A = \begin{bmatrix} 1 & 1 \\ 4 & 1 \end{bmatrix}$

Then the two eigenvalues are the roots of the equation

$$\det \begin{bmatrix} 1 - \lambda & 1 \\ 4 & 1 - \lambda \end{bmatrix}$$

$$= (1 - \lambda)^2 - 4$$
$$= \lambda^2 - 2\lambda - 3 = 0$$

or $\lambda_1 = -1$ and $\lambda_2 = 3$.

Definition: If λ is an eigenvalue of the matrix A, then an eigenvector corresponding to λ is a nonzero vector satisfying $Ax = \lambda x$.

Example: $X = \begin{bmatrix} 1 \\ 2 \end{bmatrix}$ is an eigenvector corresponding to $\lambda_2 = 3$, since

$$\begin{bmatrix} 1 & 1 \\ 4 & 1 \end{bmatrix} \begin{bmatrix} 1 \\ 2 \end{bmatrix} = 3 \begin{bmatrix} 1 \\ 2 \end{bmatrix}$$

Definition: The inverse of a matrix A is the matrix A^{-1} that satisfies $AA^{-1} = A^{-1}A = I$, where I is the identity matrix.

Example: The inverse of

$$A = \begin{bmatrix} 2 & -2 \\ 1 & 1 \end{bmatrix}$$

is

$$A^{-1} = \begin{bmatrix} \frac{1}{4} & \frac{1}{2} \\ -\frac{1}{4} & \frac{1}{2} \end{bmatrix}$$

since

$$AA^{-1} = A^{-1}A = \begin{bmatrix} 1 & 0 \\ 0 & 1 \end{bmatrix}$$

Diagonalization of a Matrix
If A is an n by n matrix with n distinct eigenvalues $\lambda_1, \lambda_2, \ldots \lambda_n$ and if

$$P = \begin{bmatrix} & | & | & & | \\ X_1 & | & X_2 & | & \cdots & | & X_n \\ & | & | & & | \end{bmatrix} = \begin{bmatrix} x_{11} & x_{12} & \cdots & x_{1n} \\ x_{21} & k_{22} & \cdots & x_{2n} \\ \cdot & \cdot & \cdot & \cdot \\ \cdot & \cdot & \cdot & \cdot \\ \cdot & \cdot & \cdot & \cdot \\ x_{n1} & x_{2n} & \cdots & x_{nn} \end{bmatrix}$$

is the matrix whose *n-th column* is the eigenvector X_n corresponding to λ_n, then

$$P^{-1}AP = \Lambda$$

where P^{-1} is the inverse of P, and Λ is the diagonal matrix

$$\Lambda = \begin{bmatrix} \lambda_1 & & & & \\ & \lambda_2 & & \bigcirc & \\ & & \cdot & & \\ & \bigcirc & & \cdot & \\ & & & & \lambda_n \end{bmatrix}$$

For example, the matrix

$$A = \begin{bmatrix} 0 & 8 \\ 2 & 0 \end{bmatrix}$$

has eigenvalues $\lambda_1 = 4$ and $\lambda_2 = -4$, with corresponding eigenvectors

$$X_1 = \begin{bmatrix} 2 \\ 1 \end{bmatrix} \quad \text{and} \quad X_2 = \begin{bmatrix} -2 \\ 1 \end{bmatrix}$$

Hence, the following product:

$$P^{-1}AP = \begin{bmatrix} 2 & -2 \\ 1 & 1 \end{bmatrix}^{-1} \begin{bmatrix} 0 & 8 \\ 2 & 0 \end{bmatrix} \begin{bmatrix} 2 & -2 \\ 1 & 1 \end{bmatrix}$$

$$= \begin{bmatrix} 1/4 & 1/2 \\ -1/4 & 1/2 \end{bmatrix} \begin{bmatrix} 0 & 8 \\ 2 & 0 \end{bmatrix} \begin{bmatrix} 2 & -2 \\ 1 & 1 \end{bmatrix}$$

is equal to the diagonal matrix

$$\Lambda = \begin{bmatrix} 4 & 0 \\ 0 & -4 \end{bmatrix}$$

(The reader can check this.)

We will now solve a simple system of two equations (along with their ICs).

Solution of the Linear System $u_t + Au_x = 0$

Consider the intitial-value problem consisting of two PDEs and two ICs

PDE 1 $\quad \dfrac{\partial u_1}{\partial t} + 8 \dfrac{\partial u_2}{\partial x} = 0$

$$-\infty < x < \infty \qquad 0 < t < \infty$$

PDE 2 $\quad \dfrac{\partial u_2}{\partial t} + 2 \dfrac{\partial u_1}{\partial x} = 0$

(29.3)

IC 1 $\quad u_1(x,0) = f(x)$

$$-\infty < x < \infty$$

IC 2 $\quad u_2(x,0) = g(x)$

This problem might correspond to finding $u_1(x,t)$ = pressure and $u_2(x,t)$ = density as a function of (a space variable) x and (time) t when we are given the initial pressure and density.

We start by writing the two PDEs in matrix form

$$\begin{bmatrix} \dfrac{\partial u_1}{\partial t} \\[2mm] \dfrac{\partial u_2}{\partial t} \end{bmatrix} + \begin{bmatrix} 0 & 8 \\ 2 & 0 \end{bmatrix} \begin{bmatrix} \dfrac{\partial u_1}{\partial x} \\[2mm] \dfrac{\partial u_2}{\partial x} \end{bmatrix} = \begin{bmatrix} 0 \\ 0 \end{bmatrix}$$

or

(29.4) $$u_t + Au_x = 0$$

where

$$A = \begin{bmatrix} 0 & 8 \\ 2 & 0 \end{bmatrix} \quad u_t = \begin{bmatrix} \dfrac{\partial u_1}{\partial t} \\[2mm] \dfrac{\partial u_2}{\partial t} \end{bmatrix} \quad u_x = \begin{bmatrix} \dfrac{\partial u_1}{\partial x} \\[2mm] \dfrac{\partial u_2}{\partial x} \end{bmatrix} \quad 0 = \begin{bmatrix} 0 \\ 0 \end{bmatrix}$$

Now, we introduce a *new* unknown (change of variables technique)

$$v = \begin{bmatrix} v_1 \\ v_2 \end{bmatrix}$$

by means of the *transformation:*

$$u = Pv$$

where P is the matrix whose columns are the eigenvectors of A (we already know P). It turns out that v will satisfy a very easy system of equations (the two equations in the new unknowns v_1 and v_2 are *independent* of each other), and, hence, we can find v_1 and v_2 easily. After we find v_1 and v_2, we then compute u_1 and u_2 from

$$u = Pv$$

First, however, let's see what system of equations v satisfies. To see this, the reader must think in terms of matrix calculus; that is, if we start with the matrix transformation $u = Pv$, we can differentiate these equations to get

(29.5)
$$\frac{\partial u}{\partial t} = P\frac{\partial v}{\partial t}$$
$$\frac{\partial u}{\partial x} = P\frac{\partial v}{\partial x}$$

(The reader should write these equations out in expanded form to make sure he or she knows what's going on.) We now substitute equations (29.5) into our system

$$u_t + Au_x = 0$$

to get

$$Pv_t + APv_x = 0$$

and if we multiply each side of this vector equation by P^{-1}, we have

$$v_t + P^{-1}APv_x = 0$$

or

(29.6)
$$v_t + \Lambda v_x = 0$$

Since we have already seen that

$$\Lambda = \begin{bmatrix} 4 & 0 \\ 0 & -4 \end{bmatrix}$$

for our matrix A, we can expand equation (29.6) to get the two equations

(29.7)
$$\frac{\partial v_1}{\partial t} + 4\frac{\partial v_1}{\partial x} = 0$$
$$\frac{\partial v_2}{\partial t} - 4\frac{\partial v_2}{\partial x} = 0$$

These are the two *uncoupled* equations that we can solve independently. In fact, the reader should know that these equations have *travelling-wave solutions*

$$v_1(x,t) = \phi(x - 4t)$$
$$v_2(x,t) = \psi(x + 4t)$$

(where ϕ and ψ are arbitrary differentiable functions).

These are the values of v_1 and v_2. To obtain our general solution u, we merely compute

$$u = Pv = \begin{bmatrix} 2 & -2 \\ 1 & 1 \end{bmatrix} \begin{bmatrix} v_1 \\ v_2 \end{bmatrix}$$

$$= \begin{bmatrix} 2 & -2 \\ 1 & 1 \end{bmatrix} \begin{bmatrix} \phi(x - 4t) \\ \psi(x + 4t) \end{bmatrix}$$

$$= \begin{bmatrix} 2\phi(x - 4t) - 2\psi(x + 4t) \\ \phi(x - 4t) + \psi(x + 4t) \end{bmatrix}$$

In other words,

(29.8)
$$u_1(x,t) = 2\phi(x - 4t) - 2\psi(x + 4t)$$
$$u_2(x,t) = \phi(x - 4t) + \psi(x + 4t)$$

Hence, we have found the general solution to the two PDEs of problem (29.3).

A typical solution (there are an infinite number) would be to let

$$\phi(\xi) = \sin \xi$$
$$\psi(\xi) = \xi^2$$
(Two arbitrary functions)

and, hence

$$u_1(x,t) = 2 \sin (x - 4t) - 2(x + 4t)^2$$
$$u_2(x,t) = \sin (x - 4t) + (x + 4t)^2$$

We now substitute general solution (29.8) into the ICs

$$u_1(x,0) = f(x)$$
$$u_2(x,0) = g(x)$$

to get

$$2\phi(x) - 2\psi(x) = f(x)$$
$$\phi(x) + \psi(x) = g(x)$$

Solving for ϕ and ψ gives

$$\phi(x) = \frac{1}{4}[f(x) + 2g(x)]$$

$$\psi(x) = \frac{1}{4}[2g(x) - f(x)]$$

and, hence, the solution to IVP (29.3) is

$$u_1(x,t) = 2\phi(x - 4t) - 2\psi(x + 4t)$$

$$= \frac{1}{2}[f(x - 4t) + 2g(x - 4t)] - \frac{1}{2}[2g(x + 4t) - f(x + 4t)]$$

(29.9)
$$u_2(x,t) = \phi(x - 4t) + \psi(x + 4t)$$

$$= \frac{1}{4}[f(x - 4t) + 2g(x - 4t)] + \frac{1}{4}[2g(x + 4t) - f(x + 4t)]$$

NOTES

1. Many numerical methods have been developed to solve *systems* of equations, and, hence, computer programs are often written to solve a system of n first-order equations. For that reason, when using these programs, it will often be necessary to write your higher order equation as a system.
2. The linear system

$$u_t + A(x,t)u_x = 0$$

can also be solved similarly to the way we solved $u_t + Au_x = 0$. The major difference is that now the matrix P of eigenvectors $A(x,t)$ may be functions of x and t.

PROBLEMS

1. Write the system of PDEs in equations (29.2) in matrix form

$$Au_t + Bu_x + Cu = 0$$

where A, B, and C are three by three matrices.
2. Find the eigenvalues and eigenvectors of the matrix

$$A = \begin{bmatrix} 1 & 1 \\ 4 & 1 \end{bmatrix}$$

3. Using the results of problem 2, what is the general solution of the system

$$\frac{\partial u_1}{\partial t} + \frac{\partial u_1}{\partial x} + \frac{\partial u_2}{\partial x} = 0$$

$$\frac{\partial u_2}{\partial t} + 4\frac{\partial u_1}{\partial x} + \frac{\partial u_2}{\partial x} = 0$$

HINT   First rewrite the system in matrix form $u_t + Au_x = 0$.

4. Verify equations (29.5) by writing them in scalar form.
5. Verify that functions u_1 and u_2 in equation (29.8) satisfy the two PDEs in equation (29.3).

OTHER READING

Introduction to Partial Differential Equations with Applications by E. C. Zachmanoglou and D. W. Thoe. Williams and Wilkins, 1976; Dover, 1986. A well-written text containing an excellent treatment of systems; more advanced than this text.

The Vibrating Drumhead (Wave Equation in Polar Coordinates)

PURPOSE OF LESSON: To show how the wave equation can describe the vibrations of a drumhead. The basic equation here is the wave equation

$$u_{tt} = c^2 \left(u_{rr} + \frac{1}{r}u_r + \frac{1}{r^2}u_{\theta\theta} \right)$$

which, if we look for fundamental *standing-wave solutions*

$$u(r,\theta,t) = R(r)\Theta(\theta)T(t)$$

is reduced to three ordinary differential equations

$$T'' + \lambda^2 c^2 T = 0 \quad \text{(Simple harmonic motion)}$$
$$r^2 R'' + rR' + (\lambda^2 r^2 - n^2)R = 0 \quad \text{(Bessel's equation)}$$
$$\Theta'' + n^2 \Theta = 0 \quad \text{(Simple harmonic motion)}$$

All products $R(r)\Theta(\theta)T(t)$ of solutions to the three ordinary differential equations describe the basic vibrations of the drumhead, while $R(r)\Theta(\theta)$ describes the shape of the vibrations. To find the vibrations of the drumhead that satisfy arbitrary initial conditions, we add the basic fundamental vibrations in such a way that the initial conditions are satisfied.

The purpose of this lesson is straightforward: to find the vibrations of a *circular drumhead* with given *boundary* and *initial* conditions. For simplicity, we let the radius of the circle be one and the boundary data be zero. Hence, the problem is to find $u(r,\theta,t)$ (which stands for the height of the drumhead from the plane) that satisfies

$$\text{PDE} \quad u_{tt} = c^2 \left(u_{rr} + \frac{1}{r}u_r + \frac{1}{r^2}u_{\theta\theta} \right) \quad 0 < r < 1$$
$$\text{BC} \quad u = 0 \quad \text{when } r = 1 \quad 0 < t < \infty$$

$$\text{ICs} \quad \begin{cases} u = f(r,\theta) \\ u_t = g(r,\theta) \end{cases} \quad \text{when } t = 0$$

See Figure 30.1.

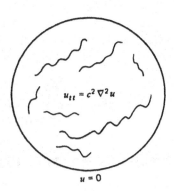

$$u_{tt} = c^2 \nabla^2 u$$

$$u = 0$$

FIGURE 30.1 Vibrating drumhead (hyperbolic IBVP).

To solve this problem, we recall the violin-string problem from Lesson 20 whose solution involved the *superposition* of an infinite number of simple vibrations. If we approach the drumhead in a similar manner, we will look for solutions of the form

$$u(r,\theta,t) = U(r,\theta)T(t)$$

This gives the *shape* $U(r,\theta)$ of the vibrations times the *oscillatory* factor $T(t)$. Carrying out this substitution (problem 1 in the problem set), we arrive at the two equations

$$\nabla^2 U + \lambda^2 U = 0 \quad \text{(Helmholtz equation)}$$
$$T'' + \lambda^2 c^2 T = 0 \quad \text{(Simple harmonic motion)}$$

where

$$\nabla^2 U = U_{rr} + \frac{1}{r}U_r + \frac{1}{r^2}U_{\theta\theta}$$

Note that we have required the separation constant to be negative (hence, we call it $-\lambda^2$), since we want $T(t)$ to be periodic.

For our next step, we want to solve the *Helmholtz* equation, but, first, it needs a boundary condition. To find it, we substitute $u(r,\theta,t) = U(r,\theta)T(t)$ into the boundary condition of the drumhead to get

$$u(1,\theta,t) = U(1,\theta)T(t) = 0 \quad 0 < t < \infty$$

or

$$U(1,\theta) = 0$$

Hence, we now have the following problem to solve in order to find the shapes $U(r,\theta)$ of the fundamental vibrations:

$$\nabla^2 U + \lambda^2 U = 0$$
$$U(1,\theta) = 0$$

This is an *elliptic eigenvalue problem* (very famous), and our purpose is to seek all λ's (if any) that yield *nonzero solutions*. The solutions $U(r,\theta)$ stand for the shapes of the fundamental modes of vibration of the drumhead, while the λ's turn out to be the roots of certain Bessel functions and are proportional to the *frequencies* of these vibrations.

So, the next step is to solve the *Helmholtz eigenvalue problem* (this is an important problem in itself). We solve it like most other linear, homogeneous PDEs with zero boundary conditions (by separation of variables).

Solution of the Helmholtz Eigenvalue Problem (Subproblem)

To solve

$$\text{PDE} \qquad \nabla^2 U + \lambda^2 U = 0$$
$$\text{BC} \qquad U(1,\theta) = 0$$

we let

$$U(r,\theta) = R(r)\Theta(\theta)$$

and plug it into the Helmholtz BVP. Doing this, we arrive at

$$r^2 R'' + rR' + (\lambda^2 r^2 - n^2)R = 0 \qquad \text{(Bessel's equation)}$$
$$R(1) = 0$$
$$\Theta'' + n^2\Theta = 0$$

(Do this yourself). Note that we have chosen the *new* separation constant n^2, $n = 0, 1, 2, \ldots$ (we get a new separation constant every time we separate variables), because we want $\Theta(\theta)$ to be periodic with period 2π (it's obvious we want the membrane (drumhead) to be periodic in θ). So, in order to solve the Helmholtz equation, we must solve the two ordinary differential equations

$$r^2 R'' + rR' + (\lambda^2 r^2 - n^2)R = 0 \qquad 0 < r < 1$$
$$R(0) < \infty \qquad \text{(Physical condition)}$$
$$R(1) = 0$$
$$\Theta'' + n^2\Theta = 0$$

Bessel's Equation

The equation

$$r^2 R'' + rR' + (\lambda^2 r^2 - n^2)R = 0$$

is well-known in ODE theory; it is called **Bessel's equation** and has two *linearly independent* solutions. They are

$$R_1(r) = AJ_n(\lambda r) \qquad n\text{-th-order Bessel function of the } first \text{ kind}$$
$$R_2(r) = BY_n(\lambda r) \qquad n\text{-th-order Bessel function of the } second \text{ kind}$$

and, hence, the general solution (all solutions) to this equation is

$$R(r) = AJ_n(\lambda r) + BY_n(\lambda r)$$

Note that the solutions depend on the n and λ in the equation. The graphs of these functions are well-known and can be found in reference 2 of Other Reading. Also see Figure 30.2.

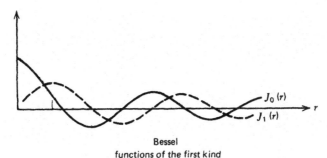

Bessel
functions of the first kind

Bessel
functions of the second kind

FIGURE 30.2 Bessel functions.

In order to find the functions $J_n(\lambda r)$ and $Y_n(\lambda r)$, we must resort to the *method of Frobenius*, which is to find solutions $R(r)$ as power series. It turns out that we can find *two* linearly independent power series $J_n(\lambda r)$ and $Y_n(\lambda r)$. Since the functions $Y_n(\lambda r)$ are unbounded at $r = 0$, we choose as our solution

$$R(r) = AJ_n(\lambda r)$$

The last step in finding $R(r)$ is to use the boundary condition $R(1) = 0$ to find λ (we're not interested in finding the coefficient A at this time). Substituting $R(1) = 0$ into $R(r) = AJ_n(\lambda r)$, we have

$$J_n(\lambda) = 0$$

In other words, in order for $R(r)$ to be *zero on the boundary* of the circle, we must pick the separation constant λ to be one of the roots of $J_n(r) = 0$; that is,

$$\lambda = k_{nm}$$

where k_{nm} is the *m*-th root of $J_n(r) = 0$. Tables of these roots are well-known, and computer programs are available to find them. A few of these roots are listed in Table 30.1.

TABLE 30.1   The *m*-th Root of $J_n(r) = 0$

		\multicolumn{5}{c}{n}
		0	1	2	3	4
m	1	2.40	3.83	5.13	6.38	7.59
	2	5.52	7.02	8.42	9.76	11.06
	3	8.65	10.17	11.62	13.02	14.37
	4	11.79	13.32	14.80	16.22	17.62
	5	14.93	16.47	17.96	19.41	20.83
	:	:	:	:	:	:

With these roots, we have just solved the Helmholtz eigenvalue problem. The eigenvalues λ are k_{nm}; and the corresponding eigenfunctions $U_{nm}(r,\theta)$ are

$$U_{nm}(r,\theta) = J_n(k_{nm}r)[A \sin(n\theta) + B \cos(n\theta)]$$

$$n = 0, 1, 2, \ldots \qquad m = 1, 2, 3, \ldots$$

We plot these functions for the different values of n and m in Figure 30.3. The general shape of $U_{nm}(r,\theta)$ is the same for different values of the constants A and B. Only the height of the vibration and placement of $\theta = 0$ is affected.

Each $U_{nm}(r,\theta)$ represents a *fundamental vibration* of the circular membrane with frequency

$$f_{nm} = k_{nm}c/2\pi \qquad \text{cycles/unit time}$$

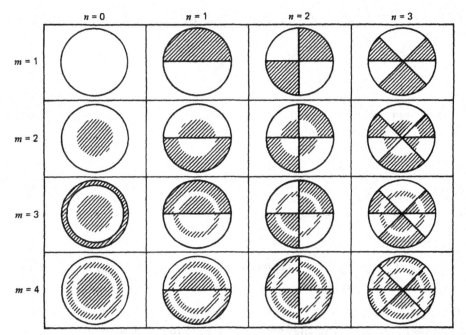

FIGURE 30.3   Fundamental vibrations of a circular drumhead showing nodal
lines.

We find these *frequencies* by solving the time equation

$$T'' + k^2_{nm}c^2T = 0$$

to get

$$T_{nm}(t) = A \sin (k_{nm}ct) + B \cos (k_{nm}ct)$$

This solution is a simple oscillatory vibration with frequency

$$f_{nm} = k_{nm}c/2\pi \qquad \text{cycles/unit time}$$

It is interesting to note that the ratio of the frequencies of $U_{nm}(r,\theta)$ to the
fundamental vibration $U_{01}(r,\theta)$ is

$$\frac{f_{nm}}{f_{01}} = \frac{k_{nm}}{k_{01}}$$

which is *not* an integer as it was in the one-dimensional wave equation. In other
words, higher vibrations for the drumhead are not pure overtones of the basic

frequencies. Note, too, that the *nodal circles* (where no vibration takes place) have radii

$$\frac{k_{n1}}{k_{nm}}, \frac{k_{n2}}{k_{nm}}, \ldots \frac{k_{nm}}{k_{nm}} = 1$$

Now that we have solved the Helmholtz equation for the basic shapes $U_{nm}(r,\theta)$, the final step is to multiply by the time factor

$$T_{nm}(t) = A \sin (k_{nm}ct) + B \cos (k_{nm}ct)$$

and add the products in such a way that the initial conditions are satisfied. That is, the solution to our problem will be

$$u(r,\theta,t) = \sum_{n=0}^{\infty} \sum_{m=1}^{\infty} J_n(k_{nm}r) \cos (n\theta)[A_{nm} \sin (k_{nm}ct) + B_{nm} \cos (k_{nm}ct)]$$

Note that $A \sin (n\theta) + B \cos (n\theta)$ was replaced by $\cos (n\theta)$ by proper choice of the angle θ. Also note that we have lumped together the constants as A_{nm} and B_{nm} (mere detail).

Rather than going through the complicated process of finding A_{nm} and B_{nm} for the general case, we will find the solution for the situation where u is independent of θ (very common). In other words, we will assume that the initial position of the drumhead depends *only* on r. In particular, we consider

$$u(r,\theta,0) = f(r)$$
$$u_t(r,\theta,0) = 0$$

(It's just as easy to do the case where $u_t \neq 0$.) With these assumptions, the solution now becomes

(30.1)
$$u(r,t) = \sum_{m=1}^{\infty} A_m J_0(k_{0m}r) \cos (k_{0m}ct)$$

and our goal is to find A_m so that

(30.2)
$$f(r) = \sum_{m=1}^{\infty} A_m J_0 (k_{0m}r)$$

To find the constants A_m, we use the orthogonality condition of the Bessel functions $\{J_0 (k_{0m}r): m = 1, 2, \ldots\}$

$$\int_0^1 rJ_0(k_{0i}r)J_0(k_{0j}r) \ dr = \begin{cases} 0 & i \neq j \\ \dfrac{1}{2}J_1^2(k_{0i}) & i = j \end{cases}$$

(the evaluation of this integral can be found in books on Bessel functions). Hence, we multiply each side of equation (30.2) by $rJ_0(k_{0j}r)$ and integrate from 0 to 1, giving

$$A_j \int_0^1 rJ_0^2(k_{0j}r) \ dr = \int_0^1 rf(r)J_0(k_{0j}r) \ dr$$

from which we can solve for A_j

(30.3) $\boxed{A_j = 2\int_0^1 rf(r)J_0(k_{0j}r) \ dr \ / \ J_1^2(k_{0j}) \qquad j = 1, 2, \ldots}$

The solution to the vibrating membrane (indepent of θ) is equation (30.1), where the coefficients are given by equation (30.3).

This solution is not so complicated as the reader might think. We can interpret it as expanding the *initial condition* $f(r)$ as the sum

$$f(r) = A_1 J_0(k_{01}r) + A_2 J_0(k_{02}r) + A_3 J_0(k_{03}r) + \cdots$$

and then inserting the oscillation factor $\cos(k_{0m}ct)$ in each term; that is,

$$u(r,t) = A_1 J_0(k_{01}r) \cos(k_{01}ct) + A_2 J_0(k_{02}r) \cos(k_{02}ct) + \cdots$$

For example, the solution to the vibrating membrane with initial conditions

$$u(r,\theta,0) = J_0(2.4r) + 0.5J_0(8.65r)$$
$$u_t(r,\theta,0) = 0$$

would be

$$u(r,t) = J_0(2.4r) \cos(2.4ct) + 0.5J_0(8.65r) \cos(8.65ct)$$

Interpretation of $J_0(k_{01}r)$, $J_0(k_{02}r)$, ...

It would be helpful for the reader to know the interpretation of $J_0(k_{01}r)$, $J_0(k_{02}r)$, ..., inasmuch as the solution is the sum of these functions. We start by drawing $J_0(r)$ (Figure 30.4)

The Vibrating Drumhead (Wave Equation in Polar Coordinates 239

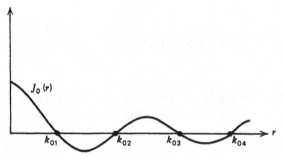

FIGURE 30.4 Zero-order Bessel function J_0 (r).

Now, in order to graph the functions $J_0(k_{01}r)$, $J_0(k_{02}r)$, ... $J_0(k_{0m}r)$ we *rescale* the r-axis so that the *m-th root* passes through $r = 1$ (Figure 30.5).

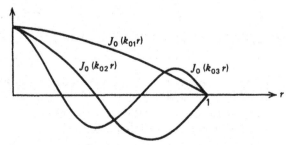

FIGURE 30.5 Graphs of $J(k_{0m}r)$ (basic building blocks of the vibrating membrane).

NOTES

1. Suppose we initially displace the circular drumhead (with $c = 1$) with the ICs

$$\text{ICs} \quad \begin{cases} u(r,\theta,0) = J_0(2.4r) + 0.10J_0(5.52r) \\ u_t(r,\theta,0) = 0 \quad \text{(Initial velocity zero)} \end{cases}$$

See Figure 30.6.

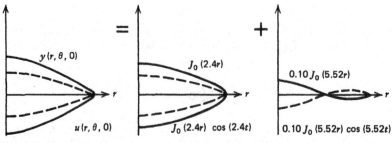

IC Decomposed into Basic Shapes

FIGURE 30.6 ICs decomposed into basic shapes.

240 Hyperbolic-Type Problems

In *general*, what we would do is expand the initial position $f(r)$ into the sum of these basic shapes $J_0(k_{0m}r)$, observe the vibration for each one, and then add them up. The basic physical principle of the vibrating membrane is that each basic shape $J_0(2.4r)$, $J_0(5.52r)$, . . . , $J_0(k_{0m}r)$, . . . gives rise to its own *simple harmonic motion* (each point on the membrane vibrates with the same frequency). Here, since our IC is the sum of two basic shapes, the vibration of the membrane will be the sum of the two simple harmonic motions caused by the shapes; that is,

$$u(r,t) = J_0(2.4r) \cos (2.4t) + 0.10J_0(5.52r) \cos (5.52t)$$

Note that these two superimposed vibrations are complicated and that the higher frequency is not a multiple of the lower.

2. For the vibrating membrane with initial conditions $u = f(r)$, $u_t = 0$, we can interpret the solution as expanding the *initial condition* $f(r)$ as a sum of basic building blocks $A_m J_0(k_{0m}r)$ and let each of them vibrate with its own frequency $\cos (k_{0m}ct)$, giving the fundamental vibration

$$A_m J_0(k_{0m}r) \cos (k_{0m}ct)$$

We then add them up to get the vibration resulting from the initial condition $f(r)$.

PROBLEMS

1. Substitute $u(r,\theta,t) = U(r,\theta)T(t)$ into the wave equation

$$u_{tt} = c^2 \left(u_{rr} + \frac{1}{r}u_r + \frac{1}{r^2}u_{\theta\theta} \right)$$

in order to separate it into the two equations

$$\nabla^2 U + \lambda^2 U = 0$$
$$T'' + \lambda^2 c^2 T = 0$$

2. Substitute $U(r,\theta) = R(r)\Theta(\theta)$ into the Helmholtz BVP

$$\nabla^2 U + \lambda^2 U = 0$$
$$U(1,\theta) = 0$$

to get

$$r^2 R'' + rR' + (\lambda^2 r^2 - n^2)R = 0$$
$$R(1) = 0$$
$$\Theta'' + n^2\Theta = 0$$

Why do we choose the separation constant to be 0, 1, 2, . . . (which we decide to call n^2)?

3. Solve

$$\text{PDE} \qquad u_{tt} = \nabla^2 u \qquad 0 < r < 1$$

$$\text{BC} \qquad u(1,\theta,t) = 0 \qquad 0 < t < \infty$$

$$\text{ICs} \qquad \begin{cases} u(r,\theta,0) = 1 - r^2 \\ u_t(r,\theta,0) = 0 \end{cases} \qquad 0 \leq r \leq 1$$

4. Solve

$$\text{PDE} \qquad u_{tt} = \nabla^2 u \qquad 0 < r < 1$$

$$\text{BC} \qquad u(1,\theta,t) = 0 \qquad 0 < t < \infty$$

$$\text{ICs} \qquad \begin{cases} u(r,\theta,0) = J_0(2.4r) \\ u_t(r,\theta,0) = 0 \end{cases} \qquad 0 \leq r \leq 1$$

5. Solve problem 4 but with the initial position of the membrane replaced by $u(r,\theta,0) = J_0(2.4r) - 0.5J_0(8.65r) + 0.25J_0(14.93r)$. What is the highest frequency in the vibration?

6. Graph the following functions for $0 \leq r \leq 1$:
   (a) $J_0(5.52r)$
   (b) $J_0(14.93r)$

OTHER READING

1. *Equations of Mathematical Physics* by A. N. Tikhonov and A. A. Samarskii. Macmillan, 1963; Dover, 1990. A comprehensive text with many examples; very good for engineers and physicists. The book solves the problem of the square drumhead; see Chapter 5.3 in particular.

2. *Tables of Functions with Formulae and Curves* by E. Jahnke and F. Emde. Stechert, 1941; Dover, 1945. One of the best known books of tables; roots of Bessel functions are tabulated here.

PART 4

Elliptic-Type
Problems

The Laplacian (an Intuitive Description)

PURPOSE OF LESSON: To present an intuitive description of the Laplacian and show what it looks like in various coordinate systems. The most common coordinate systems in two and three dimensions are:

$$\nabla^2 u = u_{xx} + u_{yy} \qquad Cartesian$$

$$\qquad\qquad\qquad\qquad\qquad\qquad\qquad \text{(Two dimensions)}$$

$$\nabla^2 u = u_{rr} + \frac{1}{r}u_r + \frac{1}{r^2}u_{\theta\theta} \qquad Polar$$

$$\nabla^2 u = u_{xx} + u_{yy} + u_{zz} \qquad Cartesian$$

$$\qquad\qquad\qquad\qquad\qquad\qquad\qquad \text{(Three dimensions)}$$

$$\nabla^2 u = u_{rr} + \frac{1}{r}u_r + \frac{1}{r^2}u_{\theta\theta} + u_{zz} \qquad Cylindrical$$

$$\nabla^2 u = u_{rr} + \frac{2}{r}u_r + \frac{1}{r^2}u_{\phi\phi} + \frac{\cot\phi}{r^2}u_\phi + \frac{1}{r^2\sin^2\phi}u_{\theta\theta} \qquad Spherical$$

Since the problem of transforming coordinates causes some students a great deal of difficulty, we will also discuss the chain rule that is used in making these transformations.

The Laplacian operator

(31.1)
$$\nabla^2 = \frac{\partial^2}{\partial x^2} + \frac{\partial^2}{\partial y^2} + \frac{\partial^2}{\partial z^2}$$

is probably the most important operator in mathematical physics. The question is, what does it mean and why should the sum of three second derivatives have anything to do with the laws of nature? The answer to this lies in the fact that the Laplacian of a function allows us to compare the function at a point with the function at neighboring points. It does what the second derivative did in one dimension and might be thought of as a second derivative generalized to higher dimensions. We now give the basic interpretation of the Laplacian that makes it so useful.

Interpretations of ∇^2 in Two Dimensions

If
1. $\nabla^2 u > 0$ at a point (x,y), then $u(x,y)$ is *smaller* than the *average* of u at its neighbors [Say on a circle around (x,y)]
2. $\nabla^2 u = 0$ at (x,y), then $u(x,y)$ is *equal* to the *average* of u at its neighbors
3. $\nabla^2 u < 0$ at (x,y), then $u(x,y)$ is *greater* then the average of u at its neighbors

Using these principles, we can now interpret some of the basic PDEs of physics.

Intuitive Meanings of Some Basic Laws of Physics

The *heat equation* $u_t = \alpha^2 \nabla^2 u$ measures temperature (or concentration) u, and the equation can be interpreted to mean that the *change* in temperature (or concentration) u_t (with respect to time) is proportional to $\nabla^2 u$. That is, the temperature at a point is *increasing* if the temperature at that point is *less* than the *average* of the temperatures on a circle around the point.

The *wave equation* $u_{tt} = \alpha^2 \nabla^2 u$ measures the displacement (among other things) of a drumhead and can be interpreted to mean that the acceleration (or force) u_{tt} of a point on the drumhead is proportional to $\nabla^2 u$. That is, the drumhead at a point is *accelerating upward* (force is up) if the drumhead at that point is *less* (in height) than the *average* of its neighbors.

Laplace's equation $\nabla^2 u = 0$ says that the solution u is always equal to the average of its neighbors. For example, a steady-state, stretched rubber membrane (not moving) satisfies Laplace's equation, hence, the height of the membrane at any point is *equal* to the *average height* of the membrane on a circle around the point. Figure 31.1 illustrates the meaning of the Laplacian.

$\nabla^2 u\,(p) = 0$

Average value of u on a small circle
around p is *equal* to $u\,(p)$

$\nabla^2 u\,(p) > 0$

Average value of u on a small circle
around p is *greater* than $u\,(p)$

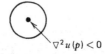
$\nabla^2 u\,(p) < 0$

Average value of u on a small circle
around p is *smaller* than $u\,(p)$

(p is a point in two or three dimensions)

FIGURE 31.1 Intuitive meaning of $\nabla^2 u$.

Poisson's equation $\nabla^2 u = f$, where f is a function that depends only on the space variables (can be a constant; can describe a number of phenomena).

1. $\nabla^2 u = -\rho$ describes the potential of an *electrostatic field* where ρ represents a constant charge density. What would be the nature of the potential field now that you know the meaning of the Laplacian?

2. $\nabla^2 u = -g(x,y)$ describes the steady state temperature $u(x,y)$ due to a heat source $g(x,y)$. If $g(x,y)$ is positive at a point, then heat is *generated* at that point. Negative $g(x,y)$ means that heat is absorbed.

3. $\nabla^2 u + \lambda u = 0$ is known as the Helmholtz equation (or the reduced-wave equation), which describes (among other things) the fundamental shapes of a stretched membrane. It's the equation we get when the time factor is separated from the wave equation or heat equation.

So far, we've talked about the intuitive meaning of the Laplacian

$$\nabla^2 u = u_{xx} + u_{yy} \qquad \text{(Two dimensions)}$$
$$\nabla^2 u = u_{xx} + u_{yy} + u_{zz} \qquad \text{(Three dimensions)}$$

However, in many problems, it is necessary to write this equation in other coordinates. For example, if the boundary of the region is a circle, then we must use *polar coordinates* (r, θ), while if we are in three dimensions and our boundary is a sphere, then we turn to *spherical coordinates* (r, θ, ϕ). So the question becomes, how do we rewrite the Laplacian in different coordinate systems?

Before we start, however, let's review briefly the five major coordinate systems in two and three dimensions; they are

Cartesian system in two dimensions
Cartesian system in three dimensions
Polar coordinates (two dimensions)
Cylindrical coordinates (three dimensions)
Spherical coordinates (three dimensions)

Polar coordinates are defined by

$$\begin{array}{ll} r^2 = x^2 + y^2 & \\ \theta = \tan^{-1}(y/x) & \end{array} \quad \text{or} \quad \begin{array}{l} x = r\cos\theta \\ y = r\sin\theta \end{array}$$

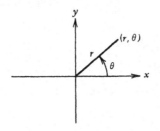

Cylindrical coordinates are defined by

$$
\begin{aligned}
r^2 &= x^2 + y^2 \\
\theta &= \tan^{-1}(y/x) \\
z &= z
\end{aligned}
\quad \text{or} \quad
\begin{aligned}
x &= r \cos \theta \\
y &= r \sin \theta \\
z &= z
\end{aligned}
$$

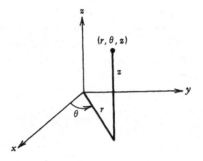

Spherical coordinates are defined by

$$
\begin{aligned}
r^2 &= x^2 + y^2 + z^2 \\
\cos \phi &= z/r \\
\tan \theta &= y/x
\end{aligned}
\quad \text{or} \quad
\begin{aligned}
x &= r \sin \phi \cos \theta \\
y &= r \sin \phi \sin \theta \\
z &= r \cos \phi
\end{aligned}
$$

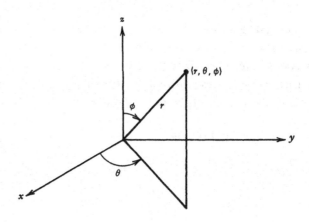

Changing Coordinates

We are now ready to change coordinates. As an illustration, we see how the two-dimensional Laplacian is transformed into *polar* coordinates. If we understand this, then we should be able to change from any coordinate system to another. So, we begin with

$$\nabla^2 u = u_{xx} + u_{yy}$$

To find the new Laplacian, we begin by drawing the diagram shown in Figure 31.2.

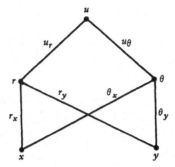

FIGURE 31.2 Functional dependence of u on (r,θ) and (x,y).

The purpose here is to illustrate exactly what depends on what. For example, the diagram in Figure 31.2 illustrates that the variable u depends on the two new variables r and θ and each, in turn, depends on x and y (lines illustrate dependence). We could also have drawn the diagram with r and θ at the bottom, but since we want to derive expressions for u_{xx} and u_{yy}, this diagram is the most appropriate. So, how do we find u_{xx} and u_{yy} in terms of r and θ?

It's really very easy; we first compute u_x and u_y. We find u_x by adding up the paths in our diagram from u to x. In this case, there are two paths, so our answer is

$$u_x = u_r r_x + u_\theta \theta_x$$
$$= u_r (\cos \theta) - u_\theta (\sin \theta / r)$$

The chain rule tells us to multiply the links of the paths together and add them up.

By a similar argument, we have

$$u_y = u_r r_y + u_\theta \theta_y$$
$$= u_r (\sin \theta) + u_\theta (\cos \theta / r)$$

Now for the second-order derivatives. We'll find u_{xx} first (u_{yy} will be an exercise for the reader). First, however, let's draw another diagram (Figure 31.3).

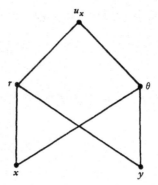

FIGURE 31.3 Functional dependence of u_x on (r,θ) and (x,y).

The reader should realize at this point that if u depends on r and θ (which, in turn, depends on x and y), then u_x will *also* depend (in general) on the same variables. So, with that, we have

$$
\begin{aligned}
u_{xx} &= (u_x)_x \\
&= (u_x)_r r_x + (u_x)_\theta \theta_x \\
&= (u_r \cos\theta - u_\theta \sin\theta/r)_r \cos\theta + (u_r \cos\theta - u_\theta \sin\theta/r)_\theta (\sin\theta/r) \\
&= (u_{rr}\cos\theta - u_{r\theta}\sin\theta/r + u_\theta \sin\theta/r^2)\cos\theta \\
&\quad + (u_{r\theta}\cos\theta - u_r \sin\theta - u_{\theta\theta}\sin\theta/r - u_\theta \cos\theta/r)(-\sin\theta/r)
\end{aligned}
$$

By a similar argument, we get

$$
\begin{aligned}
u_{yy} &= (u_{rr}\sin\theta + u_{r\theta}\cos\theta/r - \cos\theta/r^2)\sin\theta + \\
&\quad (u_{r\theta}\sin\theta + u_r \cos\theta + u_{\theta\theta}\cos\theta/r - u_\theta \sin\theta/r)(\cos\theta/r).
\end{aligned}
$$

Adding u_{xx} and u_{yy}, we get the desired result

$$
\boxed{\nabla^2 u = u_{xx} + u_{yy} = u_{rr} + \frac{1}{r}u_r + \frac{1}{r^2}u_{\theta\theta}}
$$

This is the Laplacian in *polar coordinates*. It has the same intuitive meaning as the Laplacian in cartesian coordinates, but a different form. Unfortunately, it has *variable coefficients*, so equations involving the polar Laplacian are more difficult to solve.

Carrying out a similar analysis for the three-dimensional Laplacian

$$
\nabla^2 u = u_{xx} + u_{yy} + u_{zz}
$$

Changing to *cylindrical coordinates*, we can show

$$\nabla^2 u = u_{rr} + \frac{1}{r}u_r + \frac{1}{r^2}u_{\theta\theta} + u_{zz}$$

Finally, if we write the Laplacian in *spherical coordinates*, we have

$$\nabla^2 u = u_{rr} + \frac{2}{r}u_r + \frac{1}{r^2}u_{\phi\phi} + \frac{\cot\theta}{r^2}u_\phi + \frac{1}{r^2\sin^2\phi}u_{\theta\theta}$$

NOTES

1. The Laplacian in cartesian coordinates is the only one with *constant coefficients*. This is one reason why problems in other coordinate systems are harder to solve. It is *still* possible to use the separation of variables method for these equations with variable coefficients; it's just that some of the resulting ordinary differential equations have variables coefficients. We arrive at a lot of fairly complicated equations, such as Bessel's equation, Legendre's equation, and other so-called classical equations of physics. To solve these equations, we must resort to infinite-series solutions and, in particular, the method of Frobenious.

2. There are many more coordinate systems than the ones we've mentioned. For example, the vibration of an elliptical drumhead would require writing the Laplacian in an elliptical coordinate system. Rather than do this, however, there is another approach: transform the ellipse into a region more to our liking (like a circle). We then solve the transformed problem into our new coordinates (polar coordinates) and then transform into the original coordinates.

PROBLEMS

1. What is the wave equation $u_{tt} = \alpha^2\nabla^2 u$ in spherical coordinates if you know the solution u depends only on r and t?

2. What is the wave equation $u_{tt} = \alpha^2\nabla^2 u$ in polar coordinates if the solution u depends only on r and t?

3. What is Laplace's equation $\nabla^2 u = 0$ in polar coordinates if u depends only on r? What are the solutions of this equation? These are the *circularly symmetric potentials* in two dimensions.

4. What is Laplace's equation in spherical coordinates if the solution u depends only on r? Can you find the solutions of this equation? These are the *spherically symmetric potentials* in three dimensions.

5. What can you say about the surface $u(x,y) = xy$?
6. Transform the three-dimensional Laplacian $\nabla^2 u = u_{xx} + u_{yy} + u_{zz}$ to spherical coordinates.

OTHER READING

Fourier Series and Orthogonal Functions by H. F. Davis. Allyn and Bacon, 1963; Dover, 1989. An excellent text, written from an intuitive viewpoint; gives the reader the between-the-lines description of many concepts.

General Nature of Boundary-Value Problems

PURPOSE OF LESSON: To explain how PDEs that don't involve the time derivative occur in nature. These differential equations have *no initial conditions* like the hyperbolic-wave equation and the parabolic-heat equation, but only boundary conditions. For that reason, these problems are called boundary-value problems (BVPs).

The three most common types of boundary conditions (BCs) are:
1. BCs of the first kind (Dirichlet BCs)
2. BCs of the second kind (Neumann BCs)
3. BCs of the third kind (Robin BCs)

These are explained and examples are shown to illustrate these ideas.

Until now, the problems we've discussed involved phenomena that changed over space *and* time. There are, however, many important problems whose outcomes do not change with time, but only with respect to space. These problems, for the most part, are described by *elliptic boundary-value problems*, and it is the purpose of this lesson to describe these types of problems.

There are two common situations in physical problems that give rise to PDEs that don't involve time; they are:
1. Steady-state problems
2. Problems where we factor out the time component in the solution

First, let's look at steady-state problems.

Steady-State Problems

Suppose we look for the steady-state solution (solution when $t \to \infty$) of the heat equation

$$u_t = \alpha^2 \nabla^2 u$$

It's obvious if the solution *doesn't change in time*, then $u_t = 0$, and so the heat equation is reduced to Laplace's equation

$$\nabla^2 u = 0$$

To illustrate the concept of the steady state in detail, let's consider the problem

$$\text{PDE} \quad u_t = u_{xx} + \sin{(\pi x)} \quad\quad 0 < x < 1 \quad\quad 0 < t < \infty$$

(32.1) $\quad$ BCs $\quad \begin{cases} u(0,t) = 0 \\ u(1,t) = 0 \end{cases} \quad 0 < t < \infty$

$$\text{IC} \quad u(x,0) = \sin{(3\pi x)} \quad\quad 0 \leqslant x \leqslant 1$$

To find the steady-state solution $u(x,\infty)$ (if it exists), we let $u_t = 0$ and solve the *boundary-value problem*

$$\frac{d^2u}{dx^2} = -\sin{(\pi x)} \quad\quad 0 < x < 1$$

$$u(0) = 0$$

$$u(1) = 0$$

In this case, we have the solution

$$u(x,\infty) = \frac{1}{\pi^2}\sin{(\pi x)}$$

If we solved problem (32.1), we would get a solution that was initially $\sin{(3\pi x)}$ but gradually looked more and more like

$$\frac{1}{\pi^2}\sin{(\pi x)}$$

For some problems, a zero steady-state solution may not exist, and for others, the steady state may be a sinusoidal function and, hence, it is not always valid to set $u_t,\ u_{tt},\ \ldots$ equal to zero. We should really know something about the physics of the problem.

Factoring out the Time Component in Hyperbolic and Parabolic Problems

In the circular drumhead problem

$$
\begin{array}{ll}
\text{PDE} & u_{tt} = \nabla^2 u \quad 0 \leqslant r < 1 \\
\text{BC} & u = 0 \qquad \text{On the circle} \\
\text{ICs} & \begin{cases} u(r,\theta,0) = f(r,\theta) \\ u_t(r,\dot\theta,0) = g(r,\theta) \end{cases}
\end{array}
$$

we looked for solutions of the form $u(r,\theta,t) = U(r,\theta)T(t)$, which yielded the Helmholtz boundary-value problem

$$
\begin{array}{ll}
\text{PDE} & \nabla^2 U + \lambda^2 U = 0 \\
\text{BC} & U(1,\theta) = 0
\end{array}
$$

This situation is common in PDEs where the solution represents a *shape* factor $U(r,\theta)$ multiplied by a time factor $T(t)$. As a matter of fact, we arrive at this same Helmholtz equation by factoring out the time component in the *heat equation*

$$
u_t = \nabla^2 u
$$

When studying boundary-value problems (BVPs), there are three types of BCs that are most common; we discuss these three types now.

The Three Main Types of BCs in Boundary-Value Problems

Boundary-Value Problems of the First Kind (Dirichlet Problems)

Here, the PDE holds over a given region of space, and the solution is *specified* on the boundary of the region. An example would be to find the steady-state temperature *inside* a region when the temperature is *given* on the boundary. Another situation would be to find the potential inside the region when the potential is given on the boundary; other examples follow.

Examples of Dirichlet Problems
Consider Laplace's equation *inside* a circle with the solution given on the

boundary

$$\text{PDE} \qquad u_{rr} + \frac{1}{r}u_r + \frac{1}{r^2}u_{\theta\theta} = 0 \qquad 0 < r < 1$$

$$\text{BC} \qquad u(1,\theta) = \sin\theta \qquad 0 \le \theta < 2\pi$$

See Figure 32.1

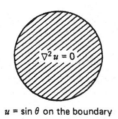

$$u = \sin\theta \text{ on the boundary}$$

FIGURE 32.1 Interior Dirichlet problem.

Another example would be an *exterior Dirichlet problem* in which we are looking for the solution of Laplace's equation *outside* the unit circle, and the solution is given *on* the circle (Figure 32.2).

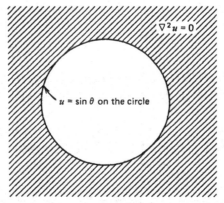

FIGURE 32.2 Exterior Dirichlet problem.

Dirichlet problems are common in electrostatics when we want to find the potential in a region with the potential given on the boundary.

Boundary-Value Problems of the Second Kind (Neumann Problems)

Here, the PDE holds in some region of space, but *now* the *outward normal derivative* $\partial u/\partial n$ (which is *proportional to the inward flux*) is specified on the boundary. These problems are common in steady-state heat flow and electrostatics, where the flux (in heat energy, electrons, and so forth) is given over the boundary.

For example, suppose the inward flow of heat varies around the circle according to

$$\frac{\partial u}{\partial r} = \sin \theta$$

The steady-state temperature inside the circle would then be given by the solution of the BVP

$$\nabla^2 u = 0 \qquad 0 < r < 1$$

$$\frac{\partial u}{\partial r} = \sin \theta \qquad r = 1 \qquad 0 \leq \theta < 2\pi$$

Here, we can see (Figure 32.3) that the *flux* of heat (cal/cm sec) across the boundary is *inward* for $0 \leq \theta < \pi$ and *outward* for $\pi \leq \theta < 2\pi$.

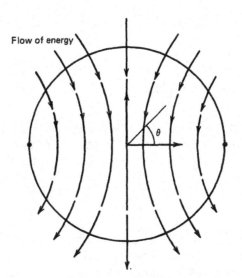

Flow of energy

(The temperature at each point is constant since the net flow in and out of that point is zero)

FIGURE 32.3 Flow of heat for the Neumann problem.

However, since the *total flux*

$$\int_0^{2\pi} \frac{\partial u}{\partial r} d\theta = \int_0^{2\pi} \sin \theta \, d\theta = 0$$

(a condition that must always be true for Neumann problems), we can say that the temperature at *each point* inside the circle does *not change* with respect to time.

In other words, Neumann problems make sense only if the net gain in heat (or whatever) across the boundary is zero. Mathematically, this says that

$$\int_C \frac{\partial u}{\partial n} = 0$$

must be true around the boundary or else the problem has no solution. For example, the interior Neumann problem

$$\nabla^2 u = 0 \qquad 0 < r < 1$$

$$\frac{\partial u}{\partial r}(1,\theta) = 1 \qquad r = 1 \qquad 0 \leqslant \theta < 2\pi$$

has no physical meaning, since the *constant* inward flux of *one* would not give rise to a steady-state solution.

The Neumann problem is somewhat different from other boundary conditions in that solutions are not unique. In other words, the Neumann problem

$$\nabla^2 u = 0 \qquad 0 < r < 1$$

$$\frac{\partial u}{\partial r}(1,\theta) = \cos(2\theta) \qquad r = 1 \qquad 0 \leqslant \theta < 2\pi$$

(note that the total flux across the boundary is zero) has an *infinite number* of solutions $u(r,\theta)$. However, once we have one solution, we can get the others just by adding a constant. For example, one solution to our Neumann problem is

$$u(r,\theta) = r^2 \cos(2\theta)$$

and it is obvious that if we add a constant to this solution, another one is obtained. For this reason, if we want to find one solution to the Neumann problem, we must have some additional information (like knowing the solution at one point).

Boundary-Value Problems of the Third Kind

These problems correspond to the PDEs being given in some region of space, but now the condition on the boundary is a mixture of the first two kinds

$$\frac{\partial u}{\partial n} + h(u - g) = 0$$

where h is a constant (input to the problem) and g a given function that can vary over the boundary. A more suggestive form of this BC would be

$$\frac{\partial u}{\partial n} = -h(u - g)$$

which says the *inward flux* across the boundary is proportional to the *difference* between the temperature u and some specified temperature g. This has the interpretation that

1. If the temperature u (on the boundary) is greater than the boundary temperature, then the flow of heat is *outward*.
2. If u is less than the boundary temperature g, then heat flows in.

This, of course, is just Newton's law of cooling, and so these types of BCs are very natural in steady-state heat flow.

Example

Suppose the temperature directly *outside* the unit circle is given by $g(\theta) = \sin \theta$. In this case, the flow of heat across the boundary would be given by

$$\frac{\partial u}{\partial r} = -h(u - \sin \theta) \qquad r = 1 \qquad 0 \le \theta < 2\pi$$

and, hence, to find the steady-state heat *inside* the circle, we would have to solve the BVP

(32.2)

$$
\begin{array}{lll}
\text{PDE} & \nabla^2 u = 0 & 0 < r < 1 \\[2mm]
\text{BC} & \dfrac{\partial u}{\partial r} + h(u - \sin \theta) = 0 & 0 \le \theta < 2\pi
\end{array}
$$

The constant h is a physical parameter and measures the amount of flux across the boundary per degree difference between u and $\sin \theta$ (it's difficult to measure, since it depends on the interface across the boundary). If h is large, then heat flow is great per temperature difference, and so the solution looks very much like the solution of the Dirichlet problem with BC $g = \sin \theta$. On the other hand, if $h = 0$, then the BC is reduced to the *insulated* BC

$$\frac{\partial u}{\partial r} = 0$$

The reader can imagine how the solutions to problem (32.2) could change as h goes from 0 to ∞. When $h = 0$, the boundary is completely insulated, and, hence, the solution $u(r, \theta)$ is a *constant* (the solution is not unique here, and any constant will work). As h gets *larger and larger*, the solution looks more and more like the solution to the Dirichlet problem with BC $u = \sin \theta$. When h

> 0, but close to *zero*, the solution will be almost identically zero (which is the *average* of $\sin \theta$ on the boundary).

PROBLEMS

1. Based on intuition, can you find the solution to the Dirichlet problem

$$\text{PDE} \qquad \nabla^2 u = 0 \qquad 0 < r < 1$$
$$\text{BC} \qquad u(1,\theta) = \sin \theta \qquad 0 \le \theta < 2\pi$$

2. Does the following Neumann problem have a solution inside the circle:

$$\text{PDE} \qquad \nabla^2 u = 0 \qquad 0 < r < 1$$
$$\text{BC} \qquad \frac{\partial u}{\partial r} = \sin^2 \theta$$

3. For different values of h, imagine the solution $u(r,\theta)$ to

$$\text{PDE} \qquad \nabla^2 u = 0 \qquad 0 < r < 1$$
$$\text{BC} \qquad \frac{\partial u}{\partial r} + h \, (u - \sin \theta) = 0$$

4. What BVP would you solve to find the steady-state solution of

$$\text{PDE} \qquad u_{tt} = u_{xx} - u_t + u \qquad 0 < x < 1 \qquad 0 < t < \infty$$
$$\text{BCs} \qquad \begin{cases} u(0,t) = 0 \\ u(1,t) = 0 \end{cases} \qquad 0 < t < \infty$$
$$\text{ICs} \qquad \begin{cases} u(x,0) = \sin (3\pi x) \\ u_t(x,0) = 0 \end{cases} \qquad 0 \le x \le 1$$

5. Now that you know the physical interpretation of the Laplacian, what is the general nature of solutions of the Helmholtz BVP

$$\text{PDE} \qquad \nabla^2 u = -\lambda^2 u \qquad 0 < r < 1$$
$$\text{BC} \qquad u(1,\theta) = 0 \qquad 0 \le \theta < 2\pi$$

6. What is the physical interpretation of the following *mixed BVP* inside the square

$$\text{PDE} \qquad u_{xx} + u_{yy} = 0 \qquad 0 < x < 1 \qquad 0 < y < 1$$

$$\text{BCs} \qquad \begin{cases} u_y(x,0) - h[u(x,0) - 2] = 0 & 0 < x < 1 \\ u(x,1) = 1 \\ u_x(0,y) = 0 \\ u_x(1,y) = 0 \end{cases} \qquad 0 < y < 1$$

OTHER READING

Partial Differential Equations of Mathematical Physics by Tyn Myint-U. Elsevier, 1973. Chapter 8 of this text contains an excellent problem set for boundary-value problems.

Interior Dirichlet Problem for a Circle

PURPOSE OF LESSON: To show how to solve the interior Dirichlet problem for the circle

$$\text{PDE} \quad u_{rr} + \frac{1}{r}u_r + \frac{1}{r^2}u_{\theta\theta} = 0 \quad 0 < r < 1$$

$$\text{BC} \quad u(1,\theta) = g(\theta) \quad 0 \leq \theta < 2\pi$$

by separation of variables. The solution can be interpreted as expanding the boundary function as

$$g(\theta) = \sum_{n=0}^{\infty} [a_n \cos(n\theta) + b_n \sin(n\theta)]$$

and then finding the solution to each of the problems

$$\nabla^2 u = 0 \qquad\qquad \nabla^2 u = 0$$
$$u(1,\theta) = \sin(n\theta) \qquad u(1,\theta) = \cos(n\theta)$$

Since these two problems have solutions

$$u(r,\theta) = r^n \sin(n\theta) \qquad u(r,\theta) = r^n \cos(n\theta)$$

the solution to the interior Dirichlet problem is

$$u(r,\theta) = \sum_{n=0}^{\infty} r^n [a_n \cos(n\theta) + b_n \sin(n\theta)]$$

After this series is obtained, some algebraic manipulations are performed to arrive at an *alternative* integral-form of the solution. This new form, *the Poisson integral solution*, brings out some interesting ideas.

This lesson presents a number of new ideas as we solve the Dirichlet problem

(33.1)

$$\text{PDE} \quad u_{rr} + \frac{1}{r}u_r + \frac{1}{r^2}u_{\theta\theta} = 0 \quad 0 < r < 1$$

$$\text{BC} \quad u(1,\theta) = g(\theta) \quad 0 \le \theta < 2\pi$$

The method of separation of variables will be the usual procedure, but after we find this series solution, we then manipulate it to get an alternative formulation (*Poisson integral formula*).

Problem (33.1) is very important in physical applications. It can be interpreted as finding the electrostatic potential inside a circle when the potential is given on the boundary. Another interpretation is the soap film model. If we start with a circular wire hoop and distort it so that the distortion is measured by $g(\theta)$ and dip it into a soap solution, a film of soap is formed within the wire. The height of the film is represented by the solution of problem (33.1), provided the displacement $g(\theta)$ is small.

The reader should be well aware of the separation-of-variables technique outlined in Figure 33.1 and should work out the details (problem 1).

A few comments on the outline in Figure 33.1

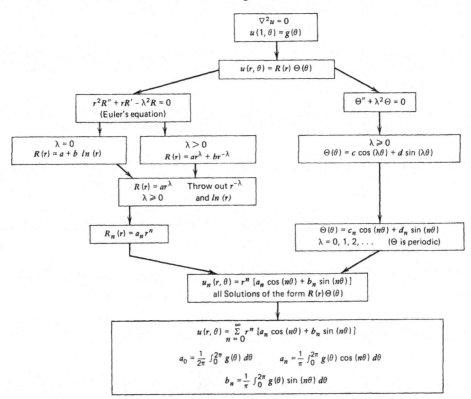

FIGURE 33.1 Outline of the solution for the interior Dirichlet problem.

1. The reader should verify that the separation constant must be *nonnegative* (that's why it's called λ^2). If the separation constant were negative, the function $\Theta(\theta)$ would not be periodic, while, on the other hand, if it were zero, we would throw out the $\ln r$ term in solution $R(r) = a + b \ln r$.
2. The reader should know how the constants a_n and b_n were obtained.

To summarize, the solution to the interior Dirichlet problem (33.1) is

(33.2) $$u(r,\theta) = \sum_{n=0}^{\infty} r^n [a_n \cos(n\theta) + b_n \sin(n\theta)]$$

Before we rewrite this solution in the alternative form, let's make some observations.

Observations on the Dirichlet Solution

1. The interpretation of our solution (33.2) is that we should expand the boundary function $g(\theta)$ as a Fourier series

$$g(\theta) = \sum_{n=0}^{\infty} [a_n \cos(n\theta) + b_n \sin(n\theta)]$$

and then solve the problem for each sine and cosine in the series. Since each of these terms will give rise to solutions $r^n \sin(n\theta)$ and $r^n \cos(n\theta)$, we can then say (by *superposition*) that

$$u(r,\theta) = \sum_{n=0}^{\infty} r^n [a_n \cos(n\theta) + b_n \sin(n\theta)]$$

2. The solution of

$$\text{PDE} \qquad \nabla^2 u = 0 \qquad 0 < r < 1$$

$$\text{BC} \qquad u(1,\theta) = 1 + \sin\theta + \frac{1}{2}\sin(3\theta) + \cos(4\theta)$$

would be

$$u(r,\theta) = 1 + r\sin\theta + \frac{r^3}{2}\sin(3\theta) + r^4\cos(4\theta)$$

Here, the $g(\theta)$ is already in the form of a Fourier series, with

$a_0 = 1$	$b_1 = 1$
$a_4 = 1$	$b_3 = 0.5$
All other a_n's $= 0$	All other b_n's $= 0$

and so we don't have to use the formulas for a_n and b_n.

3. If the radius of the circle were arbitrary (say R), then the solution would be

$$u(r, \theta) = \sum_{n=0}^{\infty} (r/R)^n [a_n \cos (n\theta) + b_n \sin (n\theta)]$$

4. Note that the constant term a_0 in solution (33.2) represents the *average* of $g(\theta)$

$$a_0 = \frac{1}{2\pi} \int_0^{2\pi} g(\theta) \, d\theta$$

This completes our discussion of the separation-of-variables solution. We now get to the interesting Poisson integral formula.

Poisson Integral Formula

We start with the separation of variables solution

$$u(r, \theta) = \sum_{n=0}^{\infty} (r/R)^n [a_n \cos (n\theta) + b_n \sin (n\theta)]$$

(we now take an arbitrary radius for the circle) and substitute the coefficients a_n and b_n. After a few manipulations involving algebra, calculus, and trigonometry, we have

$$u(r, \theta) = \frac{1}{2\pi} \int_0^{2\pi} g(\theta) \, d\theta + \frac{1}{\pi} \sum_{n=1}^{\infty} (r/R)^n \int_0^{2\pi} g(\alpha)$$

$$[\cos (n\alpha) \cos (n\theta) + \sin (n\alpha) \sin (n\theta)] d\alpha$$

$$= \frac{1}{2\pi} \int_0^{2\pi} \left\{ 1 + 2 \sum_{n=1}^{\infty} (r/R)^n \cos [n(\theta - \alpha)] \right\} g(\alpha) \, d\alpha$$

$$= \frac{1}{2\pi} \int_0^{2\pi} \left\{ 1 + \sum_{n=1}^{\infty} (r/R)^n [e^{in(\theta - \alpha)} + e^{-in(\theta - \alpha)}] \right\} g(\alpha) \, d\alpha$$

$$= \frac{1}{2\pi} \int_0^{2\pi} \left\{ 1 + \frac{re^{i(\theta - \alpha)}}{[R - re^{i(\theta - \alpha)}]} + \frac{re^{-i(\theta - \alpha)}}{[R - re^{-i(\theta - \alpha)}]} \right\} g(\alpha) \, d\alpha$$

$$= \frac{1}{2\pi} \int_0^{2\pi} \left[\frac{R^2 - r^2}{R^2 - 2rR \cos (\theta - \alpha) + r^2} \right] g(\alpha) \, d\alpha$$

This last equation is what we were looking for; it's the **Poisson Integral Formula.** So what we have is an alternative form for the solution to the interior Dirichlet problem

$$(33.3) \quad u(r,\theta) = \frac{1}{2\pi} \int_0^{2\pi} \left[\frac{R^2 - r^2}{R^2 - 2rR \cos(\theta - \alpha) + r^2} \right] g(\alpha) \, d\alpha$$

Observations on the Poisson Integral Solution

1. We can interpret the Poisson integral solution (33.3) as finding the potential u at (r,θ) as a *weighted average* of the boundary potentials $g(\theta)$ weighted by the Poisson kernel

$$\text{Poisson kernel} = \frac{R^2 - r^2}{R^2 - 2rR \cos(\theta - \alpha) + r^2}$$

This tells us something about physical systems; namely, that the potential at a point is the *weighted average* of neighboring potentials. The Poisson kernel tells just how much *weight* to assign each point (Figure 33.2).

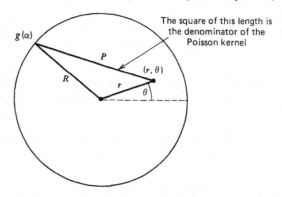

FIGURE 33.2 $u(r,\theta)$ as a weighted sum of boundary potentials.

For boundary values $g(\alpha)$ close to (r,θ), the Poission kernel gets *large*, since the denominator of the Poisson kernel is the square of the distance from (r,θ) to (R,α) (Figure 33.2). Due to this fact, the integral puts more weight on those values of $g(\alpha)$ that are closest to (r,θ). Unfortunately, if (r,θ) is extremely close to the boundary $r = R$, then the Poisson kernel gets very large for those values of α that are closest to (r,θ). For this reason, when (r,θ) is close to the boundary, the *series* solution works better for evaluating the numerical value of the solution.

2. If we evaluate the potential at the center of the circle by the Poisson integral, we find

$$u(0,0) = \frac{1}{2\pi} \int_0^{2\pi} g(\alpha)\, d\alpha$$

In other words, the potential at the center of the circle is the average of the boundary potentials.

This completes our discussion. Offhand, the reader may think that this isn't a very important problem, since the domain of the problem is so simple. It's true that Laplace's equation is easiest when the region in question is a circle, square, half plane, quarter plane, and so on, but there are a couple of important points to note.

1. In many cases, the experimenter designs the physical apparatus and has the option of shaping the boundary any way he or she likes.
2. Later on, we will study transformations known as *conformal mappings*, which allow us to transform complicated regions into simple ones (like circles). Hence, to solve the Dirichlet problem in an arbitrary region, all we have to do is transform it *conformally* into a circle, use the solutions we've found in this lesson, and transform back into the original coordinates.

NOTES

1. We can always solve the BVP (nonhomogeneous PDE)

$$\begin{array}{lll} \text{PDE} & \nabla^2 u = f & \text{Inside } D \\ \text{BC} & u = 0 & \text{On the boundary of } D \end{array}$$

by
(a) Finding any solution V of $\nabla^2 V = f$ (A particular solution)
(b) Solving the new BVP

$$\begin{array}{lll} \nabla^2 W = 0 & \text{Inside } D \\ W = V & \text{On the boundary of } D \end{array}$$

(c) Observing that $u = V - W$ is our desired solution
In other words, we can transfer the nonhomogeneity from the *PDE* to the *BC*.

2. We can solve the BVP (nonhomogeneous BC)

$$\begin{array}{lll} \text{PDE} & \nabla^2 u = 0 & \text{Inside } D \\ \text{BC} & u = f & \text{On the boundary of } D \end{array}$$

by
(a) Finding any function V that satisfies the BC $V = f$ on the boundary of D
(b) Solving the new BVP

$$\nabla^2 W = \nabla^2 V \quad \text{Inside } D$$
$$W = 0 \quad \text{On the boundary of } D$$

(c) Observing that $u = V - W$ is the solution to our problem
In other words, we can transform the nonhomogeneity from the BC to the PDE.

PROBLEMS

1. Carry out the details for the separation-of-variables solution to the interior Dirichlet problem (33.1). It's very important to know the details, since when we solve other Dirichlet problems in the next lesson, there will be some very interesting differences. You should especially know why the separation constant λ cannot be negative. Also when $\lambda = 0$, an important solution is thrown out; what is it? Look at the outline in Figure 33.1.
2. What is the solution to the interior Dirichlet problem

$$\text{PDE} \quad u_{rr} + \frac{1}{r}u_r + \frac{1}{r^2}u_{\theta\theta} = 0 \quad 0 < r < 1$$

with the follwing BCs:

(a) $u(1,\theta) = 1 + \sin\theta + \frac{1}{2}\cos\theta$
(b) $u(1,\theta) = 2$
(c) $u(1,\theta) = \sin\theta$
(d) $u(1,\theta) = \sin 3\theta$

What do the solutions look like? Do they satisfy Laplace's equation?
3. What is the solution to the following interior Dirichlet problem with radius $R = 2$:

$$\text{PDE} \quad \nabla^2 u = 0 \quad 0 < r < 2$$
$$\text{BC} \quad u(2,\theta) = \sin\theta \quad 0 \leq \theta < 2\pi$$

What does the graph look like?
4. What would be the solution of problem 3 if the BCs were changed to $u(2,\theta) = \sin(2\theta)$? What does this graph look like?

5. Solve

$$\text{PDE} \quad \nabla^2 u = 0 \quad 0 < r < 1$$

$$\text{BC} \quad u(1,\theta) = \begin{cases} \sin\theta & 0 \leq \theta < \pi \\ 0 & \pi \leq \theta < 2\pi \end{cases}$$

Roughly, what does the solution look like?

6. What does the Poisson kernel look like as a function of α: $0 \leq \alpha < 2\pi$ for $r = 3R/4$, $\theta = \pi/2$? In other words, draw the graph of the Poisson kernel.

7. Verify note 1 in the lesson.

8. Verify note 2 in the lesson.

OTHER READING

Partial Differential Equations: An Introduction by E. C. Young. Allyn and Bacon, 1972. This text has a nice discussion on the Poisson integral formula.

The Dirichlet Problem in an Annulus

PURPOSE OF LESSON: To solve the Dirichlet problem between two circles (annulus):

$$\text{PDE} \quad u_{rr} + \frac{1}{r}u_r + \frac{1}{r^2}u_{\theta\theta} = 0 \quad R_1 < r < R_2$$

$$\text{BCs} \quad \begin{aligned} u(R_1,\theta) &= g_1(\theta) \\ u(R_2,\theta) &= g_2(\theta) \end{aligned} \quad 0 \leq \theta \leq 2\pi$$

The technique used will be separation of variables, and it is similar to the *interior Dirichlet problem* except that now we don't throw out the solutions

$$\frac{1}{r^n}\sin(n\theta) \qquad \frac{1}{r^n}\cos(n\theta) \qquad \ln r$$

as we did before. Hence, our solution will be

$$u(r,\theta) = a_0 + b_0 \ln r +$$
$$\sum_{n=1}^{\infty} [(a_n r^n + b_n r^{-n})\cos(n\theta) + (c_n r^n + d_n r^{-n})\sin(n\theta)]$$

We will also discuss briefly the solution to the exterior Dirichlet problem. In this case, we throw out those terms that are unbounded at $r = \infty$. Hence, the exterior Dirichlet problem

$$\text{PDE} \quad \nabla^2 u = 0 \quad 1 < r < \infty$$
$$\text{BC} \quad u(1,\theta) = g_1(\theta) \quad 0 \leq \theta \leq 2\pi$$

has the solution

$$u(r,\theta) = \sum_{n=0}^{\infty} r^{-n}[a_n \cos(n\theta) + b_n \sin(n\theta)]$$

There are many regions of interest where we might solve the Dirichlet problem. Just to name a few, we could have the Dirichlet problem:

(a) Inside a circle (Lesson 33)
(b) In an annulus (this lesson)
(c) Outside a circle (this lesson)
(d) Inside a sphere (later)
(e) Between two spheres (later)
(f) Between two lines (in two dimensions)
(g) Between two planes (in three dimensions)

The list is endless. Our intention is to solve a representative sample of Dirichlet problems, so that the reader learns the general principles and is able to solve new ones on his or her own.

The goal in this lesson is to find the shape of a soap film between two warped hoops. Our intuition is probably not so good here as it was in the interior Dirichlet problem. The model for this problem is

(34.1)

$$\text{PDE} \qquad u_{rr} + \frac{1}{r}u_r + \frac{1}{r^2}u_{\theta\theta} = 0 \qquad R_1 < r < R_2$$

$$\text{BCs} \qquad \begin{cases} u(R_1,\theta) = g_1(\theta) \\ u(R_2,\theta) = g_2(\theta) \end{cases} \qquad 0 \le \theta \le 2\pi$$

The general picture for this problem can be seen in Figure 34.1.

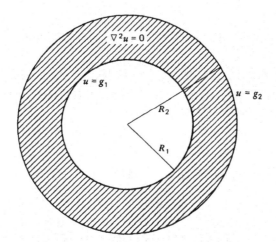

FIGURE 34.1 Laplace's equation in an annulus.

The problem here is to find the solution $u(r,\theta)$ *between* the two circles $r = R_1$ and $r = R_2$, which is given as $g_1(\theta)$ and $g_2(\theta)$ on the circles. We begin by looking for solutions of the form

$$u(r,\theta) = R(r)\Theta(\theta)$$

Substituting this into Laplace's equation, we get the two following ODEs in $R(r)$ and $\Theta(\theta)$:

$$r^2 R'' + rR' - \lambda^2 R = 0 \quad \text{(Euler's equation)}$$
$$\Theta'' + \lambda^2 \Theta = 0$$

Note that in the two equations, we required the separation constant to be greater than, or equal to, zero (we've called it λ^2), or else the solution for $\Theta(\theta)$ would not be periodic. The next step will be to solve these two ODEs and multiply their solutions together. This will give us our solutions to the PDEs that are of the form $R(r)\Theta(\theta)$.

The more interesting of the two equations is Euler's equation. Fortunately, it is one of the few ODEs with variable coefficients that can be solved fairly easily. To find the solution, it is best to consider two cases.

SOLUTION OF EULER'S EQUATION $(r^2 R'' + rR' - \lambda^2 R = 0)$

CASE 1 $(\lambda = 0)$ Here Euler's equation reduces to

$$r^2 R'' + rR' = 0$$

and it is easy to see that the general solution is

$$R(r) = a + b \ln r$$

This can be found by letting $V(r) = R'(r)$ and solving the new equation

$$rV'(r) + V(r) = 0$$

for $V(r)$. After finding $V(r) = c_1/r$ (use the ODE technique of separation of variables), we substitute back to get

$$R(r) = c_1 \ln r + c_2$$

CASE 2 $(\lambda > 0)$ Here Euler's equation is

$$r^2 R'' + rR' - \lambda^2 R = 0$$

and to solve this, we look for solutions of the form $R(r) = r^\alpha$. The goal is to find two values of α (say α_1 and α_2) so that the general solution will be

$$R(r) = ar^{\alpha_1} + br^{\alpha_2}$$

Plugging $R(r) = r^\alpha$ into Euler's equation yields $\alpha = \lambda, -\lambda$ and, hence,

$$R(r) = ar^\lambda + br^{-\lambda}$$

Using this solution to Euler's equation, we now have

$$\lambda = 0 \quad \begin{cases} R(r) = a + b \ln r \\ \Theta(\theta) = c + d\theta \end{cases}$$

$$\lambda > 0 \quad \begin{cases} R(r) = ar^\lambda + br^{-\lambda} \\ \Theta(\theta) = c \cos (\lambda\theta) + d \sin (\lambda\theta) \end{cases}$$

and using the requirement that $\Theta(\theta)$ must be *periodic* with period 2π, we have that λ must be $0, 1, 2, \ldots$. Hence, we arrive at the following solutions to Laplace's equation:

Product Solutions to Laplace's Equation

c (constants)

$c \ln r$

$cr^n \cos (n\theta)$

$cr^n \sin (n\theta)$

$cr^{-n} \cos (n\theta)$

$cr^{-n} \sin (n\theta)$

Since any sum of these solutions is also a solution, we arrive at our general solution

(34.2)
$$u(r,\theta) = a_0 + b_0 \ln r + \sum_{n=1}^{\infty} [(a_n r^n + b_n r^{-n}) \cos (n\theta) + (c_n r^n + d_n r^{-n}) \sin (n\theta)]$$

to Laplace's equation.

The only task left is to determine the constants in the sum (34.2) so that $u(r,\theta)$ satisfies the BCs

$$u(R_1,\theta) = g_1(\theta)$$
$$u(R_2,\theta) = g_2(\theta)$$

Substituting the above solution (34.2) into these BCs and integrating gives the following equations:

$$\begin{cases} a_0 + b_0 \ln R_1 = \dfrac{1}{2\pi} \int_0^{2\pi} g_1(s) \, ds \\[2mm] a_0 + b_0 \ln R_2 = \dfrac{1}{2\pi} \int_0^{2\pi} g_2(s) \, ds \end{cases} \qquad \text{(Solve for } a_0, b_0)$$

$$
(34.3) \quad
\begin{cases}
a_n R_1^n + b_n R_1^{-n} = \dfrac{1}{\pi} \displaystyle\int_0^{2\pi} g_1(s) \cos(ns)\, ds \\[4mm]
a_n R_2^n + b_n R_2^{-n} = \dfrac{1}{\pi} \displaystyle\int_0^{2\pi} g_2(s) \cos(ns)\, ds
\end{cases}
\quad \text{(Solve for } a_n, b_n)
$$

$$
\begin{cases}
c_n R_1^n + d_n R_1^{-n} = \dfrac{1}{\pi} \displaystyle\int_0^{2\pi} g_1(s) \sin(ns)\, ds \\[4mm]
c_n R_2^n + d_n R_2^{-n} = \dfrac{1}{\pi} \displaystyle\int_0^{2\pi} g_2(s) \sin(ns)\, ds
\end{cases}
\quad \text{(Solve for } c_n, d_n)
$$

From these equations, we can solve for the constants a_0, b_0, a_n, b_n, c_n, and d_n. We've now solved problem (34.1). The solution is (34.2), where the constants are determined by equations (34.3).

We will work a few simple problems in order to give the reader a feeling for this solution.

Worked Problems for the Dirichlet Problem in an Annulus

Example 1
Suppose the potential on the inside circle is zero, while the outside potential is $\sin \theta$

$$\text{PDE} \qquad \nabla^2 u = 0 \qquad 1 < r < 2$$

$$\text{BCs} \qquad \begin{cases} u(1,\theta) = 0 \\ u(2,\theta) = \sin\theta \end{cases} \qquad 0 \leq \theta \leq 2\pi$$

The first step in computing the solution is to compute the integrals in equations (34.3). Carrying out these simple calculations and solving the necessary equations for a_0, b_0, a_n, b_n, c_n, and d_n yields:

$$
\begin{aligned}
a_0 &= 0 \\
b_0 &= 0 \\
a_n &= 0 \\
b_n &= 0
\end{aligned}
\qquad n = 1, 2, \ldots
$$

$$
c_n = \begin{cases} 2/3 & n = 1 \\ 0 & \text{all other } n\text{'s} \end{cases}
$$

$$
d_n = \begin{cases} -2/3 & n = 1 \\ 0 & \text{all other } n\text{'s} \end{cases}
$$

These values give the solution as

$$u(r,\theta) = \frac{2}{3}\left(r - \frac{1}{r}\right)\sin\theta$$

We can easily check that $u(r,\theta)$ satisfies the two BCs. It's obvious that it satisfies Laplace's equation, since it is in the form of the general solution (34.2).

Example 2
Consider the problem with *constant* potentials on the boundaries

$$\text{PDE} \qquad \nabla^2 u = 0 \qquad 1 < r < 2$$

$$\text{BCs} \qquad \begin{cases} u(1,\theta) = 3 \\ u(2,\theta) = 5 \end{cases} \quad 0 \le \theta \le 2\pi$$

In this case, we can save ourselves a lot of time, since it's obvious that the solution is independent of θ (since the BCs are independent of θ). In other words, we know our solution must be of the form $a_0 + b_0 \ln r$. Using our two equations for a_0 and b_0, we obtain

$$a_0 + b_0 \ln 1 = 3$$
$$a_0 + b_0 \ln 2 = 5$$

or

$$a_0 = 3 \qquad b_0 = 2/\ln 2 = 2.9$$

Hence, the solution is

$$u(r,\theta) = 3 + 2.9 \ln r$$

The graph of this solution is given in Figure 34.2.

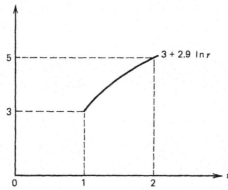

FIGURE 34.2 Radial slice of the potential inside the annulus ($1 < r < 2$).

Example 3
Another interesting problem is

$$\text{PDE} \qquad \nabla^2 u = 0 \qquad 1 < r < 2$$

$$\text{BCs} \qquad \begin{cases} u(1,\theta) = \sin\theta \\ u(2,\theta) = \sin\theta \end{cases} \qquad 0 \leqslant \theta \leqslant 2\pi$$

A quick check of the coefficients a_0, b_0, a_n, b_n, c_n, and d_n reveals that they are all zero except for c_1 and d_1. In fact, the equations for c_1 and d_1 are

$$c_1 + d_1 = \frac{1}{\pi} \int_0^{2\pi} \sin^2 s \; ds = 1$$

$$2c_1 + d_1/2 = \frac{1}{\pi} \int_0^{2\pi} \sin^2 s \; ds = 1$$

Solving for c_1 and d_1 gives $c_1 = 1/3$ and $d_1 = 2/3$. Hence, the solution is

$$u(r,\theta) = \left(\frac{1}{3}r + \frac{2}{3r} \right) \sin\theta$$

The shape of this curve for different values of θ is shown in Figure 34.3.

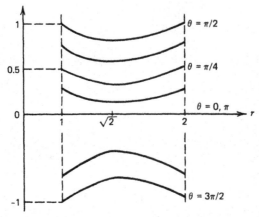

FIGURE 34.3   Soap film between $u(1,\theta) = \sin\theta$ and $u(2,\theta) = \sin\theta$.

We finish this lesson with a short discussion of the Dirichlet problem outside the circle.

Exterior Dirichlet Problem

The exterior Dirichlet problem

$$\text{PDE} \quad u_{rr} + \frac{1}{r}u_r + \frac{1}{r^2}u_{\theta\theta} = 0 \quad 1 < r < \infty$$

$$\text{BC} \quad u(1,\theta) = g(\theta) \quad 0 \leq \theta \leq 2\pi$$

is solved exactly like the interior Dirichlet problem in Lesson 33 except that now we throw out the solutions that are *unbounded* as r goes to *infinity*.

$$r^n \cos(n\theta) \qquad r^n \sin(n\theta) \qquad \ln r$$

Hence, we are left with the solution

(34.4)
$$u(r,\theta) = \sum_{n=0}^{\infty} r^{-n}[a_n \cos(n\theta) + b_n \sin(n\theta)]$$

where a_n and b_n are exactly as before

$$a_0 = \frac{1}{2\pi} \int_0^{2\pi} g(\theta)\, d\theta$$

$$a_n = \frac{1}{\pi} \int_0^{2\pi} g(\theta) \cos(n\theta)\, d\theta \qquad b_n = \frac{1}{\pi} \int_0^{2\pi} g(\theta) \sin(n\theta)\, d\theta$$

In other words, we merely expand $u(1,\theta) = g(\theta)$ as a Fourier series

$$g(\theta) = \sum_{n=0}^{\infty} [a_n \cos(n\theta) + b_n \sin(n\theta)]$$

and then insert the factor r^{-n} in each term to get the solution.
   To gain a little familiarity with this solution, two examples are given.

Examples of the Exterior Dirichlet Problem

Example 1
The Exterior Problem:

$$\text{PDE} \quad \nabla^2 u = 0 \quad 1 < r < \infty$$

$$\text{BC} \quad u(1,\theta) = 1 + \sin\theta + \cos(3\theta) \quad 0 \leq \theta \leq 2\pi$$

has the solution

$$u(r,\theta) = 1 + \frac{1}{r} \sin \theta + \frac{1}{r^3} \sin (3\theta)$$

Example 2
The Exterior Problem:

$$\text{PDE} \quad \nabla^2 u = 0 \quad 1 < r < \infty$$
$$\text{BC} \quad u(r,\theta) = \cos (4\theta) \quad 0 \leqslant \theta \leqslant 2\pi$$

has the solution

$$u(r,\theta) = \frac{1}{r^4} \cos (4\theta)$$

The reader should envision what this solution looks like.

NOTES

1. The exterior Dirichlet problem for arbitrary radius R

$$\text{PDE} \quad \nabla^2 u = 0 \quad R < r < \infty$$
$$\text{BC} \quad u(R,\theta) = g(\theta) \quad 0 \leqslant \theta \leqslant 2\pi$$

has the solution

$$u(r,\theta) = \sum_{n=0}^{\infty} (r/R)^{-n} [a_n \cos (n\theta) + b_n \sin (n\theta)]$$

2. The only solutions of the two-dimensional Laplace equation that depend only on r are *constants* and $\ln r$. The potential $\ln r$ is very important and is called the **logarithmic potential**; it will be discussed in more detail later.

PROBLEMS

1. Solve the Dirichlet problem

$$\text{PDE} \quad \nabla^2 u = 0 \quad 1 < r < 2$$
$$\text{BCs} \quad \begin{cases} u(1,\theta) = \cos \theta \\ u(2,\theta) = \sin \theta \end{cases}$$

2. What is the solution to the exterior Dirichlet problem

$$\text{PDE} \quad \nabla^2 u = 0 \quad 1 < r < \infty$$

for the following BCs:
- (a) $u(1,\theta) = 1$
- (b) $u(1,\theta) = 1 + \cos(3\theta)$
- (c) $u(1,\theta) = \sin(\theta) + \cos(3\theta)$
- (d) $u(1,\theta) = \begin{cases} 1 & 0 \leqslant \theta < \pi \\ 0 & \pi \leqslant \theta < 2\pi \end{cases}$

3. The exterior *Neumann* problem

$$\text{PDE} \qquad \nabla^2 u = 0 \qquad 1 < r < \infty$$

$$\text{BCs} \qquad \frac{\partial u}{\partial r}(1,\theta) = g(\theta) \qquad 0 \leqslant \theta \leqslant 2\pi$$

has a solution that is the same form as the Dirichlet problem

$$u(r,\theta) = \sum_{n=0}^{\infty} r^{-n}[a_n \cos(n\theta) + b_n \sin(n\theta)]$$

but now the coefficients a_n and b_n must satisfy the new BC. Substitute this solution in the BC

$$\frac{\partial u}{\partial r}(1,\theta) = \sin\theta$$

in order to obtain the solution to

$$\nabla^2 u = 0 \qquad 1 < r < \infty$$

$$\frac{\partial u}{\partial r}(1,\theta) = \sin\theta$$

Does your solution check? Of course, once you have this solution, any constant plus this solution is also a solution.

4. Substitute the general solution (34.2) into the BC

$$u(R_1,\theta) = g_1(\theta)$$
$$u(R_2,\theta) = g_2(\theta)$$

and integrate to get equations (34.3).

OTHER READING

Partial Differential Equations by Tyn Myint-U. Elsevier, 1973. An excellent discussion of the Dirichlet problem; many problems are worked out.

Laplace's Equation in Spherical Coordinates (Spherical Harmonics)

PURPOSE OF LESSON: To find particular solutions of Laplace's equation in spherical coordinates, so that they can be fitted together in various ways to solve different problems (like Dirichlet, Neumann, for example). We will also solve the *interior Dirichlet problem*

$$\text{PDE} \quad (r^2 u_r)_r + \frac{1}{\sin \phi} [\sin \phi \, u_\phi]_\phi + \frac{1}{\sin^2 \phi} u_{\theta\theta} = 0$$

$$\text{BC} \quad u(1,\theta,\phi,) = g(\phi) \quad 0 \leq \phi \leq \pi$$

for the special case where the boundary potential $g(\phi)$ depends *only* on ϕ (the angle from the north pole). Here, we expand the boundary potential $g(\phi)$ as an infinite series of *surface harmonics*

$$g(\phi) = \sum_{n=0}^{\infty} a_n P_n(\cos \phi)$$

where the surface harmonics $P_n(\cos \phi)$ (called Legendre polynomials) are all particular solutions to Laplace's equation and are *polynomials* in $\cos \phi$ of degree n. After finding this expansion, the solution is just

$$u(r,\phi) = \sum_{n=0}^{\infty} a_n r^n P_n(\cos \phi)$$

The analogous *exterior Dirichlet problem* has the solution

$$u(r,\phi) = \sum_{n=0}^{\infty} \frac{a_n}{r^{n+1}} P_n(\cos \phi)$$

An important problem in physics is to find the potential inside or outside a sphere when the potential is given on the boundary. For the *interior problem*, we must find the function $u(r,\theta,\phi)$ that satisfies

$$\text{PDE} \qquad (r^2 u_r)_r + \frac{1}{\sin \phi} [\sin \phi \, u_\phi]_\phi + \frac{1}{\sin^2 \phi} u_{\theta\theta} = 0$$

(35.1)

$$\text{BC} \qquad u(1,\theta,\phi) = g(\theta,\phi) \qquad -\pi \leqslant \theta \leqslant \pi \qquad 0 \leqslant \phi \leqslant \pi$$

Note that this spherical Laplacian is written in a different form than those we've seen before. This form is slightly more compact and easier to use.

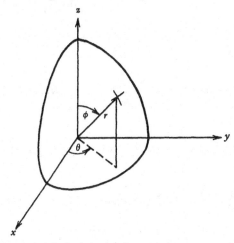

FIGURE 35.1   Dirichlet problem interior to a sphere.

A typical application of this model would be to find the temperature inside a sphere when the temperature is specified on the boundary. Quite often $g(\theta,\phi)$ has a *specific form*, so that it isn't necessary to solve the problem in its most general form.

Two important special cases are considered in this lesson. One is the case when $g(\theta,\phi)$ is *constant*, and the other is when it depends *only* on the angle ϕ (the angle from the north pole).

Special Cases of the Dirichlet Problem

Special Case 1 $g(\theta,\phi) = $ constant)

In this case, it is clear that the solution is independent of θ and ϕ, and so Laplace's equation reduces to the ODE

$$(r^2 u_r)_r = 0$$

This is a simple ODE that the reader can easily solve; the general solution is

(35.2) $$u(r) = \frac{a}{r} + b$$

In other words, constants and c/r are the only potentials that depend only on the radial distance from the origin. The potential $1/r$ is very important in physics and is called the **Newtonian potential.** We now work two problems where the potential depends only on r.

Problem 1   (Potential *interior* to a sphere)

$$\begin{array}{ll} \text{PDE} & \nabla^2 u = 0 \qquad 0 < r < 1 \\ \text{BC} & u(1,\theta,\phi) = 3 \end{array}$$

Here solution (35.2) must be $u(r,\theta,\phi) = 3$ in order to be bounded.

Problem 2   (Potential *between* two spheres each at constant potential)
Suppose we want to find the steady-state temperature between two spheres held at different temperatures.

$$\text{PDE} \qquad \nabla^2 u = 0 \qquad R_1 < r < R_2$$

$$\text{BCs} \qquad \begin{cases} u(R_1,\theta,\phi) = A \\ u(R_2,\theta,\phi) = B \end{cases}$$

Since we know the potential has the general form

$$u(r) = \frac{a}{r} + b$$

we substitute it in the BCs and solve for a and b; doing this gives

$$a = (A - B)\frac{R_1 R_2}{(R_2 - R_1)} \qquad b = \frac{R_2 B - R_1 A}{(R_2 - R_1)}$$

and, hence,

$$u(r) = \frac{(A - B)R_1 R_2}{(R_2 - R_1)r} + \frac{R_2 B - R_1 A}{(R_2 - R_1)}$$

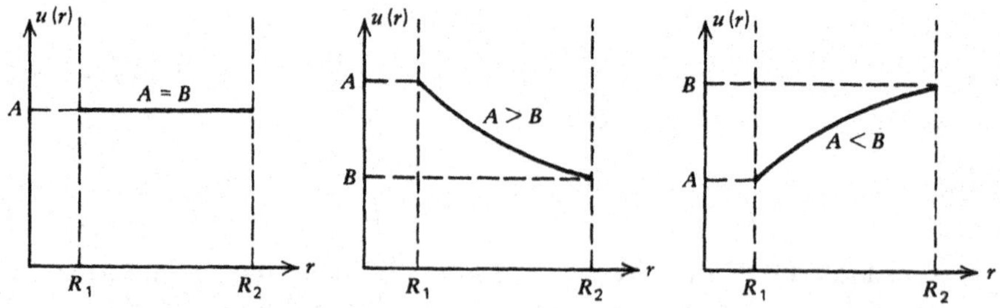

FIGURE 35.2   Potential between two concentric spheres each at constant potential A and B.

The graph of this potential is given for different values of the boundary potentials A and B in Figure 35.2.

Special Case 2 [$g(\theta,\phi)$ depends *only* on ϕ]

Here, the Dirichlet problem takes the form

$$\text{PDE} \quad (r^2 u_r)_r + \frac{1}{\sin \phi} [\sin \phi \, u_\phi]_\phi = 0 \quad 0 < r < 1$$

$$\text{BC} \quad u(1,\theta,\phi) = g(\phi) \quad 0 \leq \phi \leq \pi$$

Using separation of variables, we look for solutions of the form

$$u(r,\phi) = R(r)\Phi(\phi)$$

and arrive at the two ODEs

$$r^2 R'' + 2rR' - n(n+1)R = 0 \quad \text{(Euler's equation)}$$
$$[\sin \phi \, \Phi']' + n(n+1) \sin \phi \, \Phi = 0 \quad \text{(Legendre's equation)}$$

The separation constant is chosen to be $n(n+1)$ for convenience; the reader will see shortly why this choice is made.

We now solve Euler's equation by substituting $R(r) = r^\alpha$ in the equation and solving for α. Doing this, we get two values

$$\alpha = \begin{cases} n \\ -(n+1) \end{cases}$$

and, hence, Euler's equation has the general solution

$$R(r) = ar^n + br^{-(n+1)}$$

Legendre's equation isn't so easy; the general strategy in solving this equation is to make the substitution

$$x = \cos \phi$$

Making this change of variable gives rise to the new Legendre's equation

$$(1 - x^2)\frac{d^2\Phi}{dx^2} - 2x\frac{d\Phi}{dx} + n(n+1)\Phi = 0 \quad -1 \leq x \leq 1$$

The idea here is to solve for $\Phi(x)$ and then substitute $x = \cos \phi$ in the solution. Legendre's equation is a linear second-order ODE with variable coefficients. One of the difficulties in this equation is that the coefficient $(1 - x^2)$ of $d^2\Phi/dx^2$ is zero at the ends of the domain $-1 \leq x \leq 1$. Equations like this are called

singular differential equations and are often solved by the method of Frobenius. Without going into the details of this method (see reference 1 of the recommended reading), we arrive at a very interesting conclusion. The only bounded solutions of Legendre's equation occur when $n = 0, 1, 2, \ldots$ and these solutions are *polynomials $P_n(x)$* (Legendre polynomials)

$n = 0 \qquad P_0(x) = 1$

$n = 1 \qquad P_1(x) = x$

$n = 2 \qquad P_2(x) = \dfrac{1}{2}(3x^2 - 1)$

$n = 3 \qquad P_3(x) = \dfrac{1}{2}(5x^3 - 3x) \qquad -1 \le x \le 1$

$\begin{matrix} \cdot & & \cdot & & \cdot \\ \cdot & & \cdot & & \cdot \\ \cdot & & \cdot & & \cdot \end{matrix}$

$n \qquad P_n(x) = \dfrac{1}{2^n n!}\dfrac{d^n}{dx^n}[(x^2 - 1)^n] \qquad$ (Rodrigues' formula)

The graphs of a few Legendre polynomials are shown in Figure 35.3.

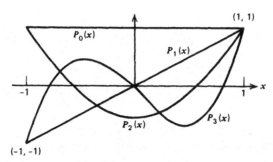

FIGURE 35.3 Legendre polynomials $P_n(x)$.

We now have that the bounded solutions of

$$r^2 R'' + 2rR' - n(n + 1)R = 0 \qquad 0 < r < 1$$
$$[\sin \phi \, \Phi']' + n(n + 1) \sin \phi \, \Phi = 0 \qquad 0 \le \phi \le \pi$$

are

$$R(r) = ar^n$$
$$\Phi(\phi) = aP_n(\cos \phi)$$

The function $P_n(\cos \phi)$ is just the n-th order Legendre polynomial with x replaced by $\cos \phi$. The final step is to form the sum

$$(35.3) \qquad u(r,\phi) = \sum_{n=0}^{\infty} a_n r^n P_n(\cos \phi)$$

in such a way that it agrees with the BC $u(1,\phi) = g(\phi)$. Substituting solution (35.3) into the BC gives

$$\sum_{n=0}^{\infty} a_n P_n(\cos \phi) = g(\phi)$$

If we *multiply* each side of this equation by $P_m(\cos \phi) \sin \phi$ and integrate ϕ from 0 to π, we get

$$\int_0^\pi g(\phi) P_m(\cos \phi) \sin \phi \, d\phi = \sum_{n=0}^{\infty} a_n \int_0^\pi P_n(\cos \phi) P_m(\cos \phi) \sin \phi \, d\phi$$

$$= \sum_{n=0}^{\infty} a_n \int_{-1}^{1} P_n(x) P_m(x) \, dx$$

$$= \begin{cases} 0 & n \neq m \\ \dfrac{2a_m}{2m + 1} & n = m \end{cases}$$

We can verify that the Legendre polynomials are orthogonal on $[-1,1]$. Hence

$$(35.4) \qquad \boxed{a_m = \frac{2m + 1}{2} \int_0^\pi g(\phi) \, P_m(\cos \phi) \sin \phi \, d\phi}$$

and so the solution to our Dirichlet problem (35.1) is

$$(35.5) \qquad \boxed{u(r,\phi) = \sum_{n=0}^{\infty} a_n r^n P_n(\cos \phi)}$$

where the coefficients a_n are given by solution (35.4).

We now give an example of a cylindrically symmetric potential.

Cylindrically Symmetric Potential   (Independent of θ)

Suppose the temperature on the surface of the sphere is given by

$$g(\phi) = 1 - \cos (2\phi) \qquad 0 \leqslant \phi \leqslant \pi$$

and suppose we would like to find the temperature *inside* the sphere. In this problem, the temperature is *constant* on circles of constant latitude (for example, the equator has temperature 2). To find u, we must solve

$$\text{PDE} \quad \nabla^2 u = 0 \quad 0 < r < 1$$
$$\text{BC} \quad u(1,\theta,\phi) = 1 - \cos(2\phi) \quad 0 \leq \phi \leq \pi$$

See Figure 35.4 for a graph of the boundary temperature.

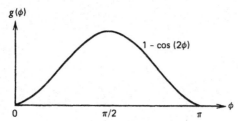

FIGURE 35.4 Temperature on latitude circles ϕ radians from the north pole.

Our goal now is to find the coefficients a_n in solution (35.5); to find them, we can either

(a) Use formula (35.4).

(b) Use existing computer programs that find the coefficients in Legendre expansions (see your computer center).

(c) Use a little common sense.

Trying the latter, consider the trigonometric identity

$$\cos(2\phi) = 2\cos^2(\phi) - 1$$

which allows us to write the boundary temperature $g(\phi)$ as

$$1 - \cos(2\phi) = 1 - [2\cos^2(\phi) - 1]$$

$$= 1 - \frac{2}{3}[3\cos^2(\phi) - 1] + \frac{1}{3}$$

$$= \frac{4}{3}P_0(\cos\phi) - \frac{4}{3}P_2(\cos\phi)$$

This gives us the expansion of $g(\phi)$ as a series of Legendre polynomials; hence, the solution to the problem is

$$u(r,\phi) = \frac{4}{3}P_0(\cos\phi) - \frac{4r^2}{3}P_2(\cos\phi)$$

$$= \frac{4}{3} - \frac{2r^2}{3}(3\cos^2\phi - 1)$$

The graphs of this solution are given in Figure 35.5 for various latitude values.

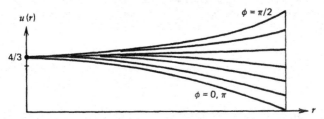

FIGURE 35.5 Temperature from the center of the sphere to the boundary.

NOTES

1. We can see that after expanding the boundary potential $g(\phi)$ as the series

$$g(\phi) = \sum_{n=0}^{\infty} a_n P_n(\cos \phi)$$

we need only multiply the n-th term by r^n to get the solution

$$u(r,\phi) = \sum_{n=0}^{\infty} a_n r^n P_n(\cos \phi)$$

2. The solution of the exterior Dirichlet problem

$$
\begin{array}{ll}
\text{PDE} & \nabla^2 u = 0 \quad 1 < r < \infty \\
\text{BC} & u(1,\theta,\phi) = g(\phi)
\end{array}
$$

is

$$u(r,\phi) = \sum_{n=0}^{\infty} \frac{b_n}{r^{n+1}} P_n(\cos \phi)$$

where

$$b_n = \frac{2n+1}{2} \int_0^{\pi} g(\phi) P_n(\cos \phi) \sin \phi \, d\phi$$

For example, the BC $g(\phi) = 3$ would yield the solution $u(r,\phi) = 3/r$. Note that in this problem, the solution goes to zero, while in *two dimensions*, the exterior solution with constant BC was *itself* a constant.

Laplace's Equation in Spherical Coordinates (Spherical Harmonics)   287

PROBLEMS

1. Substitute $R(r) = r^\alpha$ into Euler's equation

$$r^2 R'' + 2rR' - n(n + 1)R = 0$$

to find $\alpha = n, - (n + 1)$

2. Make the change of variable $x = \cos \phi$ to change the old Legendre's equation in ϕ

$$[\sin \phi \; \Phi']' + n(n + 1) \sin \phi \; \Phi = 0 \qquad 0 \leqslant \phi \leqslant \pi$$

to the new Legendre's equation in x

$$(1 - x^2) \frac{d^2\Phi}{dx^2} - 2x \frac{d\Phi}{dx} + n(n + 1)\Phi = 0 \qquad -1 \leqslant x \leqslant 1$$

3. Verify Rodrigues' formula

$$P_n(x) = \frac{1}{2^n n!} \frac{d^n}{dx^n} [(x^2 - 1)^n]$$

for the Legendre polynomials P_0, P_1, P_2, and P_3.

4. Solve the interior Dirichlet problem

$$\begin{array}{ll} \text{PDE} & \nabla^2 u = 0 \qquad 0 < r < 1 \\ \text{BC} & u(1,\phi) = \cos (3\phi) \end{array}$$

Use a trig formula to try to write $\cos (3\phi)$ in terms of $\cos \phi$, $\cos^2 \phi$, $\cos^3 \phi$, . . . and then combine them to get the expansion

$$\cos (3\phi) = a_0 P_0(\cos \phi) + a_1 P_1 (\cos \phi) + \cdots$$

5. Solve

$$\nabla^2 u = 0 \qquad 0 < r < 1$$

$$u(1,\phi) = \begin{cases} 1 & 0 \leqslant \phi \leqslant \pi/2 \\ -1 & \pi/2 < \phi \leqslant \pi \end{cases}$$

This is the problem where the boundary of the northern hemisphere is hot $(+1)$, while the southern hemisphere is cold (-1).

6. What is the solution to the exterior Dirichlet problem

$$\text{PDE} \qquad \nabla^2 u = 0 \qquad 1 < r < \infty$$
$$\text{BC} \qquad u(1,\phi) = 1 + \cos \phi$$

Does it check?

OTHER READING

Elementary Differential Equations and Boundary-Value Problems by W. Boyce and R. DiPrima. Wiley & Sons, 1969. One of the best ODE textbooks; Euler's, Bessel's, and Legendre's equations are discussed in this book.

A Nonhomogeneous Dirichlet Problem (Green's Function)

PURPOSE OF LESSON: To show how a nonhomogeneous Dirichlet problem can be solved by the Green's function approach (the impulse-response function). This important technique *resolves* the right-hand side of the equation (generally thought of as an input of some kind) into a *continuum* of impulses (delta functions or point inputs) at the different points of the domain. The *response* to each of these impulses is then found (Green's function or the impulse-response function), and then they are *summed* (integrated) to give the overall response.

A common problem in applied mathematics is to find the potential in some region of space in response to a forcing term $f(x,y)$ acting *inside* the region. In *electrostatics*, the potential (volts) in a region D is sought in response to a charge density $f(x,y)$ throughout that region. A typical example would be to find the potential inside a circle in two dimensions that satisfies (Poisson's equation with zero BC)

(36.1)

$$\text{PDE} \qquad u_{rr} + \frac{1}{r}u_r + \frac{1}{r^2}u_{\theta\theta} = f(r,\theta)$$

$$\text{BC} \qquad u(1,\theta) = 0 \qquad 0 \le \theta \le 2\pi$$

Note that we have chosen the boundary values to be zero. If we wanted to solve the general case, where both the equation and BC were nonhomogeneous, we could add the Poisson integral formula from Lesson 33 to the solution from this lesson.

In order to gain a little intuition about nonhomogeneous differential equations, let's consider graphing the solution to the following Poisson's equation:

$$\text{PDE} \qquad \nabla^2 u = -q \qquad 0 < r < 1 \qquad (q \text{ a positive constant})$$

$$\text{BC} \qquad u(1,\theta) = 0 \qquad 0 \le \theta \le 2\pi$$

Here, the potential (temperature if you like) is fixed at zero on the boundary, and the Laplacian of u is always equal to $-q$ inside the circle. Since $\nabla^2 u(p)$ measures the difference between $u(p)$ and the average of its neighbors, Poisson's equation says that the surface $u(r,\theta)$ will always be concave down, so to speak. In other words, it will look like a thin membrane fixed at the boundary that was continuously being pushed up by a stream of air from below. If the right-hand side were a function $f(x,y)$ that changed over the domain, then the concavity at each point would change.

We now get to the major part of this lesson: to introduce Green's function and solve equation (36.1).

First, however, we must introduce the notion of potential due to *point sources and sinks*.

Potentials from Point Sources and Sinks

In solving a nonhomogeneous linear equation, it is sufficient to solve the equation with a point source, since we can find the solution to the general problem by summing the responses to point sources. Our goal here is to find the potential in some region of space due to a point source (or sink). We can interpret these points in a variety of ways. In heat flow, we could think of a source as a point where heat is created and a sink as a point where it is destroyed. On the other hand, in electrostatics, a point source would be a single positive charge (proton), while a sink would be a single negative charge (electron). In any case, whatever the interpretation, we will now find the potential $u(r)$ in two dimensions that depends on a single point source (the potential in three dimensions is left as a problem).

Suppose we have a single point source of magnitude $+q$ located at the origin. It is clear that the heat (or whatever) will flow outward along radial lines, and, hence, if we compute the total outward flux across a circle of radius r, we have the situation described in figure 36.1

$$\text{Total outward flux across the circle} \quad = -\int_0^{2\pi} u_r(r) r \, d\theta$$

$$= -2\pi r u_r(r)$$

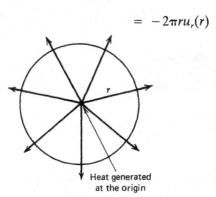

Heat generated
at the origin

FIGURE 36.1   Radial flow of heat due to a point source.

But the outward flux must be equal to the heat generated within the circle (conservation of energy), and so we have

$$-2\pi r u_r(r) = q$$

Solving this simple differential equation for $u(r)$, we get

$$u(r) = \frac{-q}{2\pi} \ln r = \frac{q}{2\pi} \ln \frac{1}{r}$$

See Figure 36.2.

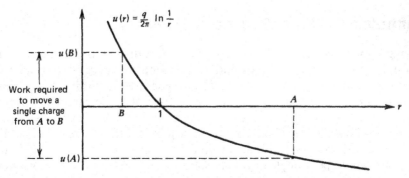

FIGURE 36.2 Potential due to a point source in two dimensions.

In terms of electrostatics, the potential difference $u(B) - u(A)$ represents the work needed to move a single positive charge from A to B (Figure 36.2). A *sink*, on the other hand, is represented by a *negative* source, and so a sink with magnitude $-q$ would give rise to a potential field

$$u(r) = \frac{-q}{2\pi} \ln \frac{1}{r}$$

This completes the discussion of potential due to point charges; we are now in position to solve the nonhomogeneous equation by means of Green's function.

Poisson's Equation inside a Circle

We will now solve the important problem

(36.2)   PDE $u_{rr} + \frac{1}{r}u_r + \frac{1}{r^2}u_{\theta\theta} = f(r,\theta)$ $0 < r < 1$

         BC $u(1,\theta) = 0$ $0 \le \theta \le 2\pi$

The Green function technique (impulse-response method) consists of two steps:
1. Finding the potential $G(r,\theta,\rho,\phi)$ at (r,θ), which we force to be zero on the boundary and which is due to a single charge (magnitude 1) at (ρ,ϕ)

2. Summing the individual responses $G(r,\theta,\rho,\phi)$ weighted by the right-hand side (charge density) $f(r,\theta)$ over all (ρ,ϕ) in the circle to get the solution

$$u(r,\theta) = \int_0^{2\pi} \int_0^1 G(r,\theta,\rho,\phi)\, f(\rho,\phi)\, \rho \, d\rho \, d\phi$$

We now find the impulse response $G(r,\theta,\rho,\phi)$ for our problem.

Finding the Potential Response $G(r,\theta,\rho,\phi)$

We first replace the right-hand side $f(r,\theta)$ by a point source of magnitude $+1$ at an arbitrary point (ρ,ϕ). Mathematically, we call a point source an impulse function (or delta function) and represent it by $\delta(r - \rho, \theta - \phi)$. We interpret this delta function as a function of r and θ that is zero for all points except at (ρ,ϕ), where the unit charge is located. In terms of forces, we could interpret the delta function as a point force of magnitude $+1$ at (ρ,ϕ). The idea now is to find the potential response (which we force to be zero on the boundary) due to a single point charge. This function is called the **impulse response function** (or **Green's function**), and it is the *response* at (r,θ) to a single *source* at (ρ,ϕ). The difficulty in finding this function is due to the fact that it must vanish on the boundary. If we didn't require zero, then the problem would be easy, since we already know that

$$\frac{1}{2\pi} \ln \frac{1}{r}$$

is the potential due to a charge at (ρ,ϕ) [where r is the distance from the charge (ρ,ϕ)].

Physically, finding $G(r,\theta,\rho,\phi)$ corresponds to one of the following:
1. Finding the equilibrium temperature inside the circle with a heat source at (ρ,ϕ) and the boundary temperature fixed at zero.
2. Finding the height of a stretched membrane fixed at zero on the boundary but pulled up to a great height at (ρ,ϕ).
3. Finding the electrostatic potential inside the circle due to a single positive charge at (ρ,ϕ) with the boundary potential grounded to zero.

We will now find Green's function; it will look something like Figure 36.3.

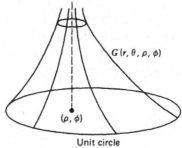

$G(r,\theta,\rho,\phi)$

(ρ,ϕ)

Unit circle

FIGURE 36.3 Green's function $G(r,\theta,\rho,\phi)$ due to a source at (ρ,ϕ).

Steps for Finding The Solution

STEP 1   Since the function

$$\frac{1}{2\pi} \ \ln \ \frac{1}{R}$$

is the potential at $P = (r,\theta)$ due to a single unit charge at $Q = (\rho,\phi)$ (where R is the *distance* from P to Q), the only thing left to do is modify the function so that it is zero on the boundary.

STEP 2   Physicists know from experiments that the potential field due to positive and negative charges placed a given distance apart give rise to *circles of constant potential* (Figure 36.4).

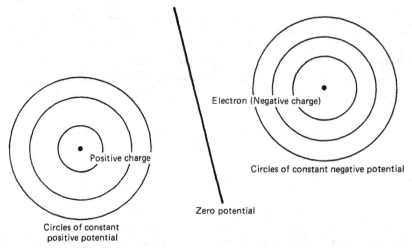

FIGURE 36.4   Potential field due to two oppositely charged particles.

So the strategy in finding Green's function is to place another charge (negative) *outside* the circle at such a point that the potential due to both is *constant* on the circle $r = 1$. We can then *subtract* this constant value to obtain a zero potential on the boundary. It is obvious now that this potential will satisfy our desired properties for $G(r,\theta,\rho,\phi)$. The big question is, of course, where do we place the *negative charge* outside the circle, so that the potential on the boundary is constant? Without going into the details, we can show rather easily that if the negative charge is placed at $\overline{Q} = (\overline{\rho},\overline{\phi}) = (1/\rho,\phi)$, then the *potential*

$$u(r,\theta) \ = \ \frac{1}{2\pi} \ \ln 1/R \ - \ \frac{1}{2\pi} \ \ln 1/\overline{R}$$

due to the two charges will be constant on the circle $r = 1$. The variables R and

$\overline{R}$ are the distances from the two charges to (r,θ). In fact, the constant potential on the circle $r = 1$ can easily be shown to be

$$\frac{-1}{2\pi} \ln \rho \quad \text{(A positive constant)}$$

See Figure 36.5

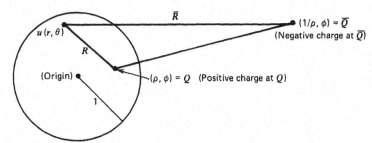

$u(r, \theta)$

$\overline{R}$

$(1/\rho, \phi) = \overline{Q}$
(Negative charge at $\overline{Q}$)

R

(Origin)

$(\rho, \phi) = Q$ (Positive charge at Q)

1

Constant potential on the circle $r = 1$

FIGURE 36.5 Charges at Q and $\overline{Q}$ giving rise to constant potential at $r = 1$.

With these steps in mind, we construct Green's function

(36.3) $\qquad G(r,\theta,\rho,\phi) = \frac{1}{2\pi} \ln 1/R - \frac{1}{2\pi} \ln 1/\overline{R} + \frac{1}{2\pi} \ln \rho$

Potential due to positive charge at Q Potential due to negative charge at $\overline{Q}$ Subtracting the constant potential on the boundary

where

$$R = \sqrt{r^2 - 2r\rho \cos(\theta - \phi) + \rho^2}$$

$$\overline{R} = \sqrt{r^2 - 2\frac{r}{\rho} \cos(\theta - \phi) + 1/\rho^2}$$

(These two formulas are just trigonometric formulas for the distance between two points in polar coordinates.) To find the solution to our original problem, we merely superimpose the impulse functions; this brings us to the final step.

STEP 3   Superposition of the impulse responses. This step is easy; we just write

$$u(r,\theta) = \int_0^{2\pi} \int_0^1 G(r,\theta,\rho,\phi)f(\rho,\phi)\rho \; d\rho \; d\phi$$

or

(36.4) $$u(r,\theta) = \frac{1}{2\pi} \int_0^{2\pi} \int_0^1 \ln (\rho\bar{R}/R)f(\rho,\phi)\rho \ d\rho \ d\phi$$

This is Green's function solution of Poisson's equation inside a circle. If we were given the charge density $f(r,\theta)$, we could evaluate this integral numerically.

NOTES

1. It is also possible to solve

$$\text{PDE} \quad \nabla^2 u = 0 \quad 0 < r < 1$$

$$\text{BC} \quad u(1,\theta) = g(\theta) \quad 0 \leqslant \theta \leqslant 2\pi$$

by means of the *Green's function* approach. In this case, the solution is

$$u(r,\theta) = \int_0^{2\pi} \frac{\partial G}{\partial r}(r,\theta,1,\phi)g(\phi) \ d\phi$$

which, if we compute $\partial G/\partial r$ (a rather tedious computation), gives

(36.5) $$u(r,\theta) = \frac{1}{2\pi} \int_0^{2\pi} \left[\frac{1 - r^2}{1 - 2r \cos (\theta - \phi) + r^2} \right] g(\phi) \ d\phi$$

which is the Poisson integral formula we found in Lesson 33.

2. The solution to the general Dirichlet problem

$$\text{PDE} \quad \nabla^2 u = f(r,\theta) \quad 0 < r < 1$$

$$\text{BC} \quad u(1,\theta) = g(\theta) \quad 0 < \theta \leqslant 2\pi$$

would be the sum of equations (36.4) and (36.5).

3. We can solve many problems in different domains by means of the Green function approach. However, we must find a new Green function for each domain and each new equation and finding Green's function is not always easy.

4. To actually evaluate solution (36.4) for most sources $f(r,\theta)$, we must resort to numerical integration on a computer.

PROBLEMS

1. Find the potential due to a point source in three dimensions.
2. Find Green's function $G(x,y,\xi,\eta)$ for Laplace's equation in the upper-half plane $y > 0$. In other words, find the potential in the upper-half plane at the point (x,y) (zero on the boundary $y = 0$) due to a point charge at (ε,η). See the following figure.

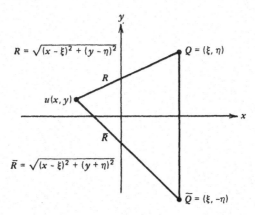

HINT If we place a negative charge at $\overline{Q} = (\xi,-\eta)$, then it's clear that the potential field on the line $y = 0$ due to the two charges at Q and $\overline{Q}$ is zero. Hence, Green's function would be the resultant field due to these two charges.

3. Using the results of problem 2, what is the solution to Poisson's equation $\nabla^2 u = -k$ in the upper-half plane with zero BC?
4. How would you go about constructing Green's function for the first quadrant $x > 0, y > 0$?
5. An alternative approach to solving Poisson's equation that works sometimes is the following; suppose you want to solve:

$$\text{PDE} \quad \nabla^2 u = 1 \quad 0 < r < 1$$
$$\text{BC} \quad u(1,\theta) = \sin \theta \quad 0 \leqslant \theta \leqslant 2\pi$$

Start by trying to find any particular solution of $\nabla^2 u = 1$ by substituting

$$u_p(r,\theta) = Ar^2$$

into the differential equation and solving for the constant.

After finding a particular solution $u_p(r,\theta)$, consider letting $u = w + u_p$ and ask the question, what boundary-value problem will $w(r,\theta)$ satisfy? After you determine this, solve for $w(r,\theta)$. Finally, what is the answer $u(r,\theta)$ of the original problem? Does it check? Look at the answer carefully; what is the interpretation of each term?

OTHER READING

1. *Introduction to Partial Differential Equations with Applications* by E. C. Zachmanoglou and D. W. Thoe. Williams and Wilkins, 1976; Dover, 1986. More advanced than this book, but well-written with several examples; could be used for extra reading by serious students.

2. *Partial Differential Equations of Mathematical Physics* by Tyn Myint-U. Elsevier, 1973. A good discussion of Green's function.

PART 5

Numerical and Approximate Methods

Numerical Solutions (Elliptic Problems)

PURPOSE OF LESSON: To show how a partial differential equation can be changed to a system of algebraic equations by replacing the *partial derivatives* in the differential equation with their *finite-difference approximations*. The system of algebraic equations can then be solved numerically by an iterative process in order to obtain an approximate solution to the PDE.

It is also pointed out that the reader can obtain an existing computer package (ELLPACK) that will solve general elliptic problems.

So far, we have studied several techniques for solving linear PDEs. However, most of the equations we've attacked were reasonably simple, had reasonably simple BCs, and had reasonably shaped domains. But many problems cannot be simplified to fit this general mold and must be solved by numerical approximations. Over the past ten years, scientists and engineers have begun to attack many more problems as a result of more computing power and more sophisticated numerical methods. Several new techniques have been developed to take advantage of high-speed computing machinery. Nonlinear problems in fluid dynamics, elasticity, and potential theory involving two and three dimensions are being solved today that were not even considered ten years ago.

There are several procedures that come under the name of numerical methods. The reader can look in reference 1 of the recommended reading for a more complete discussion of these techniques. This lesson and the next two show how the very popular *finite-difference method* can be used to solve elliptic, hyperbolic, and parabolic equations.

To begin, we introduce the idea of *finite differences*. We then show how to use these finite differences to solve a Dirichlet problem inside a square.

Finite-Difference Approximations

First, we recall the Taylor series expansion of a function $f(x)$

$$f(x + h) = f(x) + f'(x)h + \frac{f''(x)}{2!}h^2 + \ldots$$

If we *truncate* this series after two terms, we have the approximation

$$f(x + h) \cong f(x) + f'(x)h$$

Hence, we can solve for $f'(x)$

$$(37.1) \qquad f'(x) \cong \frac{f(x + h) - f(x)}{h}$$

which is called the **forward-difference approximation** to the first derivative $f'(x)$.

We could also replace h by $-h$ in the Taylor series and arrive at the **backward-difference approximation**

$$(37.2) \qquad f'(x) \cong \frac{f(x) - f(x - h)}{h}$$

or by subtracting

$$f(x - h) \cong f(x) - f'(x)h$$

from

$$f(x + h) \cong f(x) + f'(x)h$$

we can obtain the **central-difference approximation**

$$(37.3) \qquad f'(x) \cong \frac{1}{2h} [f(x + h) - f(x - h)]$$

By retaining *another term* in the Taylor series, this type of analysis can be extended to arrive at the central-difference approximation of the second derivative $f''(x)$

$$(37.4) \qquad f''(x) \cong \frac{1}{h^2} [f(x + h) - 2f(x) + f(x - h)]$$

We now extend the finite-difference approximations to *partial derivatives*. If we begin with the Taylor series expansion in two variables

$$u(x + h, y) = u(x, y) + u_x(x, y)h + u_{xx}(x, y)\frac{h^2}{2!} + \dots$$

$$u(x - h, y) = u(x, y) - u_x(x, y)h + u_{xx}(x, y)\frac{h^2}{2!} - \dots$$

we can deduce the following:

$$u_x(x, y) \cong \frac{u(x + h, y) - u(x, y)}{h} \qquad \text{(Forward difference)}$$

$$u_{xx}(x,y) \cong \frac{1}{h^2}[u(x + h,y) - 2u(x,y) + u(x - h,y)]$$

$$u_y(x,y) \cong \frac{u(x,y + k) - u(x,y)}{k}$$

$$u_{yy}(x,y) \cong \frac{1}{k^2}[u(x,y + k) - 2u(x,y) + u(x,y - k)]$$

Which approximation to use (forward, central, or backward) depends on the problem, but in this lesson, we will use the central-difference approximation.

To illustrate how to use these approximations, we consider the simple Dirichlet problem.

Dirichlet Problem Solved by the Finite-Difference Method

(37.5) PDE $u_{xx} + u_{yy} = 0$ $0 < x < 1$ $0 < y < 1$

BCs $u = 0$ On the top and sides of the square
$u(x,0) = \sin(\pi x)$ $0 \leq x \leq 1$

We begin this problem by drawing the grid system on the xy-plane shown in Figure 37.1.

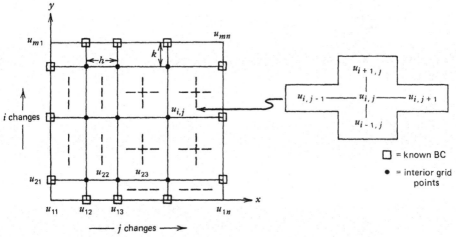

FIGURE 37.1 Grid lines for the Dirichlet problem inside a square.

It is also convenient (especially if we want to use a computer) to use the following notation:

$$u(x,y) = u_{i,j}$$

$$u(x,y + k) = u_{i+1,j}$$

$$u(x, y - k) = u_{i-1,j}$$

$$u(x + h, y) = u_{i,j+1}$$

$$u(x - h, y) = u_{i,j-1}$$

$$u_x(x,y) = \frac{1}{2h} (u_{i,j+1} - u_{i,j-1})$$

$$u_y(x,y) = \frac{1}{2k} (u_{i+1,j} - u_{i-1,j})$$

$$u_{xx}(x,y) = \frac{1}{h^2} (u_{i,j+1} - 2u_{i,j} + u_{i,j-1})$$

(Central-difference formulas)

$$u_{yy}(x,y) = \frac{1}{k^2} (u_{i+1,j} - 2u_{i,j} + u_{i-1,j})$$

Our strategy for solving this Dirichlet problem is to replace the partial derivatives in Laplace's equation

$$u_{xx} + u_{yy} = 0$$

by their finite-difference approximations. Doing this and using the compact notation $u_{i,j}$, we have the following *difference equation:*

$$\nabla^2 u = \frac{1}{h^2} (u_{i,j+1} - 2u_{i,j} + u_{i,j-1}) + \frac{1}{k^2} (u_{i+1,j} - 2u_{i,j} + u_{i-1,j}) = 0$$

By letting the two discretization sizes h and k be the same, Laplace's equation is replaced by

(37.6) $$(u_{i+1,j} + u_{i-1,j} + u_{i,j+1} + u_{i,j-1} - 4u_{i,j}) = 0$$

or solving for $u_{i,j}$

$$u_{i,j} = \frac{1}{4} (u_{i+1,j} + u_{i-1,j} + u_{i,j+1} + u_{i,j-1})$$

Note that here the $u_{i,j}$'s would stand for the solution at the *interior* grid points. This last equation says that we can approximate the solution $u_{i,j}$ by *averaging* the solution at the *four neighboring grid points.* Hence, we can devise a numerical strategy for solving the problem.

Numerical Algorithm for Solving the Dirichlet Problem (Liebmann's method)

STEP 1   Seek the solution $u_{i,j}$ at the interior grid points by setting them equal to the *average* of all the BCs (reasonable start).

STEP 2 Systematically run over all the *interior* grid points, replacing the old estimates by the average of its four neighbors. It doesn't make much difference in what order this process is carried out, but, generally, it is done in a row by row (or column by column) manner. After a few iterations, this process will converge to an approximate solution of the problem. The rate of change of this process is generally slow but can be speeded up in a number of ways; interested readers should consult reference 1 of the recommended reading.

This completes the discussion of our Dirichlet problem; the reader is asked to carry out three iterations of Liebmann's method in the problems.

NOTES

1. If we write equations (37.6) for four interior grid points (that is, $m = n = 4$), we will get the four algebraic equations:

(37.7)
$$-4u_{22} + 0 + \sin(\pi/3) + u_{23} + u_{32} = 0$$
$$-4u_{23} + u_{22} + \sin(2\pi/3) + 0 + u_{33} = 0$$
$$-4u_{32} + 0 + u_{22} + u_{33} + 0 = 0$$
$$-4u_{33} + u_{32} + u_{23} + 0 + 0 = 0$$

   from which we can solve for u_{22}, u_{23}, u_{32}, and u_{33}. The solution of these equations can be found by *iterative methods*, and Liebmann's method is one of them.

2. If we made our discretization sizes h and k smaller (so that we had more grid points), the analysis would be similar except that the system of algebraic equations (37.7) would be larger. In general, the number of *equations* will be equal to the number of *interior grid points*.

3. The system of equations (37.7) can be written in matrix form

$$\begin{bmatrix} -4 & 1 & 1 & 0 \\ 1 & -4 & 0 & 1 \\ 1 & 0 & -4 & 1 \\ 0 & 1 & 1 & -4 \end{bmatrix} \begin{bmatrix} u_{22} \\ u_{23} \\ u_{32} \\ u_{33} \end{bmatrix} = \begin{bmatrix} -\sin(\pi/3) \\ -\sin(2\pi/3) \\ 0 \\ 0 \end{bmatrix} = \begin{bmatrix} -0.86 \\ -0.86 \\ 0 \\ 0 \end{bmatrix}$$

   In general, when we have several equations (maybe 1,000) this coefficient matrix takes on a specific form with many zeros. The solution of these sparse systems of equations can be found by special numerical methods. Iterative procedures, such as Jacobi's method, Gauss Seidel, and successive over-relaxation (SOR) are commonly used (along with techniques for speeding up convergence).

4. To solve the Neumann problem where there are *derivatives* on the boundary, we must also replace these derivatives by some finite difference approximation.

5. We can also solve equations like:
   (a) $u_{xx} + u_{yy} = f(x,y)$ (Nonhomogeneous equations)
   (b) $xu_{xx} + u_{yy} + 2u = \sin(x - y)$ (Variable coefficients; non-homogeneous)
   (c) $\sin xu_{xx} + u_{xy} + 3u = 0$ (Variable coefficients)
   by the finite-difference method.
6. If the domain of the problem is an *irregularly* shaped region, we can overlay the region with grid lines and then approximate the solution at nearby grid points by interpolating the boundary conditions. After doing this, we can proceed in the usual manner. See Figure 37.2.

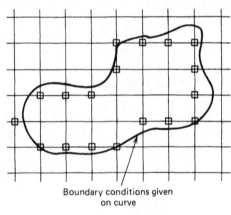

Boundary conditions given
on curve

□= New BC found by
interpolation of BC
on curve

FIGURE 37.2

7. Several journals list computer programs for solving PDEs; some of them are:
   (a) *ACM Transactions on Mathematical Software*
   (b) *Computer Journal*
   (c) *Numerische Mathematik*
   (d) *BIT*

In addition, an extensive package of programs, called ELLPACK, has recently been designed for the purpose of solving fairly general elliptic boundary-value problems. This package will solve a wide variety of problems in two or three dimensions, various coordinate systems, arbitrary boundaries, general BCs, by an assortment of different methods.*

* Anyone interested in obtaining information about this program should contact Dr. John Rice, ELLPACK User's Guide CSD-TR 226, Computer Center, Purdue University, West Lafayette, Indiana 47907.

PROBLEMS

1. Derive approximation equation (37.4) for the second derivative $f''(x)$

$$f''(x) = \frac{1}{h^2}[f(x + h) - 2f(x) + f(x - h)]$$

2. Carry out the computation for two iterations in Dirichlet problem (37.5) using the Liebmann iterative process. Is the method converging?

3. What algebraic equations must be solved when you use finite-difference approximations to solve the following Poisson equation inside the square:

$$\text{PDE} \quad u_{xx} + u_{yy} = f(x,y) \quad 0 < x < 1 \quad 0 < y < 1$$
$$\text{BC} \quad u(x,y) = g(x,y) \quad \text{On the boundary}$$

4. What algebraic equations must you solve when replacing the derivatives in

$$\text{PDE} \quad u_{xx} + u_{yy} + 2u = 0 \quad 0 < x < 1 \quad 0 < y < 1$$
$$\text{BC} \quad u(x,y) = g(x,y) \quad \text{On the boundary}$$

by their finite differences?

5. How would you solve the Neumann problem inside the square

$$\text{PDE} \quad u_{xx} + u_{yy} = 0 \quad 0 < x < 1 \quad 0 < y < 1$$

$$\text{BC} \quad \begin{cases} u = 0 & \text{On the top, bottom, and} \\ & \text{left-hand side of the square} \\ \dfrac{\partial u}{\partial x}(1,y) = 1 & 0 \leqslant y \leqslant 1 \end{cases}$$

by the finite-difference method?

6. Write a *flow diagram* to solve the Dirichlet problem inside the square

$$\text{PDE} \quad u_{xx} + u_{yy} = f(x,y) \quad 0 < x < 1 \quad 0 < y < 1$$
$$\text{BC} \quad u(x,y) = g(x,y) \quad \text{On the boundary}$$

with an arbitrary number of grid lines. If you know a computer language, write a program to carry out these computations.

OTHER READING

1. *Numerical Methods for Partial Differential Equations* by W. F. Ames. Academic Press, 1977. An up-to-date authoritative text on numerical techniques.

2. *Numerical Analysis* by S. S. Kunz. McGraw-Hill, 1964. Chapter 13 offers a clear, precise summary of some numerical methods in PDE theory.

3. *Numerical Solution of Partial Differential Equations* by G. D. Smith. Oxford University Press, 1965. A concise book describing finite difference methods in PDE theory; clearly written.

An Explicit Finite-Difference Method

PURPOSE OF LESSON: To introduce the idea of explicit finite-difference methods and show how they can be used to solve hyperbolic and parabolic problems. The basic idea is that after a PDE like

$$u_t = u_{xx}$$

is replaced by its finite-difference approximation, we can solve for the solution explicitly at *one value of time* in terms of the solution at *earlier values of time*. In this way, an initial-boundary-value problem (hyperbolic or parabolic) can be solved by consecutively finding the solution at larger and larger values of time.

A problem we face is that as we make the grid sizes *small* so that the finite differences accurately represent the derivatives, the number of computations *increases*, and so the roundoff error increases.

In the previous lesson, we solved elliptic boundary-value problems (steady-state problems) where the PDE was satisfied in a given region of space, and the solution (or its derivative) was specified on the boundary. In those types of problems, we found the approximate solution at the *interior grid points* by solving a system of algebraic equations. In other words, the solution at all the interior grid points was found *simultaneously*.

In this lesson, we will show how *time-dependent problems* can be solved by finite-difference approximations. The idea here is that if we are given the solution when time is *zero*, we can then find the solution for $t = \Delta t, 2\Delta t, 3\Delta t, \ldots$ by means of a *marching process*. Replacing both the *space* and *time* derivatives by their finite-difference approximations, we can then solve for the solution $u_{i,j}$ in the difference equation *explicitly* in terms of the solution at earlier values of time. This process is called an **explicit-type marching process**, since we find the solution at a *single* value of time in terms of the solution at earlier values of time.

To show how this method works, we consider a representative problem from heat flow.

The Explicit Method for Parabolic Equations

Consider the problem of heat flow along a rod initially at temperature zero, where the left end of the rod is fixed at temperature one, and the right-hand side experiences a heat loss (or gain) proportional to the difference between the temperature at that end and an outside temperature that is given by $g(t)$. In other words, we solve the problem

$$\text{PDE} \qquad u_t = u_{xx} \qquad 0 < x < 1 \qquad 0 < t < \infty$$

$$\text{BCs} \qquad \begin{cases} u(0,t) = 1 \\ u_x(1,t) = -[u(1,t) - g(t)] \end{cases} \qquad 0 < t < \infty$$

$$\text{IC} \qquad u(x,0) = 0 \qquad 0 \leqslant x \leqslant 1$$

To solve this problem by finite differences, we start by drawing the usual rectangular grid system with grid points:

$$\begin{aligned} x_j &= jh \qquad j = 0, 1, 2, \ldots, n \\ t_i &= ik \qquad i = 0, 1, 2, \ldots, m \end{aligned}$$

See Figure 38.1.

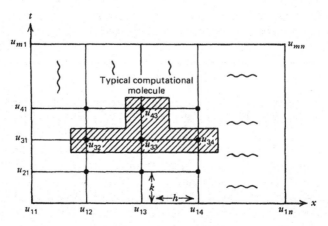

FIGURE 38.1   Grid system for a heat-flow problem.

Note that in Figure 38.1, the $u_{i,j}$ on the *left* and *bottom* are given BCs and ICs, and our job is to find the other $u_{i,j}$'s. To do this, we begin by replacing the partial derivatives u_t and u_{xx} in the heat equation with their approximations

$$u_t = \frac{1}{k}[u(x, t+k) - u(x,t)] = \frac{1}{k}(u_{i+1,j} - u_{i,j})$$

$$u_{xx} = \frac{1}{h^2} [u(x + h,t) - 2u(x,t) + u(x - h,t)] = \frac{1}{h^2} (u_{i,j+1} - 2u_{i,j} + u_{i,j-1})$$

By substituting these expressions into $u_t = u_{xx}$ and solving for the solution at the largest value of time, we have

(38.2)
$$\boxed{u_{i+1,j} = u_{i,j} + \frac{k}{h^2} [u_{i,j+1} - 2u_{i,j} + u_{i,j-1}]}$$

This is the formula we are looking for, since it gives us the solution at one value of time in terms of the solution at earlier values of time (note that the index i stands for time). Figure 38.1 shows those values of the solution that are involved in the formula.

We are now almost ready to begin the computations. First, however, we must approximate the derivative in the right-hand BC

$$u_x(1,t) = - [u(1,t) - g(t)]$$

by

(38.3)
$$\frac{1}{h} [u_{i,n} - u_{i,n-1}] = -[u_{i,n} - g_i]$$

where $g_i = g(ik)$ is given. Note that we have replaced $u_x(1,t)$ by the *backward-difference approximation*, since the forward-difference approximation would require knowing values of $u_{i,j}$ outside the domain. Solving now for $u_{i,n}$ in this BC gives us

(38.4)
$$u_{i,n} = \frac{u_{i,n-1} + hg_i}{1 + h}$$

With this equation and our explicit formula (38.2), we are ready to begin the computations.

Algorithm for the Explicit Method

STEP 1 Find the solution at the grid points for $t = \Delta t$ by using the explicit formula

$$u_{2,j} = u_{1,j} + \frac{k}{h^2} [u_{1,j+1} - 2u_{1,j} + u_{1,j-1}] \qquad j = 2, 3, \ldots, n - 1$$

See Figure 38.2.

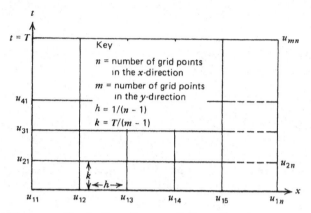

FIGURE 38.2   Diagram illustrating the explicit method.

STEP 2   Find $u_{2,n}$ from formula (38.4)

$$u_{2,n} = \frac{u_{2,n-1} + hg_2}{1 + h}$$

Steps 1 and 2 find the solution for $t = \Delta t$. To find the solution for $t = 2\Delta t$ (second row from the bottom in Figure 38.2), repeat steps 1 and 2, moving up one more row (increase i by 1) and using the values of $u_{i,j}$ just computed; for $t = 3\Delta t, 4\Delta t, \ldots$, keep repeating the same process.

In order for the reader to be able to computerize this method, we will present a fairly detailed flow diagram of the method in Figure 38.3. Those students not familiar with flow diagrams should think of them as links between computational algorithms and detailed computer programs. Flow diagrams explain in a precise manner how the computations should be carried out.

NOTES

1. There is a serious deficiency in the explicit method, for if the step size in t is large compared to the step size in x, then machine roundoff error can grow until it ruins the accuracy of the solution. The relative size of these two numbers x and t depends on the particular equation and the BCs, but, generally, the step size in t should be much smaller than the step size in x. In reference 3 of the recommended reading in Lesson 37, the author proves that we must have $k/h^2 \leq 0.5$ in order for this method to work.
2. A general rule of thumb is that as the step sizes Δt and Δx are made smaller, the *truncation error* of approximating partial derivatives by finite differences decreases. However, the smaller these grid sizes, the more computations necessary, and, hence, the *roundoff error*, as a result of rounding off our computations, will increase. Therefore, we have the phenomenon illustrated in Figure 38.5.

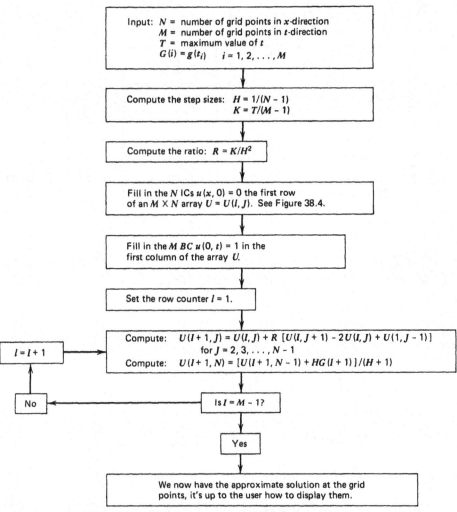

Input: N = number of grid points in x-direction
        M = number of grid points in t-direction
        T = maximum value of t
        $G(i) = g(t_i)$ $i = 1, 2, \ldots, M$

Compute the step sizes: $H = 1/(N - 1)$
                         $K = T/(M - 1)$

Compute the ratio: $R = K/H^2$

Fill in the N ICs $u(x, 0) = 0$ the first row of an $M \times N$ array $U = U(I, J)$. See Figure 38.4.

Fill in the M BC $u(0, t) = 1$ in the first column of the array U.

Set the row counter $I = 1$.

$I = I + 1$

Compute: $U(I + 1, J) = U(I, J) + R\,[U(I, J + 1) - 2U(I, J) + U(I, J - 1)]$
          for $J = 2, 3, \ldots, N - 1$
Compute: $U(I + 1, N) = [U(I + 1, N - 1) + HG(I + 1)]/(H + 1)$

Is $I = M - 1$?

No

Yes

We now have the approximate solution at the grid points, it's up to the user how to display them.

FIGURE 38.3   Flow diagram of the explicit method.

3. The hyperbolic problem

$$\text{PDE} \qquad u_{tt} = u_{xx} \qquad 0 < x < 1 \qquad 0 < t < \infty$$

$$\text{BCs} \qquad \begin{cases} u(0,t) = g_1(t) \\ u(1,t) = g_2(t) \end{cases} \qquad 0 < t < \infty$$

$$\text{ICs} \qquad \begin{cases} u(x,0) = \phi(x) \\ u_t(x,0) = \psi(x) \end{cases} \qquad 0 \leq x \leq 1$$

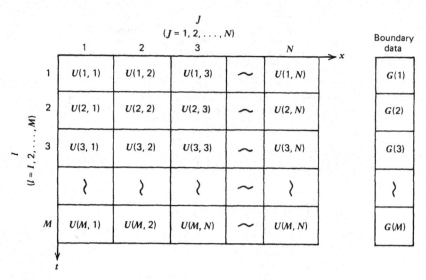

FIGURE 38.4 Arrays used in the explicit method.

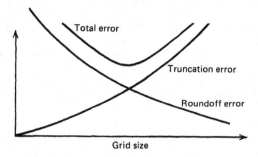

Total error

Truncation error

Roundoff error

Grid size

FIGURE 38.5 Total error as a function of grid size.

can also be solved by the explicit finite-difference method. Here, we can approximate the derivatives u_{tt} and u_{xx} by

$$u_{tt} \cong \frac{1}{k^2} [u(x,t + k) - 2u(x,t) + u(x,t - k)]$$

$$u_{xx} \cong \frac{1}{h^2} [u(x + h,t) - 2u(x,t) + u(x - h,t)]$$

and the derivative $u_t(x,0)$ in the IC by

$$u_t(x,0) \cong \frac{1}{k} [u(x,k) - u(x,0)] = \frac{1}{k} [u(x,k) - \phi(x)]$$

Hence, solving for $u(x,t + k)$ explicitly in terms of the solution at earlier values of time gives

$$u(x,t + k) = 2u(x,t) - u(x,t - k)$$
(38.5)
$$+ \left(\frac{k}{h}\right)^2 [u(x + h,t) - 2u(x,t) + u(x - h,t)]$$

From this equation, it is clear that we must already know the solution at *two* previous time steps, and, hence, we must use the initial-velocity condition

$$\frac{1}{k} [u(x,k) - \phi(x)] = \psi(x)$$

to get us started. Solving for $u(x,k)$ gives $u(x,k) = \phi(x) + k\psi(x)$, and, thus, we can find the solution for $t = \Delta t$. The solution at all later values of time can now be found by our explicit formula (38.5).

PROBLEMS

1. Find the finite-difference solution of the heat-conduction problem

$$\text{PDE} \quad u_t = u_{xx} \quad 0 < x < 1 \quad 0 < t < \infty$$

$$\text{BCs} \quad \begin{cases} u(0,t) = 0 \\ u(1,t) = 0 \end{cases} \quad 0 < t < \infty$$

$$\text{IC} \quad u(x,0) = \sin(\pi x) \quad 0 \le x \le 1$$

for $t = 0.005, 0.010, 0.015$ by the explicit method. Let $h = \Delta x = 0.1$. Plot the solution at $x = 0, 0.1, 0.2, 0.3, \ldots, 0.9, 1$ for $t = 0.015$.
2. Solve problem 1 analytically (separation of variables) and evaluate the analytical solution at the grid points: $x = 0, 0.1, 0.2, \ldots, 0.9, 1$ for $t = 0.015$. Compare these results to your numerical solution in problem 1. (You may wish to write a small computer program or use a calculator to evaluate the separation-of-variables solution.)
3. Write a flow diagram to carry out the computations of the hyperbolic problem discussed in note 3 of this lesson.
4. Do problem 1 except now *replace* the BC at $x = 1$ by

$$u_x(1,t) = -[u(1,t) - 1]$$

OTHER READING

Finite-Difference Methods for PDEs by G. F. Forsythe and W. R. Wasow. John Wiley & Sons, 1960. An excellent text with several physical examples illustrated, soil-drainage problems, oil-flow problems, and a meteorological-forecast problem are a few of the problems discussed.

An Implicit Finite-Difference Method (Crank-Nicolson Method)

PURPOSE OF LESSON: To show how time-dependent problems can be solved by another finite-difference scheme known as *implicit methods*. In this method, we again replace the partial derivatives in the problem by their finite-difference approximations, but unlike explicit methods (where we solved for $u_{i+1,j}$ explicitly in terms of earlier values), in implicit methods, we solve *a system of equations* in order to find the solution at the largest value of time. In other words, for each new value of time we solve a system of algebraic equations to find *all* the values.

Implicit methods have an advantage over explicit ones, since the step size can be made larger without worrying about excessive buildup of round-off error.

A popular implicit method known as the *Crank-Nicholson method* will be used to solve a parabolic problem.

The difficulty with the explicit methods that we discussed in the last lesson is that the step size in time must be small in order for the method to work properly. In particular, if we were to solve the simple heat-flow problem

$$\text{PDE} \qquad u_t = u_{xx} \qquad 0 < x < 1 \qquad 0 < t < \infty$$

$$(39.1) \qquad \text{BC} \quad \begin{cases} u(0,t) = g_1(t) \\ u(1,t) = g_2(t) \end{cases} \quad 0 < t < \infty$$

$$\text{IC} \qquad u(x,0) = f(x) \qquad 0 \leqslant x \leqslant 1$$

by the *explicit method*, it would be necessary for the grid sizes Δt and Δx to satisfy

$$\frac{\Delta t}{(\Delta x)^2} \leqslant 0.5$$

in order for the method to be numerically stable (the roundoff errors don't build up). See reference 1 (p. 45) of the recommended reading for details of numerical stability. In other words, if the grid size Δx in the x-direction were chosen to be $\Delta x = 0.1$, then the time increment Δt must be $\Delta t \leq 0.5\Delta x^2 = 0.005$ (hence, to go from $t = 0$ to $t = 1$ would take 200 steps).

There are, however, procedures (implicit methods) that allow us to take larger steps by doing more work per step; in these methods, we can take relatively large steps by solving a *system of algebraic equations* at each step. To illustrate how these methods work, we solve the following heat-flow problem.

The Heat-Flow Problem Solved by an Implicit Method

Consider the following problem:

$$\text{PDE} \qquad u_t = u_{xx} \qquad 0 < x < 1 \qquad 0 < t < \infty$$

(39.2) $\qquad$ BCs $\qquad \begin{cases} u(0,t) = 0 \\ u(1,t) = 0 \end{cases} \qquad 0 < t < \infty$

$$\text{IC} \qquad u(x,0) = 1 \qquad 0 \leq x \leq 1$$

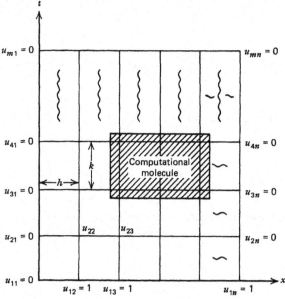

FIGURE 39.1 Grid system for implicit scheme ($\Delta x = 0.2$).

We replace the partial derivatives u_t and u_{xx} by the following approximations:

$$u_t(x,t) = \frac{1}{k}[u(x,t + k) - u(x,t)]$$

$$u_{xx}(x,t) = \frac{\lambda}{h^2}[u(x + h,t + k) - 2u(x,t + k) + u(x - h,t + k)]$$
$$+ \frac{(1 - \lambda)}{h^2}[u(x + h,t) - 2u(x,t) + u(x - h,t)]$$

where λ is a chosen number in the interval $[0,1]$. Note that our approximation for u_{xx} is a *weighted average* of the central-difference approximation to the derivative u_{xx} at time values t and $t + k$. In the special case when $\lambda = 0.5$, it is just the ordinary average of these two central differences, while if $\lambda = 0.75$, our approximation puts weights of 0.75 and 0.25 on each of the two terms (note, if $\lambda = 0$, it is the usual *explicit* finite-difference method we used in the last lesson).

If we now substitute the approximations for u_t and u_{xx} into our problem, we get the new *finite-difference problem*

$$\text{Difference equation} \quad \frac{1}{k}(u_{i+1,j} - u_{i,j})$$

$$= \frac{\lambda}{h^2}(u_{i+1,j+1} - 2u_{i+1,j} + u_{i+1,j-1}) + \frac{(1 - \lambda)}{h^2}(u_{i,j+1} - 2u_{i,j} + u_{i,j-1})$$

(39.3)

$$\text{BC} \quad \begin{cases} u_{i,1} = 0 \\ u_{i,n} = 0 \end{cases} \quad i = 1, 2, \ldots, m$$

$$\text{IC} \quad u_{1,j} = 1 \quad j = 2, \ldots, n-1$$

See Figure 39.1.

Now, if we rewrite the difference equation in (39.3), putting the $u_{i,j}$'s with the largest time subscript (*i*-subscript) on the left-hand side of the equation, we arrive at the equation

(39.4) $\quad -\lambda r u_{i+1,j+1} + (1 + 2r\lambda)u_{i+1,j} - \lambda r u_{i+1,j-1}$
$$= r(1 - \lambda)u_{i,j+1} + [1 - 2r(1 - \lambda)]u_{i,j} + r(1 - \lambda)u_{i,j-1}$$

where we have set $r = k/h^2$ for convenience. Note that for a *fixed subscript i* and for j going from 2 to $n - 1$, this is a system of $n - 2$ equations in the $n - 2$ unknowns $u_{i+1,2}, u_{i+1,3}, u_{i+1,4}, \ldots u_{i+1,n-1}$ [which are the interior grid points at $t = (i + 1)\Delta t$].

An Implicit Finite-Difference Method (Crank-Nicolson Method)   319

To help show exactly what $u_{i,j}$'s are involved in this formula, we write it in the symbolic or molecular form shown in Figure 39.2.

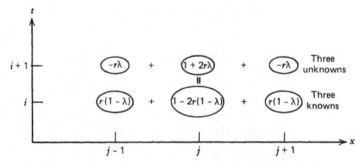

FIGURE 39.2   The molecular form of the implicit formula.

We now show how equation (39.4) can be used to find the solution of problem (39.2).

Implicit Algorithm for Heat Problem (39.2)

STEP 1   Pick some value for λ $(0 \le \lambda \le 1)$. Note that if $\lambda = 0$, then equation (39.4) is the same as the explicit formula we developed in lesson 38.

STEP 2   Pick $h = \Delta x = 0.2$ and $k = \Delta t = 0.08$ $(r = k/h^2 = 2)$. This gives *six* grid points in the x-direction (four *interior* grid points); see Figure 39.1. Also let's pick the weight parameter $\lambda = 0.5$ (which is called the Crank-Nicolson method). If we now apply our computational molecule to the first and second rows $(i = 1)$, moving it from left to right $(j = 2, 3, 4, 5)$, we get the following four equations:

$$-u_{21} + 3u_{22} - u_{23} = u_{11} - u_{12} + u_{13} = 1$$
$$-u_{22} + 3u_{23} - u_{24} = u_{12} - u_{13} + u_{14} = 1$$
$$-u_{23} + 3u_{24} - u_{25} = u_{13} - u_{14} + u_{15} = 1$$
$$-u_{24} + 3u_{25} - u_{26} = u_{14} - u_{15} + u_{16} = 1$$

which if written in matrix form, placing the four unknown interior grid points u_{22}, u_{23}, u_{24}, and u_{25} on the left-hand side of the equation, gives

(39.5)
$$\begin{bmatrix} 3 & -1 & 0 & 0 \\ -1 & 3 & -1 & 0 \\ 0 & -1 & 3 & -1 \\ 0 & 0 & -1 & 3 \end{bmatrix} \begin{bmatrix} u_{22} \\ u_{23} \\ u_{24} \\ u_{25} \end{bmatrix} = \begin{bmatrix} 1 \\ 1 \\ 1 \\ 1 \end{bmatrix}$$

This system of equations is called a **tridiagonal** system, and to solve it, we use a method that transforms a tridiagonal system of the form

$$
\begin{bmatrix}
b_1 & c_1 & 0 & 0 & \cdots & & 0 \\
a_1 & b_2 & c_2 & 0 & \cdots & & 0 \\
0 & a_2 & b_3 & c_3 & & 0 & 0 \\
\vdots & & & & & & \\
& & & & & & \\
& & & & & c_{n-1} & \\
0 & 0 & 0 & \cdots & & a_{n-1} & b_n
\end{bmatrix}
\begin{bmatrix}
x_1 \\
x_2 \\
x_3 \\
\vdots \\
\\
\\
x_n
\end{bmatrix}
=
\begin{bmatrix}
d_1 \\
d_2 \\
d_3 \\
\vdots \\
\\
\\
d_n
\end{bmatrix}
$$

into an equivalent one

$$
\begin{bmatrix}
1 & c_1^* & 0 & 0 & \cdots & & 0 \\
0 & 1 & c_2^* & 0 & \cdots & & 0 \\
0 & 0 & 1 & c_3^* & & & 0 \\
\vdots & & & & & & \\
& & & & & & \\
& & & & & c_{n-1}^* & \\
0 & 0 & \cdots & & & 0 & 1
\end{bmatrix}
\begin{bmatrix}
x_1 \\
x_2 \\
x_3 \\
\vdots \\
\\
\\
x_n
\end{bmatrix}
=
\begin{bmatrix}
d_1^* \\
d_2^* \\
d_3^* \\
\vdots \\
\\
\\
d_n^*
\end{bmatrix}
$$

where

$$
c_1^* = c_1/b_1 \qquad c_{j+1}^* = \frac{c_{j+1}}{b_{j+1} - a_j c_j^*} \qquad j = 1, 2, \ldots, n - 2
$$

and

$$
d_1^* = d_1/b_1 \qquad d_{j+1}^* = \frac{d_{j+1} - a_j d_j^*}{b_{j+1} - a_j c_j^*} \qquad j = 1, 2, \ldots, n - 1
$$

There is nothing magical about this transformation; it just involves rewriting the original system of equations in an equivalent form. The point is, once we have written the system of equations in the new form, it is easy to solve. Solving from bottom to top, we have

$$
x_n = d_n^* \qquad x_j = d_j^* - c_j^* x_{j+1} \qquad j = n - 1, \quad n - 2, \ldots, 2, 1
$$

Applying this method to our system of four equations (39.5), we get:

$$
u_{22} = 0.60
$$
$$
u_{23} = 0.80
$$
$$
u_{24} = 0.80
$$
$$
u_{25} = 0.60
$$

An Implicit Finite-Difference Method (Crank-Nicolson Method) 321

This gives us the solution (approximation) at the interior grid points for $t = \Delta t$. After finding these values, we move to the next time value and solve a new set of equations.

This implicit method takes more work at each value of time than does the explicit method, but it enables us to pick a larger Δt and still get a good approximation.

PROBLEMS

1. Derive equation (39.4) from the difference equation in (39.3).
2. Tell how you would solve the problem

$$\text{PDE} \qquad u_t = u_{xx} \qquad 0 < x < 1$$

$$\text{BCs} \qquad \begin{cases} u(0,t) = 1 \\ u_x(1,t) + u(1,t) = g(t) \end{cases} \qquad 0 < t < \infty$$

$$\text{IC} \qquad u(x,0) = 0 \qquad 0 \leqslant x \leqslant 1$$

   by the implicit finite-difference method.
3. How would you solve

$$\text{PDE} \qquad u_t = u_{xx} + u \qquad 0 < x < 1$$

$$\text{BCs} \qquad \begin{cases} u(0,t) = 0 \\ u(1,t) = 0 \end{cases} \qquad 0 < t < \infty$$

$$\text{IC} \qquad u(x,0) = 1 \qquad 0 \leqslant x \leqslant 1$$

   by the implicit method?
4. What is the molecular form of equation (39.4) when we pick $\lambda = 1$?
5. Write a flow diagram to solve heat-flow problem (39.2). Write a computer program if facilities are available. A good experiment would be to solve this problem numerically with a simple IC $u(x,0) = \sin(\pi x)$ for different values of the parameter λ. You could compare the true analytical solution, which, in this case, is

$$u(x,t) = e^{-\pi^2 t} \sin(\pi x)$$

   with the numerical solution for different values of λ.
6. Solve the system of algebraic equations (39.5) using the formulas given in the lesson.

OTHER READING

Numerical Methods in PDEs by W. F. Ames. Academic Press, 1977. An excellent book with applications to fluid dynamics and elasticity.

Analytic versus Numerical Solutions

PURPOSE OF LESSON: To discuss the relative merits and demerits of analytic and numerical solutions to PDEs. The importance of mathematical models in identifying physical quantities (parameter identification) is presented, and an important example from biology is discussed.

It's probably time we had a discussion about the relative merits and demerits of analytic and numerical solutions. First of all, let's make sure we know what we mean when we talk about these types of solutions.

Meaning of Analytic Solutions

Analytic solutions are those solutions where the unknown variable u is given as a mathematical expression in terms of the *independent variables* and *parameters* of the system which are generally infinite series or integrals.

Meaning of Numerical Solutions

Numerical solutions, on the other hand, refer to finding the solution of PDEs by replacing the differential equation with an *approximate equation* and solving the easier one. For example, the method of finite-difference approximations replaces partial derivatives with finite differences, so we approximate the solution to a PDE by solving a finite-difference equation. The result is generally a table of numbers listing the solution u for various values of the independent variables.

Now that we know the basic meaning of the two types of solutions, let's ask which is better.

Comparing Numerical and Analytic Solutions

Suppose we have the simple parabolic IBVP

$$\text{PDE} \qquad u_t = \alpha^2 u_{xx} \qquad 0 < x < 1 \qquad 0 < t < \infty$$

$$(40.1) \qquad \text{BCs} \qquad \begin{cases} u(0,t) = 0 \\ u(1,t) = 0 \end{cases} \qquad 0 < t < \infty$$

$$\text{IC} \qquad u(x,0) = 1 \qquad 0 \leqslant x \leqslant 1$$

whose solution is shown in Figure 40.1.

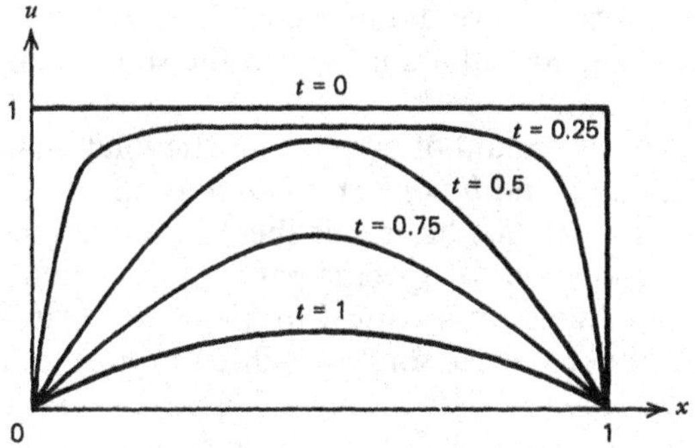

FIGURE 40.1 Solution of heat equation (40.1) for various values of time.

The question here is, would we rather be given the analytic solution

$$(40.2) \qquad \boxed{u(x,t) = \frac{2}{\pi}\left[e^{-(\pi\alpha)^2 t}\sin(\pi x) + \frac{1}{3}e^{-(3\pi\alpha)^2 t}\sin(3\pi x) + \ldots\right]}$$

to the problem or the *numerical solution* (with $\alpha = 1$)?

TABLE 40.1

	x: 0	0.1	0.2	0.3	0.4	0.5	0.6	0.7	0.8	0.9	1.
0	1	1	1	1	1	1	1	1	1	1	1
0.01	0	0.2	0.34	·	·	·	·	·	0.34	0.2	0
0.02	0	0.15	·	·	·	·	·	·	·	0.15	0
t 0.03	0	·	·	·	·	·	·	·	·	·	0
·	·	·	·	·	·	·	·	·	·	·	·
·	·	·	·	·	·	·	·	·	·	·	·
·	·	·	·	·	·	·	·	·	·	·	·

This is a good question; the answer depends on what we want to do with the solution. There are, however, some clear cut advantages to each of the two types of solutions; first the advantages.

Advantages of the Analytic Solution

1. Equation (40.2) obviously contains more information than a table of numbers. If we wanted to evaluate the solution at any specific point (x,t), we could do so with any degree of accuracy merely by adding a sufficient number of terms in the infinite series. An upper bound on the error could be found without much difficulty.
2. The analytic solution allows us to find the solution at a *single point* (x,t) without going through the entire marching process of finding the solution at all other points, as we did in the explicit and implicit methods.
3. The analytic solution allows us to find the solution at any point and not just the grid points.
4. Probably most important of all, the analytic solution tells us how *physical parameters*, *initial* and *boundary conditions* affect the solution.

Numerical solutions do not bring out these interrelationships, since we are finding the numeric solution for specific parameters, initial and boundary conditions. In many situations, it is critical to know the relationship between the parameters of the model and the solution, since our goal may be to estimate the *parameters from the solution*. For example, suppose we measure the solution u experimentally, and we know the analytic solution

$$\boxed{u \ = \ \text{function of the parameters}}$$

then we can more or less solve for the *parameters* as a function of the data via

$$\boxed{\text{parameters} \ = \ \text{function of } u \ = \ \text{function of the data}}$$

This concept is called **parameter identification**, and it is one of the major reasons for solving PDEs. Later on in this lesson, we give an important example of parameter identification in biology. First, however, let's see why numerical solutions are worthwhile.

Advantages of Numerical Solutions

There is one major advantage to numerical solutions, and it is that many problems do not have known analytic solutions. Practically all nonlinear PDEs must be solved by numerical methods, and, in fact, most realistic models in physics, chemistry, biology, and so forth, are nonlinear in nature. The linear models represent, for the most part, approximations where we have thrown out certain nonlinear components. Some very important nonlinear equations such as:

1. Nonlinear wave equation $u_{tt} = u_{xx} + f(u)$
2. Reaction-diffusion equation $u_t = u_{xx} + f(u)$

3. Hodgkin-Huxley equations
$$\begin{cases} u_t = u_{xx} + f(u,v) \\ v_t = g(u,v) \end{cases}$$

do not have known analytic solutions for all nonlinearities f and g. Hence, the general attack for most nonlinear problems (and some linear ones) involves the use of numerical solutions.

We now consider an example of how analytic solutions can be used to find important physical parameters. More details of this problem can be found in the recommended reading.

Parameter Identification (in Biology)

Suppose a biologist is trying to determine how fast potassium ions (K^+) diffuse in an exoplasm solution. By knowing the diffusion coefficient, we can tell a lot about how nerve impulses are transmitted along axons. The problem is that this coefficient is practically impossible to measure directly. What we can do, however, is find a mathematical relationship between the potassium concentration $u(x,t)$ and the diffusion coefficient D, so that by *measuring $u(x,t)$*, we *can* find D. The following example shows how this works.

Biologists Hodgkin and Keyes found that after isolating giant squid axons in a special salt solution, the concentration of radioactive potassium (^{42}K) along the axons could initially be approximated by the curve

$$u(x,0) = Ae^{-x^2/a}$$

In other words, the parameters A and a in the curve were found so that the equation *fit* the observed data points (least-squares fit). See Figure 40.2.

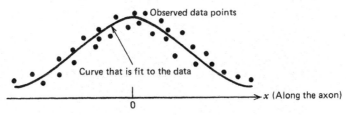

FIGURE 40.2   Initial concentration of ^{42}K.

It was also determined that *after* this initial concentration, the concentration in a normal environment flowed down the axon both by *convection* and *diffusion*. Hence, Hodgkin and Keyes assumed the concentration u could be described by the convection-diffusion IVP

(40.3) $\qquad$ PDE $\qquad u_t = Du_{xx} - Vu_x \qquad -\infty < x < \infty$
$\qquad\qquad$ IC $\qquad u(x,0) = Ae^{-x^2/a} \qquad -\infty < x < \infty$

This problem was solved in Lesson 15 by transforming into moving coordinates. If the reader remembers, the general idea of moving coordinates was to forget V and solve the pure diffusion problem to get

$$u(x,t) = \frac{A}{2\sqrt{\pi Dt}} \int_{-\infty}^{\infty} e^{-\xi^2/a} e^{-(x-\xi)^2/4Dt} \, d\xi$$

$$= \frac{A\sqrt{a}}{\sqrt{a + 4Dt}} \, e^{-[x^2/(a+4Dt)]}$$

(which is the solution to the problem with $V = 0$). Substituting $x - Vt$ for x gives the solution to problem (40.3); namely,

(40.4) $\qquad\qquad u(x,t) = \dfrac{A\sqrt{a}}{\sqrt{a + 4Dt}} \, e^{-[(x-Vt)^2/(a+4Dt)]}$

It turned out that Hodgkin and Keyes were able to measure V directly, so that equation (40.4) gives a relationship between ^{42}K concentration and the diffusion coefficient D ($A, a,$ and V are all known). By designing an experiment where they measured the potassium concentration $u(x,t)$ at some *fixed point* x_0 along the axon at different values of time, they were able to find the value of D that made the *theoretical* curve (40.4) fit the data (Figure 40.3).

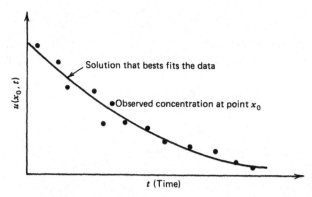

FIGURE 40.3 Fitting the curve $u(x_0, t)$ to the observed data points.

In other words, if we pick various values of the diffusion coefficient D, we get different curves $u(x_0, t)$. Hence, we pick the value of D that makes the solution to the PDE fit the experimental data. In their experiment, Hodgkin and Keyes found that the diffusion coefficient of K^+ in exoplasm was $D = 1.5 \times 10^{-5}$ cm²/sec.

PROBLEMS

1. How could you construct an experiment to estimate the parameter α in problem (40.1) using analytic solution (40.2)?
2. Least-squares approximations minimize the sum of squares (SS) between the *curve* and the *data points*. For example, the least-squares line $y(x) = a + bx$ that approximates data points (x_1, y_1), (x_2, y_2), . . . (x_n, y_n) would be the specific line that minimizes (see Figure 40.4).

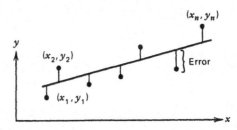

FIGURE 40.4   Least-squares approximation at data points.

$$SS = \sum_{i=1}^{n} [y_i - (a + bx_i)]^2$$

Find constants a and b in terms of the data points (x_i, y_i) so that this line is the least-squares approximation.
3. An important problem in biochemistry is determining the molecular weight of the macromolecule myoglobin. One approach is to place a certain blood solution into an ultracentrifuge (very fast centrifuge) and spin it for a given length of time. The equation that describes the concentration of the liquid in the centrifuge is known as **Lamm's equation:**

$$u_t = \frac{1}{r} \frac{\partial}{\partial r} (Dru_r - s\omega^2 r^2 u) \qquad 0 < r < 1$$

where
  r = distance from the center of the centrifuge
  D = diffusion coefficient (depends on the molecular weight of myoglobin)
  s = sedimentation coefficient (computed experimentally)
  ω = angular velocity of the centrifuge (known)
  $u(r,t)$ = concentration of the medium in the centrifuge

The approach in finding the molecular weight of myoglobin is to find the steady-state solution $u(r, \infty)$ of Lamm's equation by letting $u_t = 0$ and solving the ODE for $u(r,\infty)$. The question is, how do you design an experiment using this steady-state solution $u(r,\infty)$ to estimate the molecular weight of myoglobin?

OTHER READING

Introduction to Mathematical Biology by S. I. Rubinow. John Wiley & Sons, 1976. Several applications of PDEs to biology are given in this text.

Classification of PDEs (Parabolic and Elliptic Equations)

PURPOSE OF LESSON: To introduce a new coordinate system (ξ, η) into the second-order linear equation

$$Au_{xx} + Bu_{xy} + Cu_{yy} + Du_x + Eu_y + Fu = G$$

for the cases

$$
\begin{aligned}
B^2 - 4AC &= 0 \quad &\text{(Parabolic case)} \\
B^2 - 4AC &< 0 \quad &\text{(Elliptic case)}
\end{aligned}
$$

and in each case, show how the equation can be written in one of the two canonical forms

$$
\begin{aligned}
u_{\eta\eta} &= \phi(\xi, \eta, u, u_\xi, u_\eta) \quad &\text{(Parabolic canonical form)} \\
u_{\xi\xi} + u_{\eta\eta} &= \phi(\xi, \eta, u, u_\xi, u_\eta) \quad &\text{(Elliptic canonical form)}
\end{aligned}
$$

In Lesson 23, we classified the general second-order linear equation in two variables

(41.1) $$Au_{xx} + Bu_{xy} + Cu_{yy} + Du_x + Eu_y + Fu = G$$

as one of three basic types and, in particular, transformed the *hyperbolic equation* into its two canonical forms.

In this lesson, we will show how *parabolic and elliptic* equations can also be reduced to canonical form.

Reducing Parabolic Equations to Canonical Form

In this case, we consider equations of the form (41.1) with $B^2 - 4AC = 0$ and introduce new coordinates (ξ, η), so that the equation takes the form

$$u_{\eta\eta} = \phi(\xi, \eta, u, u_\xi, u_\eta)$$

Introducing new coordinates $\xi = \xi(x,y)$ and $\eta = \eta(x,y)$ into equation (41.1) gives us the same equation as before; namely,

$$(41.2) \qquad \overline{A}u_{\xi\xi} + \overline{B}u_{\xi\eta} + \overline{C}u_{\eta\eta} + \overline{D}u_{\xi} + \overline{E}u_{\eta} + \overline{F}u = \overline{G}$$

where:

$$
\begin{aligned}
\overline{A} &= A\xi_x^2 + B\xi_x\xi_y + C\xi_y^2 \\
\overline{B} &= 2A\xi_x\eta_x + B(\xi_x\eta_y + \xi_y\eta_x) + 2C\xi_y\eta_y \\
\overline{C} &= A\eta_x^2 + B\eta_x\eta_y + C\eta_y^2 \\
(41.3) \qquad \overline{D} &= A\xi_{xx} + B\xi_{xy} + C\xi_{yy} + D\xi_x + E\xi_y \\
\overline{E} &= A\eta_{xx} + B\eta_{xy} + C\eta_{yy} + D\eta_x + E\eta_y \\
\overline{F} &= F \\
\overline{G} &= G
\end{aligned}
$$

However, our goal now is to set $\overline{B}$ and either $\overline{A}$ or $\overline{C}$ equal to zero and solve for ξ and η (we can see in a moment that $B^2 - 4AC$ must be zero in order to carry out this plan). Here, we'll set $\overline{A}$ and $\overline{B}$ equal to zero and solve the resulting equations. First, setting $\overline{A} = 0$ and solving for $[\xi_x/\xi_y]$ we have

$$[\xi_x/\xi_y] = -B/2A$$

Hence, we can find the coordinate $\xi = \xi(x,y)$ that satisfies this equation by setting

$$\frac{dy}{dx} = -[\xi_x/\xi_y] = B/2A$$

and finding the implicit solution

$$\xi(x,y) = c$$

to this equation.

For example, if $\dfrac{dy}{dx} = B/2A = 3$, then we have $y - 3x = c$, and, hence, $\xi(x,y)$ $= y - 3x$ satisfies $\xi_x/\xi_y = -3$ (and, thus, makes $\overline{A} = 0$).

So, we are half done; we have found one coordinate $\xi = \xi(x,y)$ that makes $\overline{A} = 0$. The last part of the problem is to find $\eta(x,y)$ so that $\overline{B} = 0$.

Here's where we get a break. It turns out (since $B^2 - 4AC = 0$) that by picking ξ so that $\overline{A} = 0$, the coefficient $\overline{B}$ is automatically zero. In fact, we'll verify this right now. The coefficient $\overline{B}$ is given by

$$\overline{B} = 2A\xi_x\eta_x + B(\xi_x\eta_y + \xi_y\eta_x) + 2C\xi_y\eta_y$$

and since $B^2 - 4AC = 0$, we can write

$$\overline{B} = 2 (\sqrt{A}\xi_x + \sqrt{C}\xi_y) (\sqrt{A}\eta_x + \sqrt{C}\eta_y)$$

and since $\xi_x/\xi_y = -B/2A = -2\sqrt{AC}/2A = -\sqrt{C/A}$, we have

$$\overline{B} = 2\sqrt{A}\left[\sqrt{A}\eta_x + \sqrt{C}\eta_y\right]$$

which, of course, is zero because $\overline{A}$ is zero.

So, by our choice of ξ, *both* $\overline{A}$ and $\overline{B}$ are zero, and, hence, we can pick η any way we like (as long as it's never parallel to the ξ coordinate). Thus, we can pick something simple like $\eta = y$.

All that remains is to find the new canonical equation, and to get this, we merely substitute ξ and η into equations (41.3) to find the coefficients $\overline{A}$, $\overline{B}$, $\overline{C}$, $\overline{D}$, $\overline{E}$, $\overline{F}$, and $\overline{G}$. This finishes the parabolic case; before going onto another topic, we present a simple example.

Transforming the Parabolic Equation $u_{xx} + 2u_{xy} + u_{yy} = 0$ into Canonical Form

We begin with the simple equation

$$u_{xx} + 2u_{xy} + u_{yy} = 0$$

where $A = 1$, $B = 2$, $C = 1$, $D = E = F = G = 0$. Hence, $B^2 - 4AC = 0$ for all x and y. To find the new coordinates ξ and η and the canonical equation, we proceed as follows:

STEP 1 Write the *characteristic equation* (only one now)

$$\frac{dy}{dx} = - [\xi_x/\xi_y] = B/2A = 1$$

Solving for y (integrating) gives

$$y = x + c$$

and, hence, $\xi = y - x$ will satisfy the characteristic equation and thus make $\overline{A} = 0$.

The η coordinate can be chosen in any way, as long as it isn't parallel to the ξ coordinate; hence we choose

$$\eta = y$$

The new coordinates

(41.4)
$$\xi = y - x$$
$$\eta = y$$

are shown in Figure 41.1.

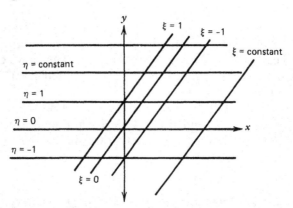

FIGURE 41.1 New coordinate system $\xi = y - x$, $\eta = y$.

STEP 2   The last step is to find the new canonical equation. Substituting ξ and η into the new coefficients $\overline{A}$, $\overline{B}$, $\overline{C}$, $\overline{D}$, $\overline{E}$, $\overline{F}$, and $\overline{G}$ gives:

$$\overline{A} = 0 \qquad \text{(Must be true; we set it equal to zero and solved for } \xi\text{)}$$
$$\overline{B} = 0 \qquad \text{(We have already shown that this is zero)}$$
$$\overline{C} = A\eta_x^2 + B\eta_x\eta_y + C\eta_y^2 = 1$$
(41.5) $\quad \overline{D} = A\xi_{xx} + B\xi_{xy} + C\xi_{yy} + D\xi_x + E\xi_y = 0$
$$\overline{E} = A\eta_{xx} + B\eta_{xy} + C\eta_{yy} + D\eta_x + E\eta_y = 0$$
$$\overline{F} = F = 0$$
$$\overline{G} = G = 0$$

Hence, our new equation

$$\overline{A}u_{\xi\xi} + \overline{B}u_{\xi\eta} + \overline{C}u_{\eta\eta} + \overline{D}u_{\xi} + \overline{E}u_{\eta} + \overline{F}u = \overline{G}$$

is simply

$$u_{\eta\eta} = 0$$

This completes the example. Before going on, let's look for a moment at this canonical form. This equation is so simple that we can find its *general solution* (all its solutions). Starting with $u_{\eta\eta} = 0$ and integrating once with respect to η gives

$$u_\eta = f(\xi)$$

where $f(\xi)$ is an arbitrary function of ξ. Integrating again gives the general solution

$$u(\xi,\eta) = \eta f(\xi) + g(\xi)$$

where $g(\xi)$ is another arbitrary function of ξ. (The reader can easily verify that any function of this form is a solution of $u_{\eta\eta} = 0$.)

Now the final step. Substituting the terms of our original coordinates x and y gives the general solution of

$$u_{xx} + 2u_{xy} + u_{yy} = 0$$

namely,

$$u(x,y) = yf(y - x) + g(y - x)$$

For example, if we pick $f(x) = \sin x$ and $g(x) = x^2$ at random, then

$$u(x,y) = y \sin (y - x) + (y - x)^2$$

should be one of the infinitely many solutions of the equation (of course, which solution actually models the physical problem depends on the initial and boundary conditions).

Reducing Elliptic Equations to Canonical Form

We start again with the general equation

$$Au_{xx} + Bu_{xy} + Cu_{yy} + Du_x + Eu_y + Fu = G$$

but now $B^2 - 4AC < 0$. Our goal is to transform this equation into the new form

$$u_{\xi\xi} + u_{\eta\eta} = \phi(\xi, \eta, u, u_\xi, u_\eta)$$

by changing the independent variables. Proceeding as we did in the two earlier cases, we are tempted to set $\overline{A} = \overline{C}$ and $\overline{B} = 0$ in the transformed equation

$$\overline{A}u_{\xi\xi} + \overline{B}u_{\xi\eta} + \overline{C}u_{\eta\eta} + \overline{D}u_\xi + \overline{E}u_\eta + \overline{F}u = \overline{G}$$

and solve for ξ and η. Unfortunately, these equations $(\overline{A} - \overline{C} = 0, \overline{B} = 0)$ do not allow us to solve for ξ and η as the earlier ones did, and for that reason, we proceed a little differently.

We find our transformation $\xi = \xi(x,y)$, $\eta = \eta(x,y)$ as a *composition* of *two* transformations; we start with the first.

Transformation 1

We first make a transformation to new coordinates ξ and η that will make our equation look like

$$u_{\xi\eta} = \psi(\xi,\eta,u,u_\xi,u_\eta)$$

It is possible to do this, but we must resort to *complex coordinates*. To find these complex coordinates ξ and η, we merely proceed as we did in the hyperbolic case by solving the characteristic equations

$$\frac{dy}{dx} = \frac{B - \sqrt{B^2 - 4AC}}{2A}$$

(Remember $B^2 - 4AC < 0$)

$$\frac{dy}{dx} = \frac{B + \sqrt{B^2 - 4AC}}{2A}$$

to get

$$\xi(x,y) = \text{constant}$$
$$\eta(x,y) = \text{constant}$$

For example, the characteristic equations

$$\frac{dy}{dx} = -\sqrt{-4x^2} = -2ix$$

$$\frac{dy}{dx} = \sqrt{-4x^2} = 2ix$$

would give us the complex conjugate coordinates

$$\xi = y + ix^2$$
$$\eta = y - ix^2$$

Transformation 2

We now make a second transformation from (ξ, η) to (α, β) via

$$\alpha = \frac{\xi + \eta}{2}$$

$$\beta = \frac{\xi - \eta}{2i}$$

and the result of this second transformation changes the first equation

$$u_{\xi\eta} = \psi(\xi, \eta, u, u_\xi, u_\eta)$$

to the final form

$$u_{\alpha\alpha} + u_{\beta\beta} = \phi(\alpha, \beta, u, u_\alpha, u_\beta)$$

where ϕ and ψ are the general names for the right hands.
Instead of showing that these two transformations (back to back) actually carry out the above result, let's apply these principles to a simple example.

Changing the Equation $y^2 u_{xx} + x^2 u_{yy} = 0$ to Canonical Form

Consider transforming the equation

$$y^2 u_{xx} + x^2 u_{yy} = 0$$

where $A = y^2$, $B = 0$, $C = x^2$, $D = E = F = G = 0$. The discriminant $B^2 - 4AC$ is equal to $-4x^2y^2$, and, hence, we will transform this equation to canonical form in the first quadrant $x > 0$, $y > 0$.

STEP 1 (First transformation)

We start by writing the two characteristic equations

$$\frac{dy}{dx} = \frac{B - \sqrt{B^2 - 4AC}}{2A} = -\frac{\sqrt{-4x^2y^2}}{2y^2} = -ix/y$$

$$\frac{dy}{dx} = \frac{B + \sqrt{B^2 - 4AC}}{2A} = \frac{\sqrt{-4x^2y^2}}{2y^2} = ix/y$$

Solving these equations by separating variables, we have the implicit relationship

$$y^2 + ix^2 = \text{constant}$$
$$y^2 - ix^2 = \text{constant}$$

and, hence,

$$\xi(x,y) = y^2 + ix^2$$
$$\eta(x,y) = y^2 - ix^2$$

(It doesn't really matter which function we call ξ and η; we could interchange the two if we wanted.) This transformation will reduce the original equation to the form

$$u_{\xi\eta} = \psi(\xi, \eta, u, u_\xi, u_\eta)$$

We don't really care about this equation (which is a *complex hyperbolic equation*), and so we continue with the second transformation.

STEP 2 (Second transformation)

Making the second transformation, we have

$$\alpha = \frac{\xi + \eta}{2} = y^2 \qquad \text{(Real part of ξ and η)}$$

$$\beta = \frac{\xi - \eta}{2i} = x^2 \qquad \text{(Complex part of ξ and η)}$$

From a notational point of view, it might be best to rename the variables (α, β) as (ξ, η) and think of our composite transformation as simply being

$$\xi(x,y) = y^2 \qquad \eta(x,y) = x^2$$

STEP 3 (Finding the new equation)

The new canonical form can be found by computing the coefficients $\overline{A}$, $\overline{B}$, $\overline{C}$, $\overline{D}$, $\overline{E}$, $\overline{F}$, and $\overline{G}$ in the equation

$$\overline{A}u_{\xi\xi} + \overline{B}u_{\xi\eta} + \overline{C}u_{\eta\eta} + \overline{D}u_\xi + \overline{E}u_\eta + \overline{F}u = \overline{G}$$

from equations (41.5) with $\xi = y^2$ and $\eta = x^2$. Doing this gives the elliptic canonical form

$$u_{\xi\xi} + u_{\eta\eta} = \frac{-\xi u_\eta - \eta u_\xi}{2\xi\eta}$$

NOTES

1. Second-order linear equations in *three or more variables* can also be classified except that we must use matrix analysis. For instance, the second-order equation in three variables

$$u_t = u_{xx} + u_{yy}$$

would be classified as a parabolic equation, while the equation

$$u_{tt} = u_{xx} + u_{yy}$$

would be a hyperbolic equation.
2. Our interest in classifying PDEs is in part due to the fact that the three basic types describe different kinds of physical phenomena, and we would like to classify mathematically these three types of physical problems.

PROBLEMS

1. Which of the following parabolic and elliptic equations are already in canonical form:

   (a) $u_t = u_{xx} - hu$
   (b) $u_{xy} + u_{xx} + 3u = \sin x$
   (c) $u_{xx} + 2u_{yy} = 0$
   (d) $u_{xx} = \sin x$

2. Transform the parabolic equation $u_{xx} + 2u_{xy} + u_{yy} + u = 2$ into canonical form.
3. Transform the elliptic equation $u_{xx} + 2u_{yy} + x^2 u_x = e^{-x^2/2}$ into canonical form.

OTHER READING

1. *Second-Order Partial Differential Equations* by M. M. Smirnov. Noordhoff, 1966. A small, well-written book; in addition to classifying second-order equations in two variables, he also classifies second-order equations in n variables.

2. *Methods of Mathematical Physics*, vol. 2 by R. Courant, and D. Hilbert. Wiley/ Interscience, 1962. One of the most famous books on PDE theory written by two of the most outstanding mathematicians of this century; more advanced than this book, but contains a wealth of information for any reader willing to spend the time.

Monte Carlo Methods (an Introduction)

PURPOSE OF LESSON: To explain the basic philosophy of Monte Carlo methods and suggest how they can be used to solve various problems. The basic idea here is that *games of chance* can be played (generally on a computer) whose outcomes approximate solutions to real-world problems. A simple example would be evaluating the integral

$$I = \int_0^1 x^2 \, dx$$

by throwing darts at the unit square $\{(x,y) : 0 < x < 1, 0 < y < 1\}$. After 100 tosses or so, we use the *fraction* of darts under the curve $y = x^2$ to approximate the integral. Generating a random game (like tossing darts) always involves generating a sequence of random numbers, and so a procedure for their generation is described.

There is an interesting technique known as the Monte Carlo method (or methods) that can be used to solve many types of problems. Here, we present a brief overview of the method and then, in the next lesson, illustrate its use in PDEs.

First of all Monte Carlo methods are procedures for solving *nonprobabilistic-type problems* (problems whose outcome does not depend on chance) by *probabilistic-type methods* (methods whose outcome depends on chance). The general philosophy of these methods is illustrated in Figure 42.1.

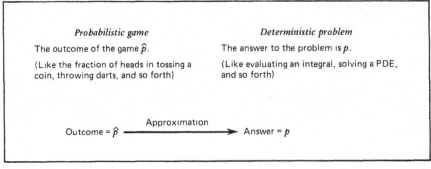

FIGURE 42.1 General philosophy of Monte Carlo method.

Evaluating an Integral

To illustrate this method, suppose we wanted to evaluate the integral

$$I = \int_a^b f(x)\, dx$$

(a nonprobabilistic problem). To use the Monte Carlo method, we would devise a game of chance whose outcome was the value of the integral (or approximates the integral). There are, of course, many games that we could devise; the actual game we used would depend on the accuracy of the approximation, simplicity of the game, and so on. An obvious game to evaluate the integral would be throwing darts at the rectangle $R = \{(x,y)\colon a \leqslant x \leqslant b,\ 0 \leqslant y \leqslant \max f(x)\}$ (Figure 42.2).

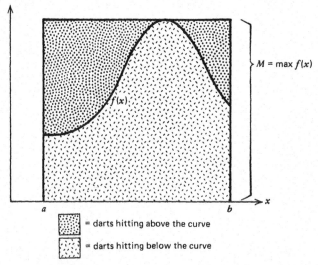

$M = \max f(x)$

$f(x)$

▦ = darts hitting above the curve

▨ = darts hitting below the curve

FIGURE 42.2   Evaluation of an integral by the Monte Carlo method.

It's fairly obvious that if we randomly toss 100 or so darts at the rectangle R enclosing the graph, then the fraction of darts hitting below the curve times the area of R will estimate the value of the integral. Hence, our outcome of the game

$$\hat{I} = [\text{fraction of tosses under } f(x)] \times (\text{area of } R)$$

is used to estimate the true value of the integral I.

To carry out the actual computation on a computer, we would have to generate the sequence of random points in some way (we'll discuss this shortly) and have the computer play the dart tossing game. Let us assume for the time being that we have a sequence of random points. The flow diagram in Figure 42.3 illustrates how the computer would attack this problem.

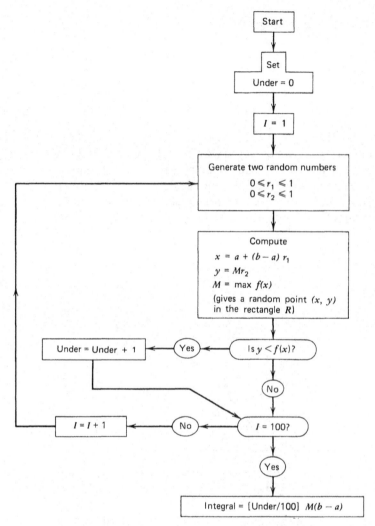

FIGURE 42.3 Flow diagram to evaluate $\int_a^b f(x)dx$ by the Monte Carlo method (100 tosses).

Random Numbers

Before going on to apply this technique to the solution of PDEs in the next lesson, we discuss the important topic of random numbers. In the integral just considered, it was necessary to generate a sequence of random points $P_i = (x_i, y_i)$ that fell inside rectangle R. In other words, the x-coordinate would have to be a random number in the interval $[a, b]$, while y_i must be in $[0, M]$. To find random numbers inside specific intervals, we start with a basic sequence of random numbers r_i (uniformly distributed) inside [0,1]. It's obvious then that if we want a random number x_i inside $[a,b]$, we just compute

$$x_i = a + (b - a)r_i$$

So everything comes down to the question, how do we generate a sequence of random numbers $\{r_i: i = 1, 2, \ldots\}$ uniformly distributed in [0,1]. The most common method in use today is the *residue* (or congruential) method. This method generates a sequence of random integers (like 2120, 1401, 177, 3013, . . .); *then*, by placing a decimal to the left of these numbers, they become numbers between zero and one (like .212, .1401, .0177, .3013, . . .).

So, to generate a sequence of *random integers* (between 0 and P, here), we use the *residue algorithm*.

Residue Algorithm for Generating Random Numbers

1. Pick the first random integer any way you like between 0 and P (P was picked in advance).
2. Multiply this random integer by some fixed integer M (picked in advance).
3. Add to that product another fixed integer K (picked in advance).
4. Divide the resulting sum by P and pick the *remainder* as the new random integer. Now go back to step 2 and repeat steps 2–4 until you have enough random integers.

This residue algorithm can be written as

$$r_{i+1} \equiv (Mr_i + K)\bmod P \qquad i = 0, 1, 2, \ldots$$

which says, if we are given a random integer r_i, then to compute a new one r_{i+1}, we multiply by M, add K, divide by P, and pick the remainder. For example, if:

$$P = 100$$
$$M = 37$$
$$K = 16$$
$$r_0 = 15$$

then we would generate the random sequence

$Mr_0 + K = 571 \longrightarrow r_1 = 71$

$Mr_1 + K = 2643 \longrightarrow r_2 = 43$

$Mr_2 + K = 1607 \longrightarrow r_3 = 7$

Note that since $P = 100$, all the integers are between zero and 100 (the so-called residue class of remainders). By placing a decimal in front of these integers, we get the sequence of random numbers between zero and one:

$$r_0 = .15$$
$$r_1 = .71$$
$$r_2 = .43$$
$$r_3 = .07 \qquad \text{(We also call these numbers } r_i\text{)}.$$

NOTES

1. Since we are dividing by $P = 100$ in our random-number generator, the remainders will be one of the integers $0, 1, 2, \ldots, 99$, and, hence, our entire process will start repeating before long. In fact, our random numbers might be

$$15, 71, \underline{43}, 7, \underline{43}, 7, \underline{43}, 7, \underline{43}, 7, \ldots \qquad \text{(Cycle of two numbers)}$$

and, hence, our method is no good. The ideal situation is to generate the entire residue class $\{0, 1, 2, \ldots 99,\}$ in a *random fashion* before starting to repeat. It can be proven mathematically that if the numbers M, K, and P are chosen according to certain rules, then no matter how we pick the first random number r_0, the algorithm will generate the entire residue class. So, if we pick P very large (like 2^{40}), we are assured that (for practical purposes) the process will never repeat.

2. Other numerical methods (such as Simpson's rule) are generally better than the Monte Carlo method for evaluating integrals unless we want to evaluate a *higher dimensional* integral like

$$I = \int_0^1 \int_0^1 \int_0^1 \int_0^1 e^{-(x^2 + y^2 + z^2 + w^2)} \, dx \, dy \, dz \, dw$$

3. It is possible to generate random samples from various statistical distributions other than the uniform distribution $f(x) = 1, 0 < x < 1$ (the usual random-number generator). Computer programs are available to generate random samples from the binomial, gamma, normal, and many other distributions. These distributions occur in nature and are of interest when it comes to modeling the real world.

PROBLEMS

1. Write a computer program to estimate the integral

$$I = \int_0^1 e^{\sin x} \, dx$$

How accurate is your answer? Plot the graph of your approximation versus the number of random tosses to see if it is converging.

2. Generate a sequence of random numbers via the algorithm

$$r_{i+1} \equiv (3r_i + 4) \bmod 7 \qquad r_0 = 0$$

3. Write a computer program (and flow diagram) to estimate

$$I = \int_0^1 \int_0^1 \int_0^1 e^{-(x^2 + y^2 + z^2)}\, dx\, dy\, dz$$

What about the accuracy?

4. How would you generate a sequence of random points inside the triangle T

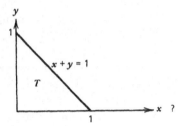

5. How would you generate a random sample from the statistical distribution given in the following diagram:

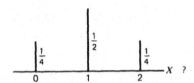

In other words, how would you generate a sequence of integers $\{0, 1, 2\}$ where 0 and 2 have probability of .25 and 1 has a probability of .5?

6. The Buffon needle problem says that a needle (of length one) tossed randomly on the American flag (the width of the red and white stripes is also one) has a probability of $2/\pi$ of crossing one of the lines between the stripes. How would you design a game (and computer program) to evaluate π?

OTHER READING

Monte Carlo Methods by J. M. Hammersley and D. C. Handscomb. Methuen and Company (London), 1964. This concise book outlines the general principles of the Monte Carlo method. Solutions of differential equations, eigenvalue problems, and integral equations are given in addition to problems in statistical mechanics, neutron diffusion, and radiation shielding. The book also contains many additional references.

LESSON **43**

Monte Carlo Solution of Partial Differential Equations

> **PURPOSE OF LESSON:** To show how random games (Monte Carlo methods) can be designed whose outcomes approximate solutions to differential equations. A specific game (tour du wino) is described whose outcome is the finite-difference approximation to a Dirichlet problem inside a square. The game is extended to include solutions to other problems as well.

In the previous lesson, we hinted at how games of chance might be designed whose expected outcomes were solutions or approximate solutions to problems in partial differential equations. This lesson shows how we can design such a game to approximate the solution of the Dirichlet problem

PDE $\quad u_{xx} + u_{yy} = 0 \quad\quad 0 < x < 1 \quad\quad 0 < y < 1$

BC $\quad u(x,y) = g(x,y) = \begin{cases} 1 & \text{On the top of the square} \\ 0 & \text{On the sides and bottom of the square} \end{cases}$

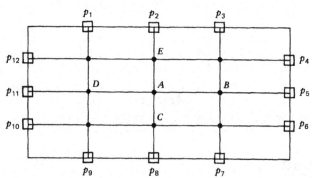

A = starting point for game $\quad\quad \bullet$ = interior grid points

$\square$ = end points p_i $\quad\quad\quad\quad g_i$ = reward for ending at p_i

FIGURE 43.1 Board for tour du wino.

(After we show how the Monte Carlo method can solve this particular problem, we will then discuss more general problems.)

To illustrate the Monte Carlo method in this problem, we introduce a game called tour du wino. To play it, we need a board on which grid lines are drawn (Figure 43.1).

Now, for the rules of the game.

How Tour du Wino is Played

STEP 1 The wino starts from an arbitrary point (point A in our case).

STEP 2 At each stage of the game, the wino staggers off randomly to one of the four neighboring grid points. (In our case, the neighbors of A are B, C, D, and E, and the probability of going to each of these neighbors is .25).

STEP 3 After arriving at a neighboring point, the wino continues this process wandering from point to point until eventually hitting a *boundary point* p_i. He then stops, and we record that point p_i. This completes one *random walk*.

STEP 4 We repeat steps 1–3 until many random walks are completed. We now compute the fraction of times the wino had ended up at each of the boundary points p_i. Table 43.1. shows a typical result after 100 random walks.

STEP 5 Suppose the wino receives a reward g_i (g_i is the value of the BC at p_i) if he ends his walk at the boundary point p_i, and suppose the goal of the game

TABLE 43.1 Probability of Random Walk Ending at p_i
(along with rewards g_i)

Boundary point p_i	$P_A(p_i)$ = fraction of times the wino ends at p_i	g_i = reward for ending at p_i
1	.04	1
2	.15	1
3	.03	1
4	.06	0
5	.17	0
6	.05	0
7	.06	0
8	.15	0
9	.03	0
10	.06	0
11	.16	0
12	.04	0

is to compute his average reward $R(A)$ for all the walks. The average reward is

$$R(A) = g_1 P_A(p_1) + g_2 P_A(p_2) + \ldots g_{12} P_A(p_{12})$$

The game is completed with the determination of $R(A)$. In our specific game with the values of g_i and $P_A(p_i)$ given in Table 43.1, we have

$$R(A) = 1(.04) + 1(.15) + 1(.03) + 0(.06) + \ldots + 0(.04)$$
$$= .22$$

We now tell the reason for playing this game.

Reason for Playing Tour du Wino

It turns out that the average reward we just obtained is the approximate solution to our Dirichlet problem at point A. This interesting observation is based on two facts.

1. Suppose the wino started at a point A that was on the *boundary* of the square. Each resulting random walk ends immediately at that point, and the wino collects the amount g_i. Thus, his average reward for starting from a boundary point is also g_i.
2. Now suppose the wino starts from an interior point. Then, the average reward $R(A)$ is clearly the average of the four average rewards of the four neighbors

$$R(A) = \frac{1}{4}[R(B) + R(C) + R(D) + R(E)]$$

Again, we ask why the wino's average reward $R(A)$ approximates the solution of the Dirichlet problem at A. We have seen that $R(A)$ satisfies the two equations

$$R(A) = \frac{1}{4}[R(B) + R(C) + R(D) + R(E)] \qquad (A \text{ an interior point})$$
$$R(A) = g_i \qquad (A \text{ a boundary point})$$

If we let g_i be the value of the boundary function $g(x,y)$ at the boundary point p_i, then our two equations are exactly the two equations we arrived at when we solved the Dirichlet problem by the finite-difference method. That is, $R(A)$ corresponds to $u_{i,j}$ in the finite-difference equations

$$u_{i,j} = \frac{1}{4}(u_{i-1,j} + u_{i+1,j} + u_{i,j-1} + u_{i,j+1}) \qquad (i,j) \text{ an interior point}$$
$$u_{i,j} = g_{i,j} \qquad g_{i,j} \text{ the solution at a boundary point } (i,j)$$

Hence, $R(A)$ will approximate the true solution of the PDE at A. The tour du wino game can be summarized in three steps

Solution of Laplace's Equation by the Monte Carlo Method

These rules give the solution at one point inside the square.

STEP 1   Generate several random walks starting at some specific point A and ending once you hit a boundary point. Keep track of how many times you hit each boundary point.

STEP 2   After completing the walks, compute the fraction of times you have ended at each point p_i. Call these fractions $P_A(p_i)$.

STEP 3   Compute the *approximate solution* $u(A)$ from the formula

$$u(A) = g_1 P_A(p_1) + g_2 P_A(p_2) + \ldots g_N P_A(p_N)$$

where g_i is the value of the function at p_i and N is the number of boundary points.

The game tour du wino can be modified to solve more complicated problems, as in the following example.

Solution to a Dirichlet Problem with Variable Coefficients

Consider the following elliptic boundary-value problem inside a square:

$$\text{PDE} \quad u_{xx} + (\sin x)u_{yy} = 0 \quad 0 < x < \pi \quad 0 < y < \pi$$

$$\text{BC} \quad u(x,y) = g(x,y) \quad \text{On the boundary of the square}$$

To solve this problem, we replace u_{xx}, u_{yy}, and $\sin x$ by

$$u_{xx} = [u_{i,j+1} - 2u_{i,j} + u_{i,j-1}]/h^2$$
$$u_{yy} = [u_{i+1,j} - 2u_{i,j} + u_{i-1,j}]/k^2 \quad \text{(Central-difference approximation)}$$
$$\sin x = \sin x_j$$

and plug them into the PDE. The grid points can be seen in Figure 43.2.

Making these substitutions and solving for $u_{i,j}$ gives

$$u_{i,j} = \frac{u_{i,j+1} + u_{i,j-1} + \sin x_j(u_{i+1,j} + u_{i-1,j})}{2(1 + \sin x_j)}$$

Look very carefully at this last equation. The coefficients of $u_{i+1,j}$, $u_{i-1,j}$, $u_{i,j+1}$, and $u_{i,j-1}$ are positive and sum to one. In other words, the solution

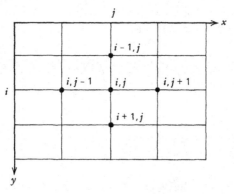

FIGURE 43.2 Grid points of the random walk.

$u_{i,j}$ is a *weighted average* of the solutions at the four neighboring points. Hence, we modify our game so that the wino doesn't stagger off to each neighbor with probability .25, but, rather, with a probability equal to the coefficient of the respective term. In other words, if the wino is at the point (i,j), he then goes to the point:

$$(i,j + 1) \text{ with probability } \frac{1}{2(1 + \sin x_j)}$$

$$(i,j - 1) \text{ with probability } \frac{1}{2(1 + \sin x_j)}$$

$$(i + 1,j) \text{ with probability } \frac{\sin x_j}{2(1 + \sin x_j)}$$

$$(i - 1,j) \text{ with probability } \frac{\sin x_j}{2(1 + \sin x_j)}$$

Other than this slight modification, the game is exactly the same as before. Modifications for other problems may be more subtle, but the ideas are similar. The reader may wish to devise his or her own games to solve other problems. The parabolic case is considered in the problems.

NOTES

1. Observe that once the fractions $P_A(p_i)$ (the fraction of times the wino ends at p_i) are computed, we can then find the solution $u(A)$ for any other boundary conditions g_i just by plugging the $P_A(p_i)$ into the formula

$$u(A) = g_1 P_A(p_1) + g_2 P_A(p_2) + \ldots + g_N P_A(p_N)$$

That is, we don't have to recompute new random walks.

2. In many cases, a researcher wants to find the solution of a PDE at only one point. If the boundary is fairly complicated and if the PDE involves three or four dimensions, then Monte Carlo methods may come to the rescue. In fact, Monte Carlo methods were originally developed to study difficult neutron-diffusion problems that were impossible to solve analytically.

PROBLEMS

1. Write a flow diagram for tour du wino to solve the problem

$$\text{PDE} \quad u_{xx} + u_{yy} = 0 \quad\quad 0 < x < 1 \quad\quad 0 < y < 1$$

$$\text{BC} \quad u(x,y) = g(x,y) \quad\quad \text{On the boundary}$$

at an interior point. Let the number of horizontal and vertical grid lines be arbitrary.

2. Write a computer program to carry out the flow diagram in problem 1.
3. How would the game tour du wino be modified to solve

$$\text{PDE} \quad u_{xx} + x^2 u_{yy} = 0 \quad\quad 0 < x < 1 \quad\quad 0 < y < 1$$

$$\text{BC} \quad u(x,y) = g(x,y) \quad\quad \text{On the boundary}$$

What is the nature of the random walk?

4. Can you devise a modified tour du wino game that will solve

$$\text{PDE} \quad u_{xx} + u_{yy} = 0 \quad\quad 0 < x < 1 \quad\quad 0 < y < 1$$

$$\text{BCs} \quad \begin{cases} u(x,1) = 0 \\ u(x,0) = 0 \\ u(0,y) = 1 \\ \dfrac{\partial u}{\partial x}(1,y) = 0 \end{cases} \quad\quad 0 < x < 1 \quad\quad 0 < y < 1$$

5. Derive the Monte Carlo game for solving the following parabolic IBVP:

$$\text{PDE} \quad u_t = \alpha^2 u_{xx} - \beta u_x - \gamma u \quad\quad 0 < x < 1 \quad\quad 0 < t < \infty$$

$$\text{BCs} \quad \begin{cases} u(0,t) = f(t) \\ u(1,t) = g(t) \end{cases} \quad\quad 0 < t < \infty$$

$$\text{IC} \quad u(x,0) = \phi(x) \quad\quad 0 \leqslant x \leqslant 1$$

HINT Replace the PDE by the finite-difference approximation from the Crank-Nicolson method; solve for $u_{i+1,j}$ in terms of its five neighbors $u_{i+1,j-1}$, $u_{i+1,j+1}$, $u_{i,j-1}$, $u_{i,j}$, $u_{i,j+1}$, and go on from there.

OTHER READING

"Monte Carlo Methods" by G. W. Brown from E. F. Beckenbach, ed., *Modern Mathematics for the Engineer*. McGraw-Hill, 1956. A short chapter illustrating the basic ideas of the Monte Carlo method.

Calculus of Variations (Euler-Lagrange Equations)

PURPOSE OF LESSON: To introduce the idea of a functional (function of a function) and explain how functionals arise naturally in physics. A very common type of functional is the integral

$$J[y] = \int_a^b F(x,y,y') \, dx$$

where the functional J is considered a function of y (a function), and the integrand $F(x,y,y')$ is assumed known. An example is the functional

$$J[y] = \int_0^1 [y^2(x) + y'^2(x)] \, dx$$

We will also show how to find the function $\bar{y}(x)$ that minimizes $J[y]$ by finding an equation (Euler-Lagrange equation) in $\bar{y}$ that must always be true when $\bar{y}$ is a minimizing function. This equation is similar to the necessary condition in calculus that states

$$\frac{df(x)}{dx} = 0$$

at those points x minimizing the function $f(x)$.

One topic that is closely related to differential equations, but which, unfortunately, many students do not study, is *calculus of variations*. This lesson, along with the next, introduces the subject and shows how PDEs can be solved by variational principles.

Calculus of variations was originally studied about the same time as calculus and deals with maximizing and minimizing *functions of functions* (called **functionals**). One of the first problems in calculus of variations was the *Brachistochrone problem* proposed by John Bernoulli in 1696, which attempts to find the path

$y(x)$ that minimizes the sliding time of a particle along a frictionless path between two points (Figure 44.1).

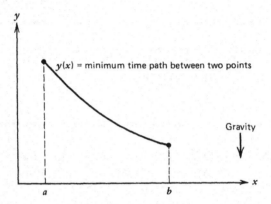

FIGURE 44.1 Brachistocrone problem (original problem in calculus of variations).

Bernoulli showed that the sliding time T could be written

$$T = \int_0^T dt = \int_0^L \frac{dt}{ds}\, ds = \int_0^L \frac{ds}{v} = \frac{1}{\sqrt{2g}} \int_0^L \frac{ds}{\sqrt{y}} = \frac{1}{\sqrt{2g}} \int_a^b \sqrt{\frac{1+y'^2}{y}}\, dx$$

and, hence, the total time $T[y]$ can be thought of as a function of a function. Since many functionals in nature are of this type, it will suffice for us to study the **general form**

(44.1)
$$J[y] = \int_a^b F(x,y,y')\, dx$$

With this motivation in mind, we now state our goal in this lesson: to find functions $y(x)$ that minimize (or maximize) functionals of the form (44.1). The strategy for this task is somewhat the same as for minimizing functions $f(x)$ in calculus. There, we found the critical points of a function by setting $f'(x) = 0$ and solving for x. In calculus of variations, things are much more subtle, since our argument is not a number, but, in fact, a function itself. However, the general philosophy is the same. We take a *functional derivative* (so to speak) with respect to the function $y(x)$ and set this to zero. This new equation is analogous to the equation

$$\frac{df(x)}{dx} = 0$$

from calculus, but now it is an ordinary differential equation, known as the **Euler-Lagrange equation**. The remainder of this lesson is devoted to finding this equation and solving it for specific problems.

Minimizing the General Functional $J[y] = \displaystyle\int_a^b F(x,y,y')\,dx$

We consider the problem of finding the function $y(x)$ that minimizes the functional

$$J[y] = \int_a^b F(x,y,y')\,dx$$

from among a class of smooth functions satisfying the boundary conditions

$$y(a) = A$$
$$y(b) = B$$

See Figure 44.2.

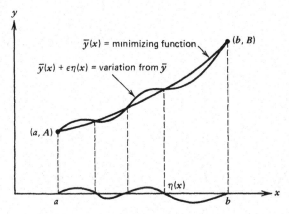

FIGURE 44.2   The variation of a function.

To find the minimizing function, call it $\bar{y}$, we introduce a small variation from $\bar{y}(x)$; namely,

$$\bar{y}(x) + \varepsilon\eta(x)$$

where ε is a small number and $\eta(x)$ is a smooth curve satisfying the BC $\eta(a) = \eta(b) = 0$; they are shown in Figure 44.2. It should be clear that if we evaluate the integral J at a *neighboring* function $\bar{y} + \varepsilon\eta$, then the functional J will be greater; that is,

$$J[\bar{y}] \leq J[\bar{y} + \varepsilon\eta]$$

for all ε. In other words, if we graph the function $\phi(\varepsilon) = J[\bar{y} + \varepsilon\eta]$ as a *function of ε*, we will have a graph something like Figure 44.3.

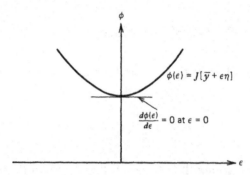

FIGURE 44.3 Graph of $J[\bar{y} + \varepsilon\eta]$ in a neighborhood of $\varepsilon = 0$.

With Figure 44.3 in mind, our strategy for finding $\bar{y}$ is to take the derivative of

$$\phi(\varepsilon) = J[\bar{y} + \varepsilon\eta]$$

with respect to ε, evaluate it at $\varepsilon = 0$, and set this equal to zero; that is,

$$\frac{d\phi(\varepsilon)}{d\varepsilon} = \frac{d}{d\varepsilon} J[\bar{y} + \varepsilon\eta]\bigg|_{\varepsilon=0}$$

$$= \int_a^b \left[\frac{\partial F}{\partial \bar{y}} \eta(x) + \frac{\partial F}{\partial \bar{y}'} \eta'(x) \right] dx$$

(The reader should carry out this step.) From integration by parts, we have

$$= \int_a^b \left\{ \frac{\partial F}{\partial \bar{y}} - \frac{d}{dx}\left[\frac{\partial F}{\partial \bar{y}'} \right] \right\} \eta(x)\, dx = 0$$

Now, since this integral is zero for *any* function $\eta(x)$ satisfying the boundary conditions $\eta(a) = \eta(b) = 0$, it is fairly obvious that the remaining portion of the integrand must be zero; that is,

(44.2) $$\frac{\partial F}{\partial \bar{y}} - \frac{d}{dx}\left[\frac{\partial F}{\partial \bar{y}'} \right] = 0 \qquad \text{(Euler-Lagrange equation)}$$

Equation (44.2) is known as the **Euler-Lagrange equation**, and although it may look complicated in its general form, once we substitute in specific functions $F(x,y,y')$, we will see immediately that it is just a second-order, ordinary differential equation in the dependent variable $\bar{y}$. In other words, we can solve this equation for the minimizing function $\bar{y}$.

So, what we have shown is that (we drop the notation $\bar{y}$ and just call it y)

If y minimizes $J[y] = \displaystyle\int_a^b F(x,y,y')\,dx$, then y *must* satisfy the equation

$$\frac{\partial F}{\partial y} - \frac{d}{dx}\left[\frac{\partial F}{\partial y'}\right] = 0$$

In order to clarify these ideas, we present an example.

Finding the Minimum of the Functional $J[y] = \displaystyle\int_0^1 [y^2 + y'^2]\,dx$

Here, we will attempt to find the function $y(x)$ that passes through the points $(0,0)$ and $(1,1)$ and minimizes

$$J[y] = \int_0^1 [y^2 + y'^2]\,dx$$

It is clear from the context of our problem that we are referring to a function that is differentiable inasmuch as the integrand depends on y'. To find $\bar{y}$, we start by writing the Euler-Lagrange equation [with BCs $y(0) = 0$ and $y(1) = 1$]

$$F_y - \frac{d}{dx}F_{y'} = 0$$

$$y(0) = 0$$
$$y(1) = 1$$

Since $F(x,y,y') = y^2 + y'^2$, we have

$$\begin{aligned}F_y &= 2y \\ F_{y'} &= 2y'\end{aligned} \quad \text{(Just differentiate } F \text{ with respect to } y \text{ and } y')$$

and, hence, the Euler-Lagrange equation is

$$2y - \frac{d}{dx}(2y') = 0$$

or just

$$y'' - y = 0$$

Solving this simple differential equation with BCs $y(0) = 0$ and $y(1) = 1$ gives

$$\bar{y}(x) = 0.42e^x - 0.42e^{-x}$$

which is shown in Figure 44.4. The claim here is that any other smooth curve $y(x)$ passing through the two boundary points will give rise to a larger $J[y]$.

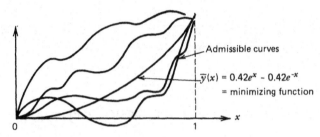

FIGURE 44.4   All possible smooth curves satisfying the BC $y(0) = 0$ and $y(1) = 1$.

NOTES

1. The Euler-Lagrange equation is analogous to setting the derivative equal to zero in the calculus. If the reader remembers, we don't always find the minimum (or maximum, for that matter) by this process. For example, the function $f(x) = x^3$ has zero derivative at $x = 0$, but this point is neither a local maximum nor minimum; the same holds true with the Euler-Lagrange equation. We should think of it as a *necessary condition* that must be true for minimizing functions but may be true for other functions as well. Quite often, however, the solution of the Euler-Lagrange equation will be a local (in fact, global) minimum by the very nature of the problem; we can often tell if this is the case.

2. The minimum $y(x)$ of the functional

$$J[y] = \int_a^b y\sqrt{1 + y'^2}\, dx$$

is the curve that gives rise to the minimum surface area of revolution when we rotate $y(x)$ about the x-axis (Figure 44.5).

The solution of the Euler-Lagrange equation with BCs $y(a) = A$ and $y(b) = B$ is part of the *hyperbolic-cosine curve* (catenary)

$$\bar{y}(x) = \alpha \cosh\left[(x - \beta)/\alpha\right]$$

where the constants α and β are determined so that the curve passes through the end points. This is a fairly difficult equation to solve, although the reader could verify that the hyperbolic cosine does satisfy the equation.

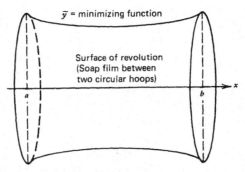

FIGURE 44.5 Minimum surface of revolution.

3. We could evaluate the functional

$$J[y] = \int_0^1 [y^2 + y'^2]\, dx$$

that was minimized in this lesson for the *minimization function*

$$\bar{y}(x) = 0.42e^x - 0.42e^{-x}$$

to see that $J[\bar{y}] = 0.46$. If we substitute any other smooth function $y(x)$ passing through the end points $(0,0)$ and $(1,1)$, we get a larger value of $J[y]$.

4. Basic principles of physics are often stated in terms of *minimizing principles* rather than differential equations. *Fermat's principle* (a light ray requires less time along its actual path than along any other path having the same end points), *Hamilton's principle* (in a conservative force field, a particle moves so as to minimize the *action integral*

$$\int_{t_1}^{t_2} (\text{kinetic energy} - \text{potential energy})\, dt$$

are examples of nature behaving in such a way that minimizes functionals.

5. Calculus of variations ideas can also be used to minimize important *multiple-integral functionals* like

$$J[u] = \int\!\!\int_D F(x,y,u,u_x,u_y)\, dx\, dy$$

where the corresponding Euler-Lagrange equation is

$$F_u - \frac{\partial}{\partial x} F_{u_x} - \frac{\partial}{\partial y} F_{u_y} = 0 \qquad \text{(A PDE)}$$

In fact, these functionals are the ones that we will be concerned with in the next lesson, since the Euler-Lagrange equation is a PDE. Our general philosophy, however, will be slightly different, since our goal is to find the function $u(x,y)$ (by some new technique) that minimizes this double-integral functional and, hence, is a solution to the PDE. In other words, we solve the PDE by *minimizing the functional*. This is in contrast to what we have been doing in this lesson with single-integral functionals, where we have been finding the minimizing function by solving the Euler-Lagrange equations. Methods for solving differential equations by finding the minimizing function of their corresponding functional $J[u]$ are called *direct methods* in the calculus of variations. Methods for finding minimizing functions of functionals $J[u]$ by solving their corresponding Euler-Lagrange equation are called *indirect methods* in the calculus of variations. The next lesson discusses the well-known direct method of Ritz.

PROBLEMS

1. Find the minimizing function $\bar{y}(x)$ of

$$J[y] = \int_0^1 \sqrt{1 + y'^2} \, dx$$

among those curves satisfying $y(0) = 0$ and $y(1) = 1$. What is your interpretation of your answer? What is the value of $J[\bar{y}]$? What is the meaning of $J[\bar{y}]$?

2. The kinetic energy of a simple vibrating mass is given by $KE = \frac{1}{2} m \dot{y}^2$ where $\dot{y} = dy/dt$, and its potential energy is given by $PE = \frac{1}{2} ky^2$.

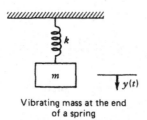

Vibrating mass at the end
of a spring

Hamilton's principle states that the motion $y(t)$ of the mass is such that the integral

$$\int_{t_1}^{t_2} [KE - PE] \, dt = \frac{1}{2} \int_{t_1}^{t_2} [m\dot{y}^2 - ky^2] \, dt$$

is minimized. If this law is valid, what differential equation must the vibrating mass satisfy?

3. Show that the minimizing function $\bar{y}(x)$ for the functional

$$J[y] = \int_0^{\pi/2} [y'^2 - y^2]\, dx$$

with BCs $y(0) = 0$ and $y(\pi/2) = 1$ is $\bar{y}(x) = \sin x$. Evaluate $J(\sin x)$.

4. Derive the Euler-Lagrange equation

$$F_u - \frac{\partial}{\partial x} F_{u_x} - \frac{\partial}{\partial y} F_{u_y} = 0$$

from the functional

$$J[u] = \int_D \int F(x, y, u, u_x, u_y)\, dx\, dy$$

OTHER READING

Methods of Mathematical Physics by R. Courant and D. Hilbert. Interscience Publishers, 1953. A classic text with many examples; see Chapters 4–6.

Variational Methods for Solving PDEs (Method of Ritz)

PURPOSE OF LESSON: To show how differential equations can be solved by interpreting the equation as the Euler-Lagrange equation of some functional and then finding the *minimizing function* of the functional (by some new method). The minimizing function will then be the solution of the PDE. The problem, of course, is to find the functional that has the original equation as its Euler-Lagrange equation. A well-known result (*minimum-energy theorem*) is presented that says finding the solution u to certain elliptic BVPs like

$$u_{xx} + y_{yy} = f \quad \text{In a region } D$$
$$u = 0 \quad \text{On the boundary of } D$$

is equivalent to finding the function u (also zero on the boundary of D) that minimizes the potential-energy functional

$$J[u] = \int_D \int [u_x^2 + u_y^2 + 2uf]\, dx\, dy$$

That is, $\nabla^2 u = f$ is the Euler-Lagrange equation of $J[u]$. An approximate minimizing function of $J[u]$ is found by the *Method of Ritz*, and so we have the solution (an approximation) to the PDE. The method of Ritz will be discussed and a functional minimized using this technique.

There is a very nice way to solve boundary-value problems (like the stretched-membrane problem) by looking for the smooth surface that minimizes the potential energy of the membrane. That is, if we think of our PDE as being the Euler-Lagrange equation of some functional $J[u]$, then we can solve the differential equation by *minimizing* the functional (since the minimizing function of the functional is also the solution of the corresponding Euler-Lagrange equation). In Lesson 44, we only considered functionals where the Euler-Lagrange equation was an ODE. This lesson considers functionals where the Euler-Lagrange equation is a PDE. for example, the functional

$$J[u] = \int_0^1 \int_0^1 [u_x^2 + u_y^2] \, dx \, dy$$

has as its Euler-Lagrange equation (proof similar to the ODE case in the last lesson)

$$u_{xx} + u_{yy} = 0$$

and so to solve the Dirichlet problem in the unit square

PDE $u_{xx} + u_{yy} = 0$ $0 < x < 1$ $0 < y < 1$

BC $u = g$ On the boundary of the square

we can alternatively find the function $u(x,y)$ that is equal to g on the boundary and minimizes $J[u]$. It probably comes as no surprise that the functional

$$J[u] = \int_0^1 \int_0^1 [u_x^2 + u_y^2] \, dx \, dy$$

represents the potential energy of the membrane, and so what we are, in fact, doing is finding the *minimum potential-energy* surface. The question, of course, is, given a differential equation, how do we find the functional $J[u]$ that represents the potential energy of the solution? The answer to this question is given in a well-known theorem (minimum-energy theorem) that states:

The solution u of the Dirichlet problem

PDE $\nabla^2 u = f$ In a region D
BC $u = 0$ On the boundary of D

is the *same function* u that minimizes (among those functions having BC $u = 0$) the energy functional

$$J[u] = \int_D \int [u_x^2 + u_y^2 + 2uf] \, dx \, dy$$

This theorem is stated for other types of BCs and is proven in the reference of Other Reading. To help understand the theorem, we present an example.

Replacing Poisson's Equation by Its Potential-Energy Functional

Consider the Dirichlet problem

(45.1)   PDE $u_{xx} + u_{yy} = f$ $0 < x < 1$ $0 < y < 1$

BC $u = 0$ On the boundary of the square

See Figure 45.1.

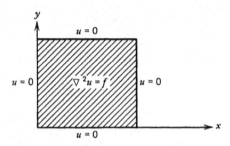

FIGURE 45.1   Poisson's equation inside a square.

Here, the energy functional $J[u]$ is

(45.2) $$J[u] = \int_0^1 \int_0^1 [u_x^2 + u_y^2 + 2uf] \, dx \, dy$$

and so to solve problem (45.1), it is sufficient to find the function $\bar{u}$ (among those that are zero on the boundary) that minimizes $J[u]$.

This completes the first part of the lesson. For the remaining part, we show how to find the minimizing function $\bar{u}$ of $J[u]$ using the method of Ritz.

Method of Ritz for Minimizing Functionals

This method is only one of several discussed in the reference of Other Reading. The idea was introduced by the mathematician W. Ritz and is quite simple; it consists of the following steps:

STEP 1   Replace the function u in the functional

$$J[u] = \int_0^1 \int_0^1 [u_x^2 + u_y^2 + 2uf] \, dx \, dy$$

by an approximating function (the user picks n)

$$u_n(x,y) = a_1\phi_1(x,y) + a_2\phi_2(x,y) + \cdots + a_n\phi_n(x,y)$$

where the functions ϕ_1, ϕ_2, ϕ_3, . . . , ϕ_n are chosen in advance so that they are all *zero on the boundary* and represent a nice enough class, so that they can reasonably approximate the solution to the problem. Typical choices for the Dirichlet problem inside the unit square would be:

$$\phi_1(x,y) = xy(1 - x)(1 - y) \quad \leftarrow \quad \text{Zero on the boundary}$$
$$\phi_2(x,y) = x\phi_1(x,y)$$
$$\phi_3(x,y) = y\phi_1(x,y)$$
$$\phi_4(x,y) = x^2\phi_1(x,y)$$
$$\phi_5(x,y) = xy\phi_1(x,y)$$
$$\phi_6(x,y) = y^2\phi_1(x,y)$$

$$
\begin{matrix}
. & . & . \\
. & . & . \\
. & . & .
\end{matrix}
$$

In other words, the first four approximations to the solution are:

$$u_1(x,y) = a_1 xy(1 - x)(1 - y)$$
$$u_2(x,y) = xy(1 - x)(1 - y)[a_1 + a_2 x]$$
$$u_3(x,y) = xy(1 - x)(1 - y)[a_1 + a_2 x + a_3 y]$$
$$u_4(x,y) = xy(1 - x)(1 - y)[a_1 + a_2 x + a_3 y + a_4 x^2]$$

$$
\begin{matrix}
. & . & . & . \\
. & . & . & . \\
. & . & . & .
\end{matrix}
$$

STEP 2 The functional

$$J[u_n] = \int_0^1 \int_0^1 \left\{ \left[\sum_{j=1}^n a_j \frac{\partial \phi_j}{\partial x} \right]^2 + \left[\sum_{j=1}^n a_j \frac{\partial \phi_j}{\partial y} \right]^2 + 2f \sum_{j=1}^n a_j \phi_j \right\} dx\, dy$$

is now a *function* of $a_1, a_2, \ldots, a_n$ and so to find these coefficients that minimize J, we set the following partial derivatives equal to zero:

$$\frac{\partial J[u_n]}{\partial a_1} = 2 \int_0^1 \int_0^1 \left\{ \sum_{j=1}^n \left[\frac{\partial \phi_j}{\partial x} \frac{\partial \phi_1}{\partial x} + \frac{\partial \phi_j}{\partial y} \frac{\partial \phi_1}{\partial y} \right] a_j + f\phi_1 \right\} dx\, dy = 0$$

$$
\begin{matrix}
. & . & . \\
. & . & . \\
. & . & .
\end{matrix}
$$

$$\frac{\partial J[u_n]}{\partial a_n} = 2 \int_0^1 \int_0^1 \left\{ \sum_{j=1}^n \left[\frac{\partial \phi_j}{\partial x} \frac{\partial \phi_n}{\partial x} + \frac{\partial \phi_j}{\partial y} \frac{\partial \phi_n}{\partial y} \right] a_j + f\phi_n \right\} dx\, dy = 0$$

This all looks pretty complicated, but if we rewrite these two equations in matrix form, we have the system of linear equations

$$Aa = b$$

where

$A = (A_{ij})$ is the $n \times n$ matrix with elements (can be computed)

$$(45.3) \qquad A_{ij} = \int_0^1 \int_0^1 \left[\frac{\partial \phi_i}{\partial x} \frac{\partial \phi_j}{\partial x} + \frac{\partial \phi_i}{\partial y} \frac{\partial \phi_j}{\partial y} \right] dx \, dy$$

$b = (b_i)$ is the vector with components

$$(45.4) \qquad b_i = - \int_0^1 \int_0^1 f(x,y) \phi_i(x,y) \, dx \, dy$$

and

$a = (a_i)$ is the unknown vector whose elements represent the coefficients in the approximate solution $u_n(x,y) = a_1\phi_1(x,y) + \ldots + a_n\phi_n(x,y)$.

STEP 3 Solve the linear system $Aa = b$ for the coefficients $a_1, a_2, \ldots, a_n$. Hence, we have the approximate minimizing function

$$u_n(x,y) = a_1\phi_1(x,y) + a_2\phi_2(x,y) + \ldots + a_n\phi_n(x,y)$$

and, hence, an *approximation* to the solution of Dirichlet problem (47.1).

NOTES

1. The *Ritz method* was used to minimize *double-integral*-type functionals in this lesson. It could also be used to minimize functionals like

$$J[y] = \int_0^1 [y^2 + y'^2] \, dx$$
$$y(0) = 0 \qquad y(1) = 1$$

The only difference here would be that approximation by functions $\phi_1, \phi_2, \phi_3, \ldots, \phi_n$ in $y_n(x) = a_1\phi_1 + a_2\phi_2 + \ldots + a_n\phi_n$ would have to satisfy the BCs

$$\phi_i(0) = 0 \qquad i = 1, 2, \ldots, n$$
$$\phi_i(1) = 1$$

2. It wasn't proven in this lesson that the Euler-Lagrange equation of the double integral

$$J[u] = \int_0^1 \int_0^1 [u_x^2 + u_y^2] \, dx \, dy$$

is *Laplace's equation*, but the proof is analogous to the arguments in Lesson 44 in the single-integral case.

3. In the reference of the recommended reading, we show how to construct the *energy functional J[u]* for many other differential equations (other than Poisson's equation) and for many other kinds of BCs (other than Dirichlet type $u = 0$). Hence, many kinds of BVPs can be solved by minimizing the corresponding energy functional.

4. The larger we pick n, the smaller the actual value of $J[u_n]$, and, hence, the more accurate the solution to the differential equation. One strategy would be to compute $u_n(x,y)$ for larger and larger values of n and see how much $J[u_n]$ is decreasing.

5. For all practical purposes, computations in the Ritz method would have to be carried out on a computer unless n was very small. The flow diagram in Figure 45.2 illustrates the computations we would make in order to solve BVP (45.1) by the Ritz method.

The user would have to provide a subroutine to define the functions ϕ_1, ϕ_2, ..., ϕ_n and evaluate them. This could be done in a subroutine like the following:

```
      SUBROUTINE BC (X,Y,PHI)
C     SUBROUTINE PROVIDED BY THE USER TO EVALUATE THE
C     FUNCTIONS PHI(1),PHI(2), ..., PHI(N)
      DIMENSION PHI(20)
      PHI(1) = X*Y*(1−X)*(1−Y)
      PHI(2) = X*PHI(1)      (these are the functions we've been using)
      PHI(3) = Y*PHI(1)
          .        .
          .        .
          .        .
      PHI(N) = (whatever it is)
      RETURN
      END
```

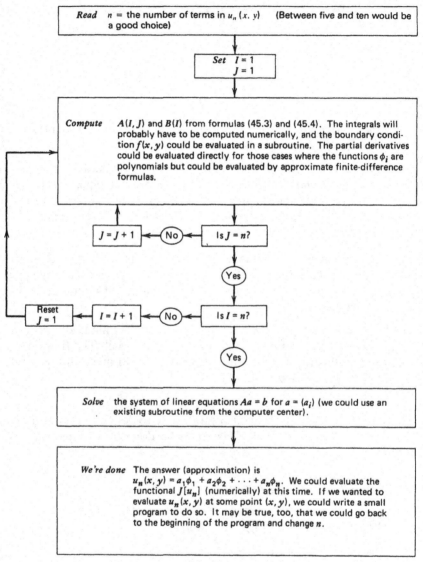

FIGURE 45.2   Flow diagram for the Ritz method.

PROBLEMS

1. What would be the *energy functional* $J[u]$ corresponding to the following problem:

$$\text{PDE} \quad u_{xx} + u_{yy} = 1 \quad 0 < x < 1 \quad 0 < y < 1$$
$$\text{BC} \quad u = 0 \quad \text{On the boundary}$$

2. How could we minimize the functional

$$J[y] = \int_0^1 [y^2 + y'^2]\, dx \qquad y(0) = 0 \qquad y(1) = 1$$

by the method of Ritz?

HINT   If we introduce a new function $z(t)$

$$z(t) = (1 - x)y(t)$$

we note that $z(t)$ satisfies $z(0) = 0$ and $z(1) = 0$.

3. Write a computer program to carry out the computation in Figure 45.2.
4. Show that the Euler-Lagrange equation of the functional $J[u]$

$$J[u] = \int_0^1 \int_0^1 [u_x^2 + u_y^2]\, dx\, dy$$

is

$$u_{xx} + u_{yy} = 0$$

5. The Dirichlet problem

$$u_{xx} + u_{yy} = \sin(\pi x) \qquad 0 < x < 1 \qquad 0 < y < 1$$
$$u = 0 \qquad \text{On the boundary of the square}$$

can be solved by the *finite sine transform* (transform the x-variable) and has the solution

$$u(x,y) = \left[Ae^{\pi y} + Be^{-\pi y} - \frac{1}{\pi^2} \right] \sin(\pi x)$$

where $A = 0.06$ and $B = 0.04$. How would we find the potential energy of the solution? It would be a good review for the reader to use the sine transform and find this solution $u(x,y)$.

OTHER READING

Variational Methods in Mathematical Physics by S. G. Miklin. Macmillan, 1964. This book discusses many types of methods for minimizing functionals and how these results can be applied to solving PDEs. In addition to the Ritz Method, the well-known method of Galerkin is discussed; see Chapters 3–4.

Perturbation Methods for Solving PDEs

PURPOSE OF LESSON: To show how different problems (nonlinear, equations with variable coefficients, irregular domains) can be solved by perturbing easier problems. In other words, to show how the solutions to easy problems can be modified so that they approximate solutions to hard problems.

For example, we will show how the solution to the nonlinear problem

$$\begin{array}{lll} \text{PDE} & \nabla^2 u + u^2 = 0 & 0 < r < 1 \\ \text{BC} & u(1,\theta) = \cos\theta & 0 \leqslant \theta \leqslant 2\pi \end{array}$$

can be approximated by perturbing the solution $u(r,\theta) = r\cos\theta$ to the *linear* Dirichlet problem

$$\begin{array}{lll} \text{PDE} & \nabla^2 u = 0 & 0 < r < 1 \\ \text{BC} & u(1,\theta) = \cos\theta & 0 \leqslant \theta \leqslant 2\pi \end{array}$$

Quite often a problem whose solution is known can be *continuously* changed so that nearby problems can be solved by a gradual modification of the solution of the first one. That is, we continuously modify (perturb) the original problem (and its solution) in a gradual way so that nearby problems are solved.

For example, Laplace's equation

$$\nabla^2 u = 0$$

can be continuously modified via the *family* of problems

$$\nabla^2 u + \varepsilon u^2 = 0 \qquad 0 \leqslant \varepsilon \leqslant 1$$

so that we arrive at the new nonlinear equation

$$\nabla^2 u + u^2 = 0$$

In this way, we can attempt to solve nonlinear problems by modifying the solution to Laplace's equation (Figure 46.1).

$$\nabla^2 u = 0 \quad\longrightarrow\quad \nabla^2 u + \varepsilon u^2 = 0 \longrightarrow \nabla^2 u + u^2 = 0$$
$$(\varepsilon = 0) \qquad\qquad (0 < \varepsilon < 1) \qquad (\varepsilon = 1)$$

Solution   Solution   Solution

$$u_0 \quad\longrightarrow\quad u = \sum_{k=0}^{\infty} \varepsilon^k u_k \quad\longrightarrow\quad u = \sum_{k=0}^{\infty} u_k$$

FIGURE 46.1 Schematic diagram of perturbation methods.

In seeking the solution to the perturbed Laplace equations

$$\nabla^2 u + \varepsilon u^2 = 0$$

we modify the solution u_0 of Laplace's equation by adding on small perturbations. It seems reasonable that the perturbed equation

$$\nabla^2 u + \varepsilon u^2 = 0$$

should have a solution of the form

(46.1) $$u = u_0 + \varepsilon u_1 + \varepsilon^2 u_2 + \ldots$$

Note that power series (46.1) in ε agrees with Laplace's solution $u = u_0$ when $\varepsilon = 0$ and will be the solution of the *nonlinear problem* (if it converges)

$$\nabla^2 u + u^2 = 0$$

when $\varepsilon = 1$. In other words, equation (46.1) acts as a *vehicle* that allows us to go from Laplace's equation to the nonlinear one. The problem is to find the functions $u_0, u_1, u_2, \ldots$ in power series (46.1). For the remaining part of this lesson, we will show how different boundary-value problems can be solved by this general principle.

A Perturbation Solution of the Nonlinear Equation $\nabla^2 u + u^2 = 0$

Suppose we would like to solve the following nonlinear Dirichlet problem

(46.2) PDE $\nabla^2 u + u^2 = 0 \qquad 0 < r < 1$

BC $u(1,\theta) = \cos\theta \qquad 0 \le \theta \le 2\pi$

The idea here is to think of this nonlinear problem as a *perturbation* of the *linear* one

(46.3)   PDE $\nabla^2 u = 0$ $0 < r < 1$

                BC $u(1,\theta) = \cos\theta$ $0 \leq \theta \leq 2\pi$

which has the solution $u_0(r,\theta) = r\cos\theta$ and ask how to *modify* $u_0(r,\theta)$, so that it satisfies problem (46.2)? As we mentioned earlier, we introduce a class of problems $\nabla^2 u + \varepsilon u^2 = 0$ and look for solutions of each of these equations of the form

(46.4) $u(r,\theta) = u_0(r,\theta) + \varepsilon u_1(r,\theta) + \varepsilon^2 u_2(r,\theta) + \ldots$

In this way, we find the solution of our nonlinear problem (46.2) by letting $\varepsilon = 1$. Substituting our perturbed candidate (46.4) in the perturbed problem

$$\text{PDE}\qquad \nabla^2 u + \varepsilon u^2 = 0$$
$$\text{BC}\qquad u(1,\theta) = \cos\theta$$

we get

$$\nabla^2(u_0 + \varepsilon u_1 + \varepsilon^2 u_2 + \ldots) + \varepsilon(u_0 + \varepsilon u_1 + \varepsilon^2 u_2 + \ldots)^2 = 0$$
$$u_0(1,\theta) + \varepsilon u_1(1,\theta) + \varepsilon^2 u_2(1,\theta) + \ldots = \cos\theta$$

Performing a little algebra and setting the coefficients of the powers of ε equal to each other, we arrive at the following *sequence* of problems $P_0, P_1, P_2, \ldots$ from which we can solve for our unknown functions $u_0, u_1, \ldots$ (note that the problem are *linear* and *nonhomogeneous*):

$$P_0 \quad \begin{cases} \nabla^2 u_0 = 0 \quad 0 < r < 1 \\ u_0(1,\theta) = \cos\theta \end{cases} \quad u_0(r,\theta) = r\cos\theta$$

$$P_1 \quad \begin{cases} \nabla^2 u_1 = -u_0^2 \\ u_1(1,\theta) = 0 \end{cases}$$

$$P_2 \quad \begin{cases} \nabla^2 u_2 = -2u_0 u_1 \\ u_2(1,\theta) = 0 \end{cases}$$

$$\begin{matrix} \cdot & \cdot & \cdot \\ \cdot & \cdot & \cdot \\ \cdot & \cdot & \cdot \end{matrix}$$

Let's start by finding $u_1(r,\theta)$ from problem P_1 (we already know u_0); we have

$$P_1 \qquad \begin{cases} \nabla^2 u_1 = -r^2 \cos^2\theta = -\dfrac{r^2}{2}[1 + \cos(2\theta)] = -\dfrac{r^2}{2} - \dfrac{r^2}{2}\cos(2\theta) \\ u_1(1,\theta) = 0 \end{cases}$$

To solve this *nonhomogeneous* problem, we resort to a technique that the reader may recall from ODE theory and that consists of the following:
1. Finding the general solution y_h of the homogeneous equation
2. Finding a particular solution y_p of the nonhomogeneous equation
3. Substituting $y_h + y_p$ into the IC and solving for the constants

This technique will work for this specific problem. In our problem P_1, the general form of the homogeneous solution of $\nabla^2 u = 0$, $0 < r < 1$ is chosen as the separation of variables solution

$$u_h(r,\theta) = \sum_{n=0}^{\infty} r^n[a_n \cos(n\theta) + b_n \sin(n\theta)]$$

Now, to find a particular solution (just one solution) of the nonhomogeneous equation

$$\nabla^2 u = -\frac{r^2}{2} - \frac{r^2}{2}\cos(2\theta)$$

we try

$$u_p(r,\theta) = Ar^4 + Br^4 \cos(2\theta)$$

[Inputs of r^n, $r^n \cos(n\theta)$, $r^n \sin(n\theta)$ give rise to solutions of the form Ar^{n+2}, $Br^{n+2} \cos(n\theta)$, $Cr^{n+2} \sin(n\theta)$, respectively]. Substituting $u_p(r,\theta)$ in the non-homogeneous equation gives

$$A = -\frac{1}{32} \qquad B = -\frac{1}{24}$$

Thus, we have

$$u_p(r,\theta) = -\frac{r^4}{32} - \frac{r^4}{24}\cos(2\theta)$$

Our final step in solving P_1 is to substitute the general solution $u(r,\theta) = u_h(r,\theta) + u_p(r,\theta)$ in the BC $u(1,\theta) = 0$ and solve for the coefficients a_n and b_n. Doing this, we get

$$\sum_{n=0}^{\infty} [a_n \cos(n\theta) + b_n \sin(n\theta)] - \frac{1}{32} - \frac{1}{24}\cos(2\theta) = 0$$

and, hence, $a_0 = 1/32$, $a_2 = 1/24$, and all the other a's and b's are zero. In other words, the solution of problem P_1 is

$$u_1(r,\theta) = \frac{1}{32} + \frac{1}{24} r^2 \cos(2\theta) - \frac{r^4}{24} \cos(2\theta) - \frac{r^4}{32}$$

$$= -\frac{(r^4 - 1)}{32} - \frac{(r^4 - r^2)}{24} \cos(2\theta)$$

The function $u_1(r,\theta)$ is called the **first perturbation** of $u_0(r,\theta)$ and by adding it to $u_0(r,\theta)$, we have the new approximation

(46.5) $\qquad u = u_0 + u_1 = r \cos\theta - \dfrac{(r^4 - 1)}{32} - \dfrac{(r^4 - r^2)}{24} \cos(2\theta)$

to problem (46.2). To find the next perturbation $u_2(r,\theta)$, we must solve

$$P_2 \qquad \begin{cases} \nabla^2 u_2 = -2u_0 u_1 \\ u_2(1,\theta) = 0 \end{cases}$$

where $u_0(r,\theta)$ and $u_1(r,\theta)$ are substituted in the right-hand side of this equation. Needless to say, without the help of a computer to carry out these algebraic manipulations, this would be a significant problem in itself. Fortunately, equation (46.5) is a fairly accurate solution to our problem. In fact, if we substituted this approximation in the left-hand side of our nonlinear equation

$$\nabla^2 u + u^2$$

we could see that it is *almost zero* everywhere inside the circle $0 < r < 1$.

In addition to solving nonlinear equations, perturbation theory can also be applied to solving problems with *irregular boundaries* (as long as they're not too irregular); we now present a simple example.

An Example of a Boundary Perturbation

PDEs aren't the only things that can be perturbed. We can find the solution of Laplace's equation inside a *deformed circle* by perturbing the solution of Laplace's equation inside a circle. For example, suppose we want to find the potential inside the region $r = 1 + \frac{1}{4} \sin\theta$ with the potential u given *on* this boundary. In other words, solve

(46.6) $\qquad \nabla^2 u = 0 \qquad 0 < r < 1 + \dfrac{1}{4} \sin\theta$

$$u\left(1 + \frac{1}{4} \sin\theta, \theta\right) = \cos\theta \qquad 0 \leq \theta \leq 2\pi$$

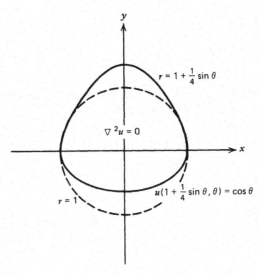

FIGURE 46.2 Laplace's equation inside a deformed circle.

See Figure 46.2. We could think of this problem as describing a soap film inside a deformed circle where the height of the wire is $\cos \theta$.

Since the general philosophy of perturbation procedures is to rewrite a difficult problem in terms of an easy one, we think of the BC on the deformed circle

$$u\left(1 + \frac{1}{4} \sin \theta, \theta\right) = \cos \theta$$

as just the *extreme* BC in the family of BCs

$$u(1 + \varepsilon \sin \theta, \theta) = \cos \theta \qquad 0 \leq \varepsilon \leq \frac{1}{4}$$

See Figure 46.3. This idea leads us to think in terms of Taylor's series. Using a Taylor series like

$$f(x + h) = f(x) + f'(x)h + f''(x)\frac{h^2}{2!} + \ldots$$

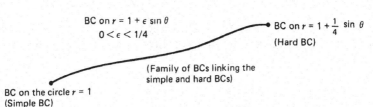

FIGURE 46.3 Schematic illustration of perturbation procedures.

we can expand the hard BC in terms of the easy one; that is,

$$u(1 + \varepsilon \sin \theta, \theta) = u(1,\theta) + u_r(1,\theta)(\varepsilon \sin \theta) + u_{rr}(1,\theta)\frac{(\varepsilon \sin \theta)^2}{2!} + \ldots$$

Substituting this in our original problem gives us the new equivalent problem (we will, of course, let $\varepsilon = 1/4$ when we get done)

(46.7)
$$\nabla^2 u = 0 \qquad 0 < r < 1 + \varepsilon \sin \theta$$

$$u(1,\theta) + u_r(1,\theta)(\varepsilon \sin \theta) + u_{rr}(1,\theta)\frac{(\varepsilon \sin \theta)^2}{2!} + \ldots = \cos \theta$$

This, of course, doesn't look like an easy problem either, but we can now divide the problem into a sequence of problems where we can separately find the functions $u_0, u_1, u_2, \ldots$ in the desired solution

(46.8)
$$u = u_0 + \varepsilon u_1 + \varepsilon^2 u_2 + \ldots$$

If we substitute this series into problem (46.7), we get the following sequence of problems from which we can find $u_0, u_1, \ldots$

$$P_0 \quad \begin{cases} \nabla^2 u_0 = 0 \qquad 0 < r < 1 \qquad \text{(Inside a } circle\text{)} \\ u_0(1,\theta) = \cos \theta \end{cases} \qquad u_0(r,\theta) = r \cos \theta$$

$$P_1 \quad \begin{cases} \nabla^2 u_1 = 0 \qquad 0 < r < 1 \qquad \text{(Inside a } circle\text{)} \\ u_1(1,\theta) = - \sin \theta \, \dfrac{\partial u_0(1,\theta)}{\partial r} = - \sin \theta \cos \theta \end{cases}$$

$$\begin{matrix} \vdots & \vdots & \vdots \\ \vdots & \vdots & \vdots \end{matrix}$$

Hence, we can solve each of these Dirichlet problems (inside the circle) for u_0, $u_1, u_2, \ldots$ and so obtain the solution

$$u = u_0 + \frac{1}{4}u_1 + \left(\frac{1}{4}\right)^2 u_2 + \ldots$$

(note that we have let $\varepsilon = 1/4$) of the Dirichlet problem in the *deformed* region. The reader will get a chance to find the first perturbation u_1 and check the approximation

$$u_0 + \frac{1}{4}u_1$$

to see how well it satisfies equation (46.6).

NOTES

Nonlinear problems are not the only types of PDE problems that can be solved by perturbation processes. We could solve the initial-value problem

(46.10)
$$u_t = (1 + x)u_{xx} \qquad -\infty < x < \infty$$
$$u(x,0) = \phi(x)$$

by introducing the parameterized equation

(46.11)
$$u_t = (1 + \varepsilon x)u_{xx}$$

and look for a solution of the form

$$u = u_0 + \varepsilon u_1 + \varepsilon^2 u_2 + \ldots$$

Substituting this expression in problem (46.11) gives us the following sequence of problems:

$$P_0 \quad \begin{cases} \dfrac{\partial u_0}{\partial t} = \dfrac{\partial^2 u_0}{\partial x^2} \\[2ex] u_0(x,0) = \phi(x) \end{cases}$$

$$P_1 \quad \begin{cases} \dfrac{\partial u_1}{\partial t} - \dfrac{\partial^2 u_1}{\partial x^2} = x\dfrac{\partial^2 u_0}{\partial x^2} \\[2ex] u_1(x,0) = 0 \end{cases}$$

$$\begin{matrix} \cdot & \cdot & \cdot \\ \cdot & \cdot & \cdot \\ \cdot & \cdot & \cdot \end{matrix}$$

Note that the two displayed problems have *constant* coefficients. The reader should be aware that ε in the perturbation equation must be small or else the infinite series will not converge.

PROBLEMS

1. Substitute equation (46.8) in problem (46.7) to find the sequence of problems $P_0, P_1, P_2, \ldots$

2. Show that the *nonlinear* problem

$$\nabla^2 u + u^2 = 0 \qquad 0 < r < 1$$
$$u(1,\theta) = \cos\theta$$

gives rise to the sequence of linear problems $P_0, P_1, P_2, \ldots$ as shown in the lesson.

3. Substitute equation (46.5) in the nonlinear problem

$$\nabla^2 u + u^2 = 0$$
$$u(1,\theta) = \cos\theta$$

to check its accuracy.

4. Solve problem P_1 in the boundary-perturbation problem and check how well $u(r,\theta) = u_0(r,\theta) + \frac{1}{4} u_1(r,\theta)$ satisfies

$$\nabla^2 u = 0$$
$$u(1 + \frac{1}{4}\sin\theta,\theta) = \cos\theta$$

OTHER READING

Partial Differential Equations: Theory and Technique by G. F. Carrier and C. E. Pearson. Academic Press, 1976. Chapters 8 and 15 will give the reader a good idea of many of the facets of perturbation theory; an excellent reference text.

Conformal-Mapping Solutions of PDEs

PURPOSE OF LESSON: To show how certain boundary-value problems in two dimensions can be transformed into new ones by means of a *conformal change of variables*. For example, we can solve Laplace's equation

$$\phi_{xx} + \phi_{yy} = 0$$

inside a complicated domain of the xy-plane (with some type of BC) by transforming the problem into a new Laplace equation

$$\phi_{uu} + \phi_{vv} = 0$$

inside a *simple* domain of a new uv-plane.

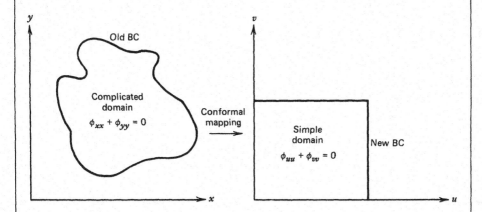

If the mapping is *conformal*, then Laplace's equation $\phi_{xx} + \phi_{yy} = 0$ in the original coordinates (x, y) will be transformed into Laplace's equation $\phi_{uu} + \phi_{vv} = 0$ in the new (u, v) coordinates (instead of some other equation, like $\phi_{uu} + 2\phi_{uv} + \phi_{vv} = 0$). We can then solve this Laplace equation for $\phi(u, v)$ in the simple domain (like a circle, half plane, square) and substitute our transformation $u = u(x,y)$, $v = v(x,y)$ in this solution to obtain our solution $\phi[u(x,y), v(x,y)]$ in terms of the original x,y coordinates.

This lesson will show the reader:

1. The nature of conformal mappings
2. How to construct them (to a certain extent)
3. Examples of problems solved using this technique

One of the major difficulties in solving boundary-value problems is due to *complicated boundaries*; even relatively simple boundaries can often cause a problem to be very difficult. One way, of course, to attack irregular boundaries is by means of perturbation methods, but this generally works only if the boundary is *close* to a simple boundary.

There is a way, however, to solve Laplace's equation in two dimensions when the boundary is fairly general, and that is by applying *conformal mappings*.

However, before we get into the actual application of this technique, we must spend a little time talking about the concept of a conformal mapping and about *complex numbers* in general.

Conformal Mappings and Complex Functions

This lesson presents only a few ideas from complex variables that will be needed later. We can imagine a complex number $z = x + iy$ as a *point* in a complex xy-plane (Figure 47.1).

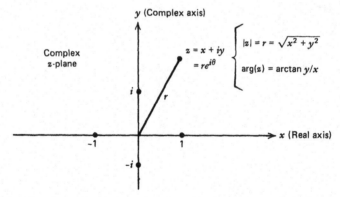

FIGURE 47.1 Complex z-plane and useful formulas.

In order to introduce the idea of a conformal mapping, we must introduce the idea of a *function of a complex variable*

$$w = f(z)$$

Here, z is a complex variable (that has some domain in the complex z-plane), and $w = u + iv$ is a new complex variable [that will run around according to the formula $f(z)$ in a new complex w-plane]. For example, the complex function $w = z^2$ defined in the *first quadrant* of the complex z-plane will map this domain onto the entire upper half $v > 0$ of the complex w-plane. In particular, $w = z^2$ will send the following points of the z-plane onto the corresponding points of the w-plane:

TABLE 47.2

z-plane	$z \rightarrow z^2$	w-plane
0	$\longrightarrow$	0
i	$\longrightarrow$	-1
$1 + i$	$\longrightarrow$	$2i$
x-axis	$\longrightarrow$	Positive real axis
First quadrant	$\longrightarrow$	Upper-half plane

See Figure 47.2.

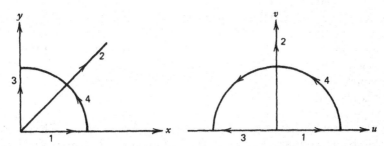

FIGURE 47.2 Mapping the first quadrant of the z-plane onto the upper w-plane.

In other words, when we talk about functions of a *complex variable* $w = f(z)$ (in contrast to functions of a real variable, which the reader is already familiar with), we discuss how *curves* in the z-plane map onto *curves* in the w-plane. Before going on, the reader should know how *complex mappings* like $w = z^2$ can be written in an equivalent *real form*. The real form tells us how the original coordinates (x,y) in the z-plane map onto the new coordinates (u,v) in the w-plane. To find the real form of the mapping, we merely write $w = z^2$ as

$$u + iv = (x + iy)^2 = x^2 - y^2 + 2ixy$$

and, hence, setting the real parts equal to the real parts and the imaginary parts equal to the imaginary parts, we have

$$\begin{aligned} u &= x^2 - y^2 \\ v &= 2xy \end{aligned} \quad \text{(Real form of } w = z^2)$$

This form will be useful later. In this particular case, this form is interesting because it says that hyperbolas in the z-plane map onto the *coordinate lines u = constant, v =* constant, in the complex w-plane.

We now get to the special kind of functions, functions of a complex variable known as conformal functions (or conformal mappings).

Definition of a Conformal Mapping

We say that a mapping $w = f(z)$ between the complex z- and w-planes is *conformal* at a point z_0 in the z-plane if the *derivative* $f'(z_0) \neq 0$. What's more, we say that $f(z)$ is conformal everywhere in a region D of the z-plane if $f'(z) \neq 0$ everywhere in D.

For example, $f(z) = z^2$ is conformal everywhere except at $z = 0$, since $f'(z) = 2z \neq 0$ for all $z \neq 0$. On the other hand, e^z is conformal everywhere in the z-plane, since $f'(z) = e^z \neq 0$ everywhere. The question, of course, is *why* conformal mappings are useful? The answer is that in solving Laplace's equation $\phi_{xx} + \phi_{yy} = 0$ inside some domain of the xy-plane, we can think of the xy-plane as the complex z-plane and introduce a complex mapping $w = f(z)$ from the z-plane to a new w-plane. The w-plane has coordinates (u,v), and so the original Laplace equation $\phi_{xx} + \phi_{yy} = 0$ will be transformed into a new PDE in coordinates u and v. The point is that if the mapping $w = f(z)$ is *conformal* everywhere in the domain of $\phi_{xx} + \phi_{yy} = 0$, then the new transformed PDE in the new coordinates u,v will still be Laplace's equation (in the new coordinates u,v). That is, $\phi_{xx} + \phi_{yy} = 0$ is transformed into the equation $\phi_{uu} + \phi_{vv} = 0$. So, the idea is to find a conformal mapping that changes the original complicated boundary to a simple one (remember $w = z^2$ changed the boundary of the first quadrant of the z-plane to the real axis of the w-plane).

We now present a few examples to show how conformal mappings can be used to solve PDEs.

Laplace's Equation in the Upper-Half Plane

Suppose we wish to solve the following Dirichlet problem in the upper-half plane (Figure 47.3):

(47.1)

$$\text{PDE} \quad \phi_{xx} + \phi_{yy} = 0 \quad -\infty < x < \infty \quad 0 < y < \infty$$

$$\text{BC} \quad \phi(x,0) = \begin{cases} 0 & |x| > 1 \\ 1 & |x| \leq 1 \end{cases}$$

One way to solve this problem would be to apply the Fourier transform to the x-variable, but an easier method is to conformally map the upper-half plane y

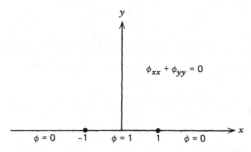

FIGURE 47.3 Boundary-value problem in the upper-half plane.

> 0 to a new region of the uv-plane. With a little effort, the reader can see that the conformal mapping

$$w = \log\left\{\frac{z - 1}{z + 1}\right\}$$

has the following properties (see Figure 47.4):

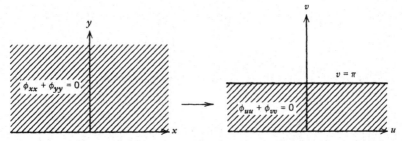

FIGURE 47.4 Conformal mapping of a hard problem into an easy one.

1. It maps the *upper half* of the z-plane conformally onto the region $-\infty < u < \infty$, $0 < v < \pi$ of the w-plane.
2. The line segment $y = 0$, $-1 < x < 1$ in the z-plane (where the potential ϕ is one) maps onto the line $v = \pi$, $-\infty < u < \infty$ in the w-plane.
3. The two segments $x > 1$, $y = 0$ and $x < -1$, $y = 0$ in the z-plane map onto the positive and negative u-axis of the w-plane, respectively.

The reader should make sure that he or she can verify these statements. The importance of these properties lies in the fact that the original problem (47.1) is now transformed into an easy one

$$\text{PDE} \qquad \phi_{uu} + \phi_{vv} = 0 \qquad -\infty < u < \infty \qquad 0 < v < \pi$$

(47.2)

$$\text{BC} \qquad \begin{cases} \phi(u,0) = 0 \\ \phi(u,\pi) = 1 \end{cases}$$

which has the obvious solution $\phi(u,v) = \dfrac{1}{\pi} v$. Hence, to find the solution of our original problem (47.1), we merely find the coordinate v in terms of x and y and substitute into $\phi(u,v) = v/\pi$. Doing this, we have

$$w = u + iv = \log\left[\frac{z - 1}{z + 1}\right]$$

$$= \log\left|\frac{z - 1}{z + 1}\right| + i\,\mathrm{arg}\left[\frac{z - 1}{z + 1}\right]$$

and, hence,

$$v = \mathrm{arg}\left[\frac{z - 1}{z + 1}\right] = \mathrm{arg}\left[\frac{x + iy - 1}{x + 1 + iy}\right]$$

$$= \mathrm{arg}\left[\frac{x^2 + y^2 - 1 + i2y}{(x + 1)^2 + y^2}\right]$$

$$= \arctan\left[\frac{2y}{x^2 + y^2 - 1}\right]$$

[from the general rule $\mathrm{arg}\,(x + iy) = \arctan(y/x)$ and we take that part of the arctan between 0 and π]

$$\phi(x,y) = \frac{1}{\pi}\tan^{-1}\left[\frac{2y}{x^2 + y^2 - 1}\right]$$

The reader should check the graph of this function along various lines $y = c$ to get an idea of what it looks like.

For our second example, we transform a region between two nonconcentric circles into an annulus.

Dirichlet Problem Between Two Nonconcentric Circles

Suppose we wish to find the potential $\phi(x,y)$ between the two circles

$$x^2 + y^2 = 1$$
$$(x - 1)^2 + y^2 = 9$$

where the potential on the inside circle is one and two on the outside circle (Figure 47.5).

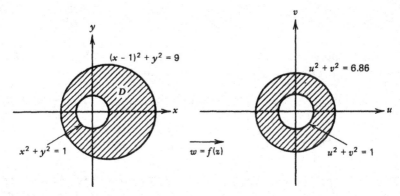

FIGURE 47.5 Conformal map onto an annulus.

In other words, our original problem is

(47.3)   PDE $\phi_{xx} + \phi_{yy} = 0$ Inside D

BC $\begin{cases} \phi(x,y) = 1 & \text{On } x^2 + y^2 = 1 \\ \phi(x,y) = 2 & \text{On } (x - 1)^2 + y^2 = 9 \end{cases}$

The problem now is to find a conformal mapping that will transform this domain into a nice easy one where our problem will be simple. In this case, it seems obvious to seek a transformation that maps our domain into an annulus. This is not a well-known transformation, but there are sources that will provide us with just the proper mapping. One of the more elaborate sources, *The Dictionary of Conformal Mappings* by H. Kober (Dover Publications), will provide the reader with hundreds of mappings between the common domains. Assuming that the reader is willing to spend a little time getting familar with the tables, he or she can find the desirable conformal transformation for the problem (47.3). In our specific case, it turns out to be the mapping

(47.4) $$w = 2t \left[\frac{z - s}{z - t} \right]$$

where $s = -0.146$ and $t = -6.85$. In the equivalent real form, it is

(47.5) $$u = \gamma \left[(x - s)(x - t) - y^2 \right]$$
$$v = \gamma \left[y(x - t) \right]$$

where

$$\gamma = 2t/[(x - t)^2 + y^2]$$

Under this conformal transformation, the original nonconcentric circles map onto the new circles $u^2 + v^2 = 1$ and $u^2 + v^2 = 6.86$ (the reader could check this).

If we now apply this conformal maping to our original problem (47.3), we arrive at the new Dirichlet problem inside an annulus

(47.6)

$$\text{PDE} \qquad \phi_{uu} + \phi_{vv} = 0 \qquad \text{In the new annulus}$$

$$\text{BCs} \qquad \begin{cases} \phi(u,v) = 1 & \text{On } u^2 + v^2 = 1 \\ \phi(u,v) = 2 & \text{On } u^2 + v^2 = 6.68 \end{cases}$$

The answer to this problem is not difficult, since we can see that the solution will be radially symmetric and we know that all radially symmetric solutions to Laplace's equation are of the form

$$\phi(r) = a \ln r + b \qquad r = u^2 + v^2$$

Substituting this equation in the BCs of (47.6), we get

$$\phi(u,v) = 0.57 \ln (u^2 + v^2) + 1$$

and so if we substitute back to our original coordinates x and y by means of the transformation (47.5), the answer to problem (47.3) is $\phi(x,y)$

$$\phi(x,y) = 0.57 \ln (u^2 + v^2) + 1$$

where u and v are given by equations (47.5).

NOTES

1. The *Schwarz-Christoffel transformation* is a procedure for constructing conformal mappings from fairly general regions of the z-plane to the upper half of the w-plane (of course, once the problem is transformed into the upper-half plane, we could transform again into yet another region). This method is described in reference 1 of the recommended reading.
2. The conformal-mapping method is somewhat restrictive because it applies only to *Laplace's equation in two dimensions* (although, with a little modification, it can be applied to Poisson's equation).

PROBLEMS

1. Where is the mapping

$$w = \log \left[\frac{z - 1}{z + 1} \right]$$

conformal? Convince yourself that it maps the upper-half z-plane onto the strip $-\infty < u < \infty$, $0 < v < \pi$ of the w-plane.

2. What is the image of the first quadrant under the mapping $w = z^3$? It might be helpful to write the complex numbers in polar form

$$z = re^{i\theta}$$

3. Solve the Dirichlet problem in the first quadrant

$$\text{PDE} \qquad \phi_{xx} + \phi_{yy} = 0 \qquad 0 < x < \infty \qquad 0 < y < \infty$$

BCs
$$\begin{cases} \phi(x,0) = \begin{cases} 1 & 0 < x < 1 \\ 0 & 1 \leqslant x < \infty \end{cases} \\[2ex] \phi(0, y) = \begin{cases} 1 & 0 < y < 1 \\ 0 & 1 \leqslant y < \infty \end{cases} \end{cases}$$

by means of the conformal mapping $w = z^2$. See the following figure.

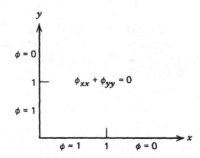

4. Solve the following mixed *Dirichlet-Neumann problem* inside a 45° wedge:

$$\text{PDE} \qquad \phi_{rr} + \frac{1}{r}\phi_r + \frac{1}{r^2}\phi_{\theta\theta} = 0 \qquad 0 < r < 1 \qquad 0 < \theta < \pi/4$$

BCs
$$\begin{cases} \phi(r,0) = 0 \\ \phi(r,\pi/4) = 1 \\ \phi_r(1,\theta) = 0 \end{cases}$$

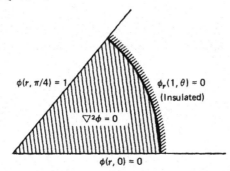

HINT   The complex function $w = \log z = \log |z| + i \arg(z)$ maps the ray $\theta = c_1$ of the z-plane onto the line $v = c_1$ of the w-plane and the curve $r = c_2$ onto the line $u = \log c_2$ (log, of course, always means natural logarithm). See the figure that follows.

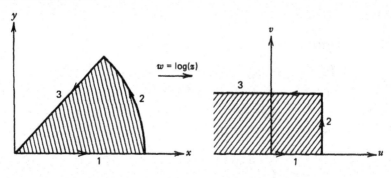

OTHER READING

1. *Complex Variables and Applications,* 3d ed. R. V. Churchill, J. W. Brown, and R. F. Verhey. McGraw-Hill, 1977. This book is one of the best and contains a good chapter on the subject of conformal mappings, which can be thought of as a part of complex variables, along with applications of PDE.

2. *Dictionary of Conformal Mappings* by H. Kober. Dover, 1960. This paperback will provide the reader with a list of conformal mappings.

ANSWERS TO SELECTED PROBLEMS

LESSON 1

1. (a) Second order, linear, nonhomogeneous, parabolic, constant coefficient coefficients, two independent variables.
   (b) Second order, linear, nonhomogeneous, parabolic, constant coefficient coefficients, two independent variables.
   (c) Second order, linear, nonhomogeneous, hyperbolic, constant coefficient coefficients, two independent variables.
   (d) Fourth order, nonlinear, two independent variables.
3. Yes, if $G = 0$.
4. $u_x = 0$ implies $u(x, y) = g(y)$.
5. $u_{xy} = 0$ implies $u(x, y) = f(x) + g(y)$.

LESSON 2

1. $u(x, t)$ approaches zero for all x as $t \to \infty$. Try $u(x, t) = e^{\alpha^2 t} \sin(\pi x)$.

2. $U(x) = -\dfrac{1}{2\alpha^2}(x^2 - x) + x$

3. $U(x) = \dfrac{1}{\sinh\sqrt{\beta/\alpha^2}} [\sinh\sqrt{\beta/\alpha^2}\, x + \sinh\sqrt{\beta/\alpha 2}\,(1 - x)]$

LESSON 3

2. Heat flows into the rod at the right end at a constant rate while the left-hand side is kept at zero degrees (zero can stand for any temperature). Keeping the left hand of the rod at zero actually means it will be a heat *sink*.
3. Both ends of the rod are insulated so the total heat energy inside the rod will eventually become distributed uniformly.

LESSON 5

3. $1 = \dfrac{4}{\pi}\left[\sin(\pi x) + \dfrac{1}{3}\sin(3\pi x) + \dfrac{1}{5}\sin(5\pi x) + \cdots\right]$

4. $u(x, t) = \dfrac{4}{\pi}\left\{e^{-\pi^2 t}\sin(\pi x) + \dfrac{1}{3}e^{-(3\pi)^2 t}\sin(3\pi x) + \cdots\right\}$

5. $u(x,t) = e^{-(2\pi)^2 t} \sin(2\pi x) + \dfrac{1}{3} e^{-(4\pi)^2 t} \sin(4\pi x) + \dfrac{1}{5} e^{-(6\pi)^2 t} \sin(6\pi x)$

6. $u(x,t) = \dfrac{8}{\pi^3} \left\{ e^{-\pi^2 t} \sin(\pi x) + \dfrac{1}{27} e^{-(3\pi)^2 t} \sin(3\pi x) + \cdots \right\}$

LESSON 6

1. $u(x,t) = x + e^{-(\pi\alpha)^2 t} \sin(\pi x)$

2. $u(x,t) = x - \dfrac{8}{\pi^3} \left\{ e^{-\pi^2 t} \sin(\pi x) + \dfrac{1}{27} e^{-(3\pi)^2 t} \sin(3\pi x) + \cdots \right\}$

LESSON 7

1. $u(x,t) = a_1 e^{-(\pi/2)^2 t} \sin(\pi x/2) + a_3 e^{-(3\pi/2)^2 t} \sin(3\pi x/2) + \cdots$

   where $a_{2n+1} = 2 \displaystyle\int_0^1 x \sin[(2n+1)\pi x/2]\, dx$

2. $\lambda_n = (n\pi/2)^2, \quad X_n(x) = \sin(n\pi x/2) \quad (n = 1,3,5,\cdots)$

3. $u(x,t) = \dfrac{1}{2} - \dfrac{4}{\pi^2} \displaystyle\sum_{n=1}^{\infty} \dfrac{1}{n^2} e^{-(n\pi)^2 t} \cos(n\pi x)$

LESSON 8

1. $u(x,t) = \dfrac{4}{\pi} e^{-\frac{1}{2}(x-t/2)} \displaystyle\sum_{n=1}^{\infty} \dfrac{1}{(2n-1)} \sin[(2n-1)\pi x]$

2. $u(x,t) = x - \dfrac{2}{\pi} e^{-t} \left\{ e^{-\pi^2 t} \sin(\pi x) - \dfrac{1}{2} e^{-(2\pi)^2 t} \sin(2\pi x) + \cdots \right\}$

3. $u(x,t) = e^{-(\pi^2+1)t} \sin(\pi x)$

LESSON 9

2. $u(x,t) = \dfrac{1}{\pi^2}[1 - e^{-\pi^2 t}] \sin(\pi x) + \dfrac{1}{(2\pi)^2}[1 - e^{-(2\pi)^2 t}] \sin(2\pi x).$

3. $u(x,t) = \left\{ \dfrac{4}{\pi} e^{-\pi^2 t} + \dfrac{1}{\pi^2}[1 - e^{-\pi^2 t}] \right\} \sin(\pi x)$

   $+ \dfrac{4}{\pi} \displaystyle\sum_{n=0}^{\infty} \dfrac{1}{2n+1} e^{-[(2n+1)\pi]^2 t} \sin[(2n+1)\pi x].$

4. $u(x,t) = \dfrac{1}{\lambda_1}[1 - e^{-\lambda_1 t}] \sin(\lambda_1 x)$ where λ_1 is the smallest root of $\tan\lambda = -\lambda$.

5. $u(x,t) = x \cos t + \displaystyle\sum_{n=1}^{\infty} T_n(t) \sin n\pi x$

    where $T_n(t) = (-1)^n \dfrac{2}{n\pi} \displaystyle\int_0^1 e^{-(n\pi)^2(t-\tau)} \sin \tau \, d\tau$

LESSON 10

3. $u(x,t) = \dfrac{2}{\pi} \displaystyle\int_0^{\infty} \dfrac{1}{\omega}(1 - \cos\omega)e^{-\omega^2 t} \sin(\omega x) \, d\omega$

LESSON 11

1. $f(x) = \dfrac{4}{\pi} \left\{ \sin(\pi x) + \dfrac{1}{3}\sin(3\pi x) + \dfrac{1}{5}\sin(5\pi x) + \cdots \right\}$

4. $F(\xi) = \sqrt{\dfrac{2}{\pi}} \dfrac{\sin\xi}{\xi}, \ |F(\xi)| = \sqrt{\dfrac{2}{\pi}} \left|\dfrac{\sin\xi}{\xi}\right|$

LESSON 12

3. $u(x,t) = \dfrac{1}{\sqrt{4\alpha^2+1}} e^{-x^2/(4\alpha^2 t+1)}$

LESSON 13

2. $u(x,t) = e^{-\alpha^2 t} \sin x$

3. $u(x,t) = \sin t * \dfrac{\alpha x}{2\sqrt{\pi t^3}} e^{-\alpha^2 x/4t}$

4. $U(x,s) = \dfrac{A}{s}\left(\dfrac{\cosh\sqrt{s}\,x}{\cosh\sqrt{s}} - 1\right)$

LESSON 14

4. $w(x,t) = x + \dfrac{2}{\pi} \displaystyle\sum_{n=1}^{\infty} \dfrac{(-1)^n}{n} e^{-(n\pi)^2 t} \sin(n\pi x)$

LESSON 15

1. $u(x,t) = e^{-t} \sin(x - 2t)$
2. $u(x,t) = e^{x-2t} \sin(x)$
3. $u(x,t) = e^{-(x-2t)^2}$

4. $u(x,t) = \dfrac{1}{\sqrt{1+4Dt}} e^{-(x-Vt)^2/(1+4Dt)}$

LESSON 16

2.   The wave initially looks like $\sin \pi x$ but vibrates to zero because of the friction term u_t.

4. $u(x,t) = c_1 e^{a(x-t)} + c_2 e^{a(x+t)}$

LESSON 17

3. $u(x,t) = \dfrac{1}{2}[e^{-(x-t)^2} + e^{-(x+t)^2}]$

4. $u(x,t) = \dfrac{1}{4}\{e^{-(x-t)^2} - e^{-(x+t)^2}\}$

LESSON 20

1. $u(x,t) = \sin(\pi x/L)\cos(\pi \alpha t/L) + \dfrac{1}{2}\sin(3\pi x/L)\cos(3\pi \alpha t/L)$

2. $u(x,t) = \left\{\dfrac{L}{3\pi\alpha}\right\}\sin(3\pi x/L)\sin(3\pi\alpha t/L)$

4. $u(x,t) = \sin(3\pi x/L)[\sin(3\pi\alpha t/L) + \cos(3\pi\alpha t/L)]$

5. $u(x,t) = \dfrac{4h}{\pi^2}\sin\pi x \cos\pi t - \dfrac{4h}{9\pi^2}\sin 3\pi x \cos 3\pi t + \cdots$

LESSON 21

2. $u(x,t) = \sin(\pi x)\left[\cos(\pi^2 t) + \dfrac{1}{\pi^2}\sin(\pi^2 t)\right]$

3. $u(x,t) = \displaystyle\sum_{n=1}^{\infty} b_n \cos(n^2\pi^2 t)\sin(n\pi x)$

   $b_n = 2\displaystyle\int_0^1 (1-x^2)\sin(n\pi x)\,dx$

LESSON 22

4.   The velocity is 1 with respect to the time scale $\tau = \alpha t$.

5. $\xi = x/v$

LESSON 23

1.   (a), (c), (d), and (e) are elliptic, (b) is hyperbolic.

3. $\xi = y - \dfrac{1}{2}x, \quad \eta = y - 3x$

4. $u_{\xi\eta} = 0$

5. $u_{\alpha\alpha} - u_{\beta\beta} = 0$

LESSON 24

2. $u(x,t) = xt$

5. $u(x,t) = \dfrac{1}{2c} \displaystyle\int_{x-ct}^{x+ct} u_t(s,0)\, ds$

LESSON 25

1. $u(x,t) = 1 + e^{-\pi^2 t} \cos(\pi x) + \dfrac{1}{2} e^{-(3\pi)^2 t} \cos(3\pi x)$

2. $u(x,t) = \displaystyle\sum_{n=1}^{\infty} B_n \sin(n\pi x)$

$B_n(t) = e^{-(n\pi\alpha)^2 t} \displaystyle\int_0^t e^{(n\pi\alpha)^2 s} F_n(s)\, ds, \quad F_n(s)$ is the sine transform of $f(x,t)$

4. All B_n's are zero except $B_1 = 1$ and $B_3 = 1/2$.

6. $u(x,t) = \displaystyle\sum_{n=1}^{\infty} B_n(t) \sin(n\pi x)$ where $B_n(t)$ is found from

$B_n'(t) + (n\pi)^2 B_n(t) = \begin{cases} 8\pi + 1 & n = 3 \\ (-1)^{n+1}(2\pi n) & n \neq 3 \end{cases}$

$B_1(0) = 1, \quad B_n(0) = 0 \quad n = 2, 3, \cdots$

LESSON 26

2. $u(x,t) = \dfrac{1}{(3\pi)^2}[1 - \cos(3\pi t)]\sin(3\pi x) + \sin(\pi x)\cos(\pi t)$

3. $u_1 + u_2$ is a solution for (a) and (c), not a solution for (b) and (d).

4. In each of four subproblems let one of the four equations be nonhomogeneous, the others homogeneous.

5. $U_n(t) = e^{-(n\pi)^2 t} \displaystyle\int_0^t e^{(n\pi)^2 s} F_n(s)\, ds$

6. Yes.

LESSON 27

1. $u(x,t) = \cos(x - t)$

2. $u(x,t) = \dfrac{1}{t^2} \sin(x/t)$

3. $u(x,y,t) = e^{-[(x-at/c)^2 + (y-bt/c)^2] - dt/c}$
4. $u(x,t) = F(x-t)e^{-t^2/2}$
5. $u(x,t) = F(2x - y, 2x - y)e^{-2(y-x)}$

LESSON 28

1. $x_0 < 0$ has characteristics $x = x_0$, $x_0 > 0$ has characteristics $t = \dfrac{1}{x_0}(x - x_0)$

2. $f(u) = \dfrac{1}{3}u^3$

3. $u(x,t) = \phi(x-t)e^{-kt}$
4. $u(x,t) = -\ln(t+1) + \phi(x-t)$

LESSON 29

1. $A = \begin{bmatrix} 0 & 0 & 0 \\ 1 & 0 & 0 \\ 0 & 0 & 1 \end{bmatrix}$ $B = \begin{bmatrix} 1 & 0 & 0 \\ 0 & 0 & 0 \\ 0 & -c^2 & 0 \end{bmatrix}$ $C = \begin{bmatrix} 0 & -1 & 0 \\ 0 & 0 & -1 \\ -b & 0 & -a \end{bmatrix}$

2. $\lambda_1 = 3$, $X_1 = \begin{bmatrix} 1 \\ 2 \end{bmatrix}$, $\lambda_2 = -1$, $X_2 = \begin{bmatrix} 1 \\ -2 \end{bmatrix}$

3. $u_1(x,t) = \phi(x-3t) + \psi(x+t)$
   $u_2(x,t) = 2\phi(x-3t) - 2\psi(x+t)$

LESSON 30

3. $u(r,t) = \displaystyle\sum_{m=1}^{\infty} A_m J_0(k_{0m}r) \cos(k_{0m}t)$

   $A_m = \dfrac{2}{J_1^2(k_{0m})} \displaystyle\int_0^1 r(1-r^2)J_0(k_{0m}r)\, dr$

4. $u(r,t) = J_0(2.4r)\cos(2.4t)$

5. $u(r,t) = J_0(2.4r)\cos(2.4t) - \dfrac{1}{2}J_0(8.65r)\cos(8.65t)$

   $+ \dfrac{1}{4}J_0(14.93r)\cos(14.93t)$

LESSON 31

1. $u_{tt} = \alpha^2 \left[u_{rr} + \dfrac{2}{r}u_r \right]$

2. $u_{tt} = \alpha^2 \left[u_{rr} + \dfrac{1}{r} u_r \right]$

3. $\dfrac{d^2 u}{dr^2} + \dfrac{1}{r} \dfrac{du}{dr} = 0, \quad u(r) = A + B \ln r$

4. $u(r) = A + \dfrac{B}{r}$

5. Since $u_{xx} + u_{yy} = 0$ at each point (x, y) the height of the surface $u(x, y) = xy$ is approximately equal to the average of the values on a small circle around (x, y).

LESSON 32

1. $u(r, \theta) = r \sin \theta$ (good guess)
2. No, the net flux is not zero.
4. $d^2 u / dx^2 + u = 0, \ u(0) = u(1) = 0$

LESSON 33

2. (a) $u(r, \theta) = 1 + r \sin \theta + \dfrac{r}{2} \cos \theta$

   (b) $u(r, \theta) = 2$
   (c) $u(r, \theta) = r \sin \theta$
   (d) $u(r, \theta) = r^3 \sin (3\theta)$

3. $u(r, \theta) = \dfrac{r}{2} \sin \theta$

4. $u(r, \theta) = \dfrac{r^2}{4} \sin (2\theta)$

5. $u(r, \theta) = \dfrac{r}{2} \sin \theta + \dfrac{2}{\pi} \left\{ \dfrac{1}{2} - \dfrac{r^2}{3} \cos 2\theta - \dfrac{r^4}{15} \cos 4\theta - \dfrac{r^6}{35} \cos 6\theta - \cdots \right\}$

6. $\dfrac{7R}{16P^2}$

LESSON 34

1. $u(r, \theta) = \left\{ -\dfrac{r}{3} + \dfrac{4}{3r} \right\} \cos \theta + \left\{ \dfrac{2r}{3} - \dfrac{2}{3r} \right\} \sin \theta$

2. (a) $u(r, \theta) = 1$

   (b) $u(r, \theta) = 1 + \dfrac{1}{r^3} \cos (3\theta)$

   (c) $u(r, \theta) = \dfrac{1}{r} \sin \theta + \dfrac{1}{r^3} \cos (3\theta)$

(d) $u(r, \theta) \doteq \dfrac{1}{2} + \dfrac{2}{\pi} \left\{ \dfrac{1}{r} \sin (\theta) + \dfrac{1}{3r^3} \sin (3\theta) + \dfrac{1}{5r^5} \sin (5\theta) + \cdots \right\}$

3. $u(r, \theta) = -\dfrac{1}{r} \sin \theta$

LESSON 35

4. $u(r, \phi) = \dfrac{8}{5} r^3 P_3(\cos \phi) - \dfrac{3r}{5} P_1(\cos \phi)$

   $= \dfrac{4}{5} r^3 [5 \cos^3 \phi - 3 \cos \phi] - \dfrac{3}{5} r \cos \phi$

5. $u(r, \phi) = \displaystyle\sum_{n=0}^{\infty} a_n r^n P_n(\cos \phi)$

   $a_n = \dfrac{2n+1}{2} \displaystyle\int_0^{\pi/2} P_n(\cos \phi) \sin \phi \, d\phi - \dfrac{2n+1}{2} \displaystyle\int_{\pi/2}^{\pi} P_n(\cos \phi) \sin \phi \, d\phi$

6. $u(r, \theta) = \dfrac{1}{r} + \dfrac{1}{r^2} \cos \phi$

LESSON 36

1. $u(r) = \dfrac{q}{4\pi r}$

2. $G(x, y, \xi, \eta) = \dfrac{1}{2\pi} \ln \left[\dfrac{\overline{R}}{R} \right]$

3. $u(x, y) = \dfrac{-k}{2\pi} \displaystyle\int_0^{\infty} \int_{-\infty}^{\infty} \ln \left[\dfrac{\overline{R}}{R} \right] d\xi \, d\eta$

4. $G(x, y, \xi, \eta) = \dfrac{1}{2\pi} \ln \left[\dfrac{R_2 R_4}{R_1 R_3} \right]$

5. $u(r, \theta) = \dfrac{1}{4} (r^2 - 1) + r \sin \theta$

LESSON 37

3. $u_{i,j} = \dfrac{1}{2} [u_{i+1,j} + u_{i-1,j} + u_{i,j+1} + u_{i,j-1}] - \dfrac{h^2}{4} f_{i,j}$

4. $u_{i,j} = \dfrac{-1}{2(h-2)} [u_{i,j+1} + u_{i,j-1} + u_{i+1,j} + u_{i-1,j}]$

LESSON 38

2. $u(x, t) = e^{-\pi^2 t} \sin (\pi x)$

LESSON 39

2. $u_{i,1} = 0$, $\dfrac{u_{i,n} - u_{i,n-1}}{h} + u_{i,n} = g_i$

LESSON 40

2. Solve for a and b from the equations $\dfrac{\partial(SS)}{\partial a} = 0$ and $\dfrac{\partial(SS)}{\partial b} = 0$.

LESSON 41

1. (a) Parabolic, canonical, (b) hyperbolic, not canonical, (c) elliptic, not canonical, (d) parabolic, canonical.

2. $16u_{\eta\eta} + u = 2$

3. $u_{\xi\xi} + u_{\eta\eta} + \left[\dfrac{1 + 2\eta}{2\eta}\right] u_\eta = e^{-\eta}$

LESSON 44

1. $\bar{y}(x) = x$
2. $m\ddot{y} + ky = 0$

LESSON 45

1. $J[u] = \displaystyle\int_0^1 \int_0^1 [u_x^2 + u_y^2 + 2u]\, dx\, dy$

2. $J[z] = \displaystyle\int_0^1 \left\{ \left[\dfrac{z}{1-x}\right]^2 + \dfrac{z'}{1-x} + \dfrac{z}{(1-x)^2} \right\} dx$

LESSON 47

1. Conformal except at $z = \pm 1$.
2. The first three quadrants.

3. $\phi(x, y) = \dfrac{1}{\pi} \tan^{-1} \left\{ \dfrac{4xy}{(x^2 - y^2)^2 + 4x^2y^2 - 1} \right\}$

4. $\phi(r, \theta) = \dfrac{4\theta}{\pi}$

APPENDIX 1

Integral Transform Tables

Table A: Exponential Fourier transform
Table B: Fourier sine transform
Table C: Fourier cosine transform
Table D: Finite Fourier sine transform
Table E: Finite Fourier cosine transform
Table F: Laplace transform

Definitions of Functions in Tables

$\delta(x)$ = delta function

$$H(x - a) = \begin{cases} 0 & x < a \\ 1 & x \geq a \end{cases} \quad \text{Heaviside function}$$

$$H(a - x) = \begin{cases} 1 & x \leq a \\ 0 & x > a \end{cases} \quad \text{Reflected Heaviside function}$$

TABLE A   Exponential Fourier Transform

$$f(x) = \mathscr{F}^{-1}[F] = \frac{1}{\sqrt{2\pi}} \int_{-\infty}^{\infty} F(\omega)e^{i\omega x}\, d\omega \qquad F(\omega) = \mathscr{F}[f] = \frac{1}{\sqrt{2\pi}} \int_{-\infty}^{\infty} f(x)e^{-i\omega x}\, dx$$

1. $f'(x)$	$i\omega F(\omega)$
2. $f''(x)$	$-\omega^2 F(\omega)$
3. $f''(x)$ (nth derivative)	$(i\omega)^n\, F(\omega)$
4. $f(ax)$ $a > 0$	$\dfrac{1}{a}F\left(\dfrac{\omega}{a}\right)$

TABLE A (cont.)

$$f(x) = \mathscr{F}^{-1}[F] = \frac{1}{\sqrt{2\pi}} \int_{-\infty}^{\infty} F(\omega)e^{i\omega x}\, d\omega \qquad\qquad F(\omega) = \mathscr{F}[f] = \frac{1}{\sqrt{2\pi}} \int_{-\infty}^{\infty} f(x)e^{-i\omega x}\, dx$$

5. $f(x - a)$ $\qquad\qquad\qquad\qquad\qquad e^{-ia\omega} F(\omega)$

6. $e^{-a^2 x^2}$ $\qquad\qquad\qquad\qquad\qquad \dfrac{1}{a\sqrt{2}} e^{-\omega^2/4a^2}$

7. $e^{-a|x|}$ $\qquad\qquad\qquad\qquad\qquad \sqrt{\dfrac{2}{\pi}}\, \dfrac{a}{a^2 + \omega^2}$

8. $\begin{cases} 1 & |x| < a \\ 0 & |x| > a \end{cases}$ $\qquad\qquad\qquad \sqrt{\dfrac{2}{\pi}}\, \dfrac{\sin a\omega}{\omega}$

9. $\begin{cases} 1 & |x| < 1 \\ 0 & |x| > 1 \end{cases}$ $\qquad\qquad\qquad \sqrt{\dfrac{2}{\pi}}\, \dfrac{i}{\omega} \left(\cos \omega - \dfrac{1}{\omega}\sin \omega\right)$

10. $\delta(x - a)$ $\qquad\qquad\qquad\qquad\qquad \dfrac{1}{\sqrt{2\pi}} e^{-ia\omega}$

11. $f(x) * g(x)$ $\qquad\qquad\qquad\qquad\qquad \dfrac{1}{\sqrt{2\pi}} F(\omega)G(\omega)$

12. $(1 + x^2)^{-1}$ $\qquad\qquad\qquad\qquad\qquad \sqrt{\dfrac{\pi}{2}} e^{-|\omega|}$

13. $xe^{-a|x|} \qquad a > 0$ $\qquad\qquad\qquad -2\sqrt{\dfrac{2}{\pi}}\, \dfrac{ia\omega}{(\omega^2 + a^2)^2}$

14. $H(x + a) - H(x - a)$ $\qquad\qquad\qquad \sqrt{\dfrac{2}{\pi}}\, \dfrac{\sin (a\omega)}{\omega}$

15. $\dfrac{a}{x^2 + a^2}$ $\qquad\qquad\qquad\qquad\qquad \sqrt{\dfrac{\pi}{2}}\, e^{-a|\omega|}$

16. $\dfrac{2ax}{(x^2 + a^2)^2}$ $\qquad\qquad\qquad\qquad -i\sqrt{\dfrac{\pi}{2}}\, \omega e^{-a|\omega|}$

17. $\begin{cases} \cos (ax) & |x| < \pi/2a \\ 0 & |x| > \pi/2a \end{cases}$ $\qquad\qquad \sqrt{\dfrac{2}{\pi}}\, \dfrac{a}{a^2 - \omega^2}\cos (\pi\omega/2a)$

TABLE A (cont.)

$$f(x) = \mathcal{F}^{-1}[F] = \frac{1}{\sqrt{2\pi}} \int_{-\infty}^{\infty} F(\omega)e^{i\omega x} \, d\omega \qquad F(\omega) = \mathcal{F}[f] = \frac{1}{\sqrt{2\pi}} \int_{-\infty}^{\infty} f(x)e^{-i\omega x} \, dx$$

18.	$\begin{array}{ll} 1 - |x| & |x| < 1 \\ 0 & |x| > 1 \end{array}$	$2\sqrt{\dfrac{2}{\pi}} \left[\dfrac{\sin(\omega/2)}{\omega} \right]^2$
19.	$\cos(ax)$	$\sqrt{\dfrac{\pi}{2}} [\delta(\omega + a) + \delta(\omega - a)]$
20.	$\sin(ax)$	$i\sqrt{\dfrac{\pi}{2}} [\delta(\omega + a) - \delta(\omega - a)]$

TABLE B Fourier Sine Transform

$$f(x) = \int_0^\infty F(\omega) \sin(\omega x) \, d\omega \qquad\qquad F(\omega) = \frac{2}{\pi} \int_0^\infty f(x) \sin(\omega x) \, dx$$

$$0 < x < \infty \qquad\qquad\qquad\qquad\qquad 0 < \omega < \infty$$

1.	$f''(x)$	$-\omega^2 F(\omega) + \dfrac{2}{\pi}\omega f(0)$
2.	$f(ax)$	$\dfrac{1}{a}F\left(\dfrac{\omega}{a}\right)$
3.	e^{-ax}	$\dfrac{2\omega}{\pi(a^2 + \omega^2)}$
4.	$x^{-1/2}$	$\left[\dfrac{2}{\pi\omega}\right]^{1/2}$
5.	$H(a - x)$	$\dfrac{2}{\pi\omega}[1 - \cos(\omega a)]$
6.	x^{-1}	1
7.	$\dfrac{x}{x^2 + a^2}$	$e^{-a\omega}$

TABLE B Fourier Sine Transform

$$f(x) = \int_0^\infty F(\omega) \sin(\omega x)\, d\omega \qquad\qquad F(\omega) = \frac{2}{\pi} \int_0^\infty f(x) \sin(\omega x)\, dx$$

$$0 < x < \infty \qquad\qquad\qquad\qquad 0 < \omega < \infty$$

8. $\dfrac{x}{x^4 + 4}$ $\qquad\qquad\qquad\qquad\qquad \dfrac{1}{2} e^{-\omega} \sin \omega$

9. $\tan^{-1} \dfrac{a}{x}$ $\qquad\qquad\qquad\qquad\quad \dfrac{1 - e^{-a\omega}}{\omega}$

10. $-x^2 f(x)$ $\qquad\qquad\qquad\qquad\quad \dfrac{2}{\pi} F''(\omega)$

11. $erfc(x/2\sqrt{a})$ $\qquad\qquad\qquad \dfrac{2}{\pi}\left[\dfrac{1 - e^{-a\omega^2}}{w}\right] \qquad a > 0$

TABLE C Fourier Cosine Transform

$$f(x) = \int_0^\infty F(\omega) \cos(\omega x)\, d\omega \qquad\qquad F(\omega) = \frac{2}{\pi} \int_0^\infty f(x) \cos(\omega x)\, dx$$

$$0 < x < \infty \qquad\qquad\qquad\qquad 0 < \omega < \infty$$

1. $f''(x)$ $\qquad\qquad\qquad\qquad\qquad -\omega^2 F(\omega) - \dfrac{2}{\pi} f'(0)$

2. $f(ax)$ $\qquad\qquad\qquad\qquad\qquad \dfrac{1}{a} F(\omega/a)$

3. e^{-ax} $\qquad\qquad\qquad\qquad\qquad \dfrac{2a}{\pi(a^2 + \omega^2)}$

4. $\delta(x)$ $\qquad\qquad\qquad\qquad\qquad \dfrac{2}{\pi}$

5. $x^{-1/2}$ $\qquad\qquad\qquad\qquad\qquad \left[\dfrac{2}{\pi\omega}\right]^{1/2}$

6. $H(a - x)$ $\qquad\qquad\qquad\qquad \dfrac{2}{\pi\omega} \sin(a\omega)$

7. e^{-ax^2} $\qquad\qquad\qquad\qquad\qquad \dfrac{1}{\sqrt{\pi a}} e^{(-\omega^2/4a)}$

TABLE C   Fourier Cosine Transform

$$f(x) = \int_0^\infty F(\omega) \cos (\omega x)\, d\omega \qquad\qquad F(\omega) = \frac{2}{\pi} \int_0^\infty f(x) \cos (\omega x)\, dx$$

$$0 < x < \infty \qquad\qquad\qquad\qquad\qquad 0 < \omega < \infty$$

8.	$\dfrac{\sin (ax)}{x}$	$H(a - \omega)$
9.	$\dfrac{a}{x^2 + a^2}$	$e^{-a\omega}$
10.	$-x^2 f(x)$	$\dfrac{2}{\pi} F''(\omega)$

TABLE D   Finite Sine Transform

The finite sine transform transforms a function $f(x)$, $0 \le x \le \pi$ into a sequence S_n, $n = 1, 2, \ldots$ by means of the formula

$$S_n = \frac{2}{\pi} \int_0^\pi f(x) \sin (nx)\, dx$$

It is convenient, insofar as tables are concerned, for the range of the variable x to lie in the interval $[0,\pi]$. The reader can transform any other interval $[a,b]$ into $[0,\pi]$ via the transformation

$$y = \frac{\pi}{b - a}(x - a)$$

There is nothing mysterious about the finite sine transform; the members of the sequence S_n are the coefficients of $\sin (nx)$ in the Fourier sine series of $f(x)$. That is,

$$f(x) = \sum_{n=1}^\infty S_n \sin (nx)$$

$$f(x) = \sum_{n=1}^\infty S_n \sin (nx) \qquad\qquad S_n = \frac{2}{\pi} \int_0^\pi f(x) \sin (nx)\, dx$$

$$0 \le x \le \pi \qquad\qquad\qquad\qquad n = 1, 2, \ldots$$

1.	$f''(x)$	$-n^2 S_n + \dfrac{2n}{\pi}\, [f(0) - (-1)^n f(\pi)]$

TABLE D (cont.)

$$f(x) = \sum_{n=1}^{\infty} S_n \sin (nx) \qquad\qquad S_n = \frac{2}{\pi} \int_0^{\pi} f(x) \sin (nx)\, dx$$

$$0 \le x \le \pi \qquad\qquad\qquad n = 1, 2, \ldots$$

	$f(x)$	S_n
2.	$\sin (mx) \qquad m = 1, 2, 3, \ldots$	$\begin{cases} 1 & n = m \\ 0 & n \ne m \end{cases}$
3.	$\displaystyle\sum_{n=1}^{\infty} a_n \sin (nx)$	a_n
4.	$\pi - x$	$\dfrac{2}{n}$
5.	x	$\dfrac{2}{n}(-1)^{n+1}$
6.	1	$\dfrac{2}{n\pi}[1 - (-1)^n]$
7.	$\begin{cases} -x & x \le a \\ \pi - x & x > a \end{cases}$	$\dfrac{2}{n}\cos (na) \qquad 0 < a < \pi$
8.	$\begin{cases} (\pi - a)x & x \le a \\ (\pi - x)a & x > a \end{cases}$	$\dfrac{2}{n^2}\sin (na) \qquad 0 < a < \pi$
9.	$\dfrac{\pi}{2}e^{ax}$	$\dfrac{n}{n^2 + a^2}[1 - (-1)^n e^{a\pi}]$
10.	$\dfrac{\sinh a(\pi - x)}{\sinh a\pi}$	$\dfrac{2n}{\pi(n^2 + a^2)}$

TABLE E   Finite Cosine Transform

The finite cosine transform transforms a function $f(x)$, $0 \le x \le \pi$ into a sequence C_n, $n = 0, 1, 2, \ldots$ by means of the formula

$$C_n = \frac{2}{\pi} \int_0^{\pi} f(x) \cos (nx)\, dx$$

It is convenient, insofar as the tables are concerned, for the range of the variable x to lie in the interval $[0,\pi]$. The reader can transform any other interval $[a,b]$ into $[0,\pi]$ via the transformation

$$y = \frac{\pi}{b - a}(x - a)$$

There is nothing mysterious about the finite cosine transform; the members of the sequence C_n are the coefficients of cos (nx) in the Fourier cosine series of $f(x)$. That is,

$$f(x) = \frac{C_0}{2} + \sum_{n=1}^{\infty} C_n \cos (nx)$$

$f(x) = \dfrac{C_0}{2} + \sum_{n=1}^{\infty} C_n \cos (nx)$	$C_n = \dfrac{2}{\pi} \int_0^{\pi} f(x) \cos (nx)\, dx$
$0 \leqslant x \leqslant \pi$	$n = 0, 1, 2, \ldots$

1. $f''(x)$

$-n^2 C_n - \dfrac{2}{\pi} [f'(0) - (-1)^n f'(\pi)]$

2. $\dfrac{a_0}{2} + \sum_{n=1}^{\infty} a_n \cos (nx)$

a_n

3. $f(\pi - x)$

$(-1)^n \dfrac{2}{\pi} C_n$

4. 1

$\begin{cases} 2 & n = 0 \\ 0 & n = 1, 2, \ldots \end{cases}$

5. $\cos (mx) \quad m = 1, 2, \ldots$

$\begin{cases} 1 & n = m \\ 0 & n \neq m \end{cases}$

6. x

$\begin{cases} \pi & n = 0 \\ \dfrac{2}{\pi n^2} [(-1)^n - 1]. & n = 1, 2, \ldots \end{cases}$

7. x^2

$\begin{cases} 2\pi^2/3 & n = 0 \\ \dfrac{4}{n^2} (-1)^n & n = 1, 2, \ldots \end{cases}$

8. $-\log (2 \sin x/2)$

$\begin{cases} 0 & n = 0 \\ 1/n & n = 1, 2, \ldots \end{cases}$

9. $\dfrac{1}{a} e^{ax}$

$\dfrac{2}{\pi} \left[\dfrac{(-1)^n e^{a\pi} - 1}{n^2 + a^2} \right]$

10. $\begin{cases} 1 & 0 < x < a \\ -1 & a < x < \pi \end{cases}$

$\begin{cases} \dfrac{2}{\pi} (2a - \pi) & n = 0 \\ \\ \dfrac{4}{n\pi} \sin (na) & n = 1, 2, \ldots \end{cases}$

TABLE F   Laplace Transforms

$f(t) = \mathcal{L}^{-1}[F(s)]$	$F(s) = \mathcal{L}[f(t)]$
1. 1	$\dfrac{1}{s}$ $s > 0$
2. e^{at}	$\dfrac{1}{s - a}$ $s > a$
3. $\sin at$	$\dfrac{a}{s^2 + a^2}$ $s > 0$
4. $\cos at$	$\dfrac{s}{s^2 + a^2}$ $s > 0$
5. $\sinh at$	$\dfrac{a}{s^2 - a^2}$ $s > |a|$
6. $\cosh at$	$\dfrac{s}{s^2 - a^2}$ $s > |a|$
7. $e^{at} \sin bt$	$\dfrac{b}{(s - a)^2 + b^2}$ $s > a$
8. $e^{at} \cos bt$	$\dfrac{s - a}{(s - a)^2 + b^2}$ $s > a$
9. t^n n = positive integer	$\dfrac{n!}{s^{n+1}}$ $s > 0$
10. $t^n e^{at}$	$\dfrac{n!}{(s - a)^{n+1}}$ $s > a$
11. $H(t - a)$	$\dfrac{e^{-as}}{s}$ $s > 0$
12. $H(t - a)f(t - a)$	$e^{-as}F(s)$
13. $e^{at}f(t)$	$F(s - a)$
14. $f(t)*g(t)$	$F(s)G(s)$
15. $f^n(t)$ (nth derivative)	$s^n F(s) - s^{n-1} f(0) - \ldots - f^{n-1}(0)$
16. $f(at)$	$\dfrac{1}{a} F\left(\dfrac{s}{a}\right)$ $a > 0$

TABLE F   Laplace Transforms

$f(t) = \mathcal{L}^{-1}[F(s)]$	$F(s) = \mathcal{L}[f(t)]$
17. $\displaystyle\int_0^t f(\tau)\, d\tau$	$\dfrac{1}{s}\, F(s)$
18. $erf(t/2a)$	$\dfrac{1}{s}\, e^{a^2 s^2}\, erfc(as)$
19. $erfc(a/2\sqrt{t})$	$\dfrac{1}{s}\, e^{-a\sqrt{s}}$
20. $J_0(at)$	$(s^2 + a^2)^{-1/2}$
21. $\delta(t - a)$	e^{-sa}
22. $\dfrac{1}{\sqrt{\pi t}}\, exp\!\left(\dfrac{-a^2}{4t}\right)$	$\dfrac{e^{-a s}}{\sqrt{s}} \qquad a \geqslant 0$
23. $\dfrac{1}{\sqrt{\pi t}} - ae^{a^2 t}\, erfc(a\sqrt{t})$	$\dfrac{1}{\sqrt{s} + a}$

PDE Crossword Puzzle

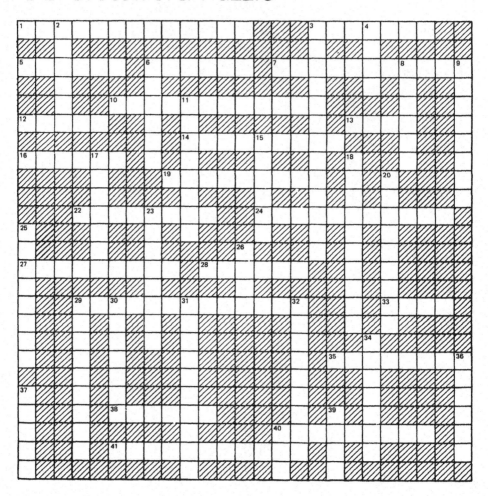

Across

1. A PDE of the form $A(x,y)u_{xx} + B(x,y)u_{yy} = f(x,y)$ is called an _____ equation.
3. A person's name associated with a method for solving initial-boundary-value problems with difficult BCs in terms of solutions of problems with simple BCs.
5. Types of equations where the principle of superposition holds.

6. The author of a famous text on PDE.
7. In order for separation of variables to work, the PDE must be _____.
10. A first-order equation of the form $u_t + f_x = 0$ is called a _____ equation.
12. When we solve Laplace's equation in polar coordinates by separation of variables, the ODE in r is named after this man.
13. The equation $u_{tt} = u_{xx}$ is commonly called the _____ equation.
14. The equation $u_{xx} + u_x + u_y - 0$ is a _____ differential equation.
16. A useful way to solve the equation $u_t = Du_{xx} - vu_x$ is by _____ coordinates.
19. The equation $u_t = u_{xx}$ is commonly called the _____ equation.
22. The Crank-Nicolson method is an _____ method.
24. The _____ of a BVP quite often are proportional to the natural frequencies of the problem.
27. When solving Laplace's equation in spherical coordinates by separation of variables, the ODE in ϕ is called _____ equation.
28. The potential due to a point charge is called _____ function.
29. The most basic idea behind the solution to linear problems is _____.
33. A method of minimizing functionals by replacing the functions by polynomials is due to _____.
35. If the dependent variable is found explicitly as a function of the independent variables of the problem, we have an _____ solution.
38. A famous mathematician/physicist responsible for the discovery of calculus and some differential equations.
40. The quantity that measures the difference between a function at a point and its surrounding points.
41. The basic vibrations of a membrane are called the _____ solutions.

Down

2. An integral transform in which the kernel is a Bessel function.
3. PDE can be simplified by changing the variables of the problem to _____ variables.
4. A famous mathematician who died at the age of 27 and had nothing to do with PDEs.
6. Initial-boundary-value problems defined on bounded domains can be solved by _____ integral transforms.

8. The PDE $r^2R'' + rR' - \lambda^2R = 0$ is called _____ equation.

9. Waves of the form $[\cos(n\pi t)\sin(n\pi x)]$ are called _____ waves.

11. A numerical method in which the time increment must be kept small.

15. The ODE in r we must solve when solving the vibrating circular membrane is due to _____.

17. The BVP of the second kind is called the _____ problem.

18. The u_x term in the equation $u_t = Du_{xx} - vu_x$ is related to the _____ of the material.

20. Responsible for a very elegant solution of the one-dimensional wave equation in free space.

23. A transform that changes derivatives to multiplication is called an _____ transformation.

25. The equation $e^x u_{xx} + u_{yy} = 0$ is a PDE of the _____ type.

26. The PDE $u_t = u_{xx}$ is called the _____ equation.

29. If the solutions of a PDE are standing waves, then we might call the PDE a _____ equation.

30. The PDE $u_{xx} + u_{yy} = f(x,y)$ is named after _____.

31. The Legendre polynomials $\{P_n(x)\}$ are _____ on the interval $[0,1]$.

32. Most nonlinear PDEs must be solved by _____ methods.

34. An integral transform that we generally use on the time variable is due to _____.

36. The PDE problem with only IC is called a _____ problem.

37. A BVP in which the BC relates the flux to the solution at the boundary is called a BVP of the _____ kind.

39. The normal derivative at the boundary of a region is related to the _____ of material across the boundary.

40. The director of the Courant Institute of Mathematical Sciences is Professor _____.

APPENDIX 3

Laplacian in Different Coordinate Systems

$\nabla^2 u = u_{xx} + u_{yy}$ Two dimensional cartesian Laplacian

$\nabla^2 u = u_{rr} + \dfrac{1}{r} u_r + \dfrac{1}{r^2} u_{\theta\theta}$ Two dimensional polar Laplacian

$\nabla^2 u = u_{xx} + u_{yy} + u_{zz}$ Three dimensional cartesian Laplacian

$\nabla^2 u = u_{rr} + \dfrac{1}{r} u_r + \dfrac{1}{r^2} u_{\theta\theta} + u_{zz}$ Three dimensional cylindrical Laplacian

$\nabla^2 u = u_{rr} + \dfrac{2}{r} u_r + \dfrac{1}{r^2} u_{\theta\theta} + \dfrac{\cot\theta}{r^2} u_\theta + \dfrac{1}{r^2 \sin\theta} u_{\phi\phi} = 0$ Three dimensional spherical Laplacian

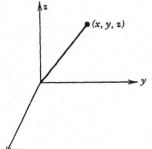

Cartesian coordinate
system

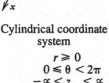

Cylindrical coordinate
system

$r \geq 0$
$0 \leq \theta < 2\pi$
$-\alpha < z < \alpha$
$x = r \cos\theta$

$y = r \sin\theta$

$z = z$

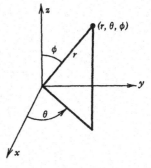

Spherical coordinate
system

$r \geq 0$
$0 \leq \theta < 2\pi$
$0 \leq \phi \leq \pi$
$x = r \sin\phi \cos\theta$

$y = r \sin\phi \sin\theta$

$z = r \cos\phi$

Types of Partial Differential Equations

Elliptic Partial Differential Equations

$\nabla^2 u = 0$ Laplace's equation

$\nabla^2 u + \lambda^2 u = 0$ Helmholtz's equation

$\nabla^2 u = k$ Poisson's equation

$\nabla^2 u + k(E - V)u = 0$ Schrödinger's equation

Hyperbolic Partial Differential Equations

$u_{tt} = c^2 u_{xx}$ One dimensional vibrating string

$u_{tt} = c^2 u_{xx} - h u_t$ Vibrating string with friction

$u_{tt} = c^2 u_{xx} - h u_t - k u$ Transmission line equation

$u_{tt} = c^2 u_{xx} + f(x,t)$ Wave equation with forced vibrations

$u_{tt} = c^2 \nabla^2 u$ Wave equation in higher dimensions

$u_{tt} = c^2 \nabla^2 u - h u_t$ Wave equation with friction

Parabolic Partial Differential Equations

$u_t = \alpha^2 u_{xx}$ One-dimensional diffusion equation

$u_t = \alpha^2 u_{xx} - h u_x$ Diffusion-convection equation

$u_t = \alpha^2 u_{xx} - k u$ Diffusion with lateral heat-concentration loss

$u_t = \alpha^2 u_{xx} + f(x,t)$ Diffusion with heat source (or loss)

Index

Astronomy

BURNHAM'S CELESTIAL HANDBOOK, Robert Burnham, Jr. Thorough guide to the stars beyond our solar system. Exhaustive treatment. Alphabetical by constellation: Andromeda to Cetus in Vol. 1; Chamaeleon to Orion in Vol. 2; and Pavo to Vulpecula in Vol. 3. Hundreds of illustrations. Index in Vol. 3. 2,000pp. 6⅛ x 9¼.

Vol. I: 0-486-23567-X
Vol. II: 0-486-23568-8
Vol. III: 0-486-23673-0

EXPLORING THE MOON THROUGH BINOCULARS AND SMALL TELE-SCOPES, Ernest H. Cherrington, Jr. Informative, profusely illustrated guide to locating and identifying craters, rills, seas, mountains, other lunar features. Newly revised and updated with special section of new photos. Over 100 photos and diagrams. 240pp. 8¼ x 11. 0-486-24491-1

THE EXTRATERRESTRIAL LIFE DEBATE, 1750–1900, Michael J. Crowe. First detailed, scholarly study in English of the many ideas that developed from 1750 to 1900 regarding the existence of intelligent extraterrestrial life. Examines ideas of Kant, Herschel, Voltaire, Percival Lowell, many other scientists and thinkers. 16 illustrations. 704pp. 5⅜ x 8½. 0-486-40675-X

THEORIES OF THE WORLD FROM ANTIQUITY TO THE COPERNICAN REVOLUTION, Michael J. Crowe. Newly revised edition of an accessible, enlightening book recreates the change from an earth-centered to a sun-centered conception of the solar system. 242pp. 5⅜ x 8½. 0-486-41444-2

A HISTORY OF ASTRONOMY, A. Pannekoek. Well-balanced, carefully reasoned study covers such topics as Ptolemaic theory, work of Copernicus, Kepler, Newton, Eddington's work on stars, much more. Illustrated. References. 521pp. 5⅜ x 8½. 0-486-65994-1

A COMPLETE MANUAL OF AMATEUR ASTRONOMY: TOOLS AND TECHNIQUES FOR ASTRONOMICAL OBSERVATIONS, P. Clay Sherrod with Thomas L. Koed. Concise, highly readable book discusses: selecting, setting up and maintaining a telescope; amateur studies of the sun; lunar topography and occultations; observations of Mars, Jupiter, Saturn, the minor planets and the stars; an introduction to photoelectric photometry; more. 1981 ed. 124 figures. 25 halftones. 37 tables. 335pp. 6½ x 9¼. 0-486-40675-X

AMATEUR ASTRONOMER'S HANDBOOK, J. B. Sidgwick. Timeless, comprehensive coverage of telescopes, mirrors, lenses, mountings, telescope drives, micrometers, spectroscopes, more. 189 illustrations. 576pp. 5⅝ x 8¼. (Available in U.S. only.) 0-486-24034-7

STARS AND RELATIVITY, Ya. B. Zel'dovich and I. D. Novikov. Vol. 1 of *Relativistic Astrophysics* by famed Russian scientists. General relativity, properties of matter under astrophysical conditions, stars, and stellar systems. Deep physical insights, clear presentation. 1971 edition. References. 544pp. 5⅝ x 8¼. 0-486-69424-0

Chemistry

THE SCEPTICAL CHYMIST: THE CLASSIC 1661 TEXT, Robert Boyle. Boyle defines the term "element," asserting that all natural phenomena can be explained by the motion and organization of primary particles. 1911 ed. viii+232pp. 5⅜ x 8½.
0-486-42825-7

RADIOACTIVE SUBSTANCES, Marie Curie. Here is the celebrated scientist's doctoral thesis, the prelude to her receipt of the 1903 Nobel Prize. Curie discusses establishing atomic character of radioactivity found in compounds of uranium and thorium; extraction from pitchblende of polonium and radium; isolation of pure radium chloride; determination of atomic weight of radium; plus electric, photographic, luminous, heat, color effects of radioactivity. ii+94pp. 5⅜ x 8½. 0-486-42550-9

CHEMICAL MAGIC, Leonard A. Ford. Second Edition, Revised by E. Winston Grundmeier. Over 100 unusual stunts demonstrating cold fire, dust explosions, much more. Text explains scientific principles and stresses safety precautions. 128pp. 5⅜ x 8½. 0-486-67628-5

THE DEVELOPMENT OF MODERN CHEMISTRY, Aaron J. Ihde. Authoritative history of chemistry from ancient Greek theory to 20th-century innovation. Covers major chemists and their discoveries. 209 illustrations. 14 tables. Bibliographies. Indices. Appendices. 851pp. 5⅜ x 8½. 0-486-64235-6

CATALYSIS IN CHEMISTRY AND ENZYMOLOGY, William P. Jencks. Exceptionally clear coverage of mechanisms for catalysis, forces in aqueous solution, carbonyl- and acyl-group reactions, practical kinetics, more. 864pp. 5⅜ x 8½.
0-486-65460-5

ELEMENTS OF CHEMISTRY, Antoine Lavoisier. Monumental classic by founder of modern chemistry in remarkable reprint of rare 1790 Kerr translation. A must for every student of chemistry or the history of science. 539pp. 5⅜ x 8½. 0-486-64624-6

THE HISTORICAL BACKGROUND OF CHEMISTRY, Henry M. Leicester. Evolution of ideas, not individual biography. Concentrates on formulation of a coherent set of chemical laws. 260pp. 5⅜ x 8½. 0-486-61053-5

A SHORT HISTORY OF CHEMISTRY, J. R. Partington. Classic exposition explores origins of chemistry, alchemy, early medical chemistry, nature of atmosphere, theory of valency, laws and structure of atomic theory, much more. 428pp. 5⅜ x 8½. (Available in U.S. only.) 0-486-65977-1

GENERAL CHEMISTRY, Linus Pauling. Revised 3rd edition of classic first-year text by Nobel laureate. Atomic and molecular structure, quantum mechanics, statistical mechanics, thermodynamics correlated with descriptive chemistry. Problems. 992pp. 5⅜ x 8½. 0-486-65622-5

FROM ALCHEMY TO CHEMISTRY, John Read. Broad, humanistic treatment focuses on great figures of chemistry and ideas that revolutionized the science. 50 illustrations. 240pp. 5⅜ x 8½. 0-486-28690-8

Engineering

DE RE METALLICA, Georgius Agricola. The famous Hoover translation of greatest treatise on technological chemistry, engineering, geology, mining of early modern times (1556). All 289 original woodcuts. 638pp. 6¾ x 11.   0-486-60006-8

FUNDAMENTALS OF ASTRODYNAMICS, Roger Bate et al. Modern approach developed by U.S. Air Force Academy. Designed as a first course. Problems, exercises. Numerous illustrations. 455pp. 5⅜ x 8½.   0-486-60061-0

DYNAMICS OF FLUIDS IN POROUS MEDIA, Jacob Bear. For advanced students of ground water hydrology, soil mechanics and physics, drainage and irrigation engineering and more. 335 illustrations. Exercises, with answers. 784pp. 6⅛ x 9¼.   0-486-65675-6

THEORY OF VISCOELASTICITY (Second Edition), Richard M. Christensen. Complete consistent description of the linear theory of the viscoelastic behavior of materials. Problem-solving techniques discussed. 1982 edition. 29 figures. xiv+364pp. 6⅛ x 9¼.   0-486-42880-X

MECHANICS, J. P. Den Hartog. A classic introductory text or refresher. Hundreds of applications and design problems illuminate fundamentals of trusses, loaded beams and cables, etc. 334 answered problems. 462pp. 5⅜ x 8½.   0-486-60754-2

MECHANICAL VIBRATIONS, J. P. Den Hartog. Classic textbook offers lucid explanations and illustrative models, applying theories of vibrations to a variety of practical industrial engineering problems. Numerous figures. 233 problems, solutions. Appendix. Index. Preface. 436pp. 5⅜ x 8½.   0-486-64785-4

STRENGTH OF MATERIALS, J. P. Den Hartog. Full, clear treatment of basic material (tension, torsion, bending, etc.) plus advanced material on engineering methods, applications. 350 answered problems. 323pp. 5⅜ x 8½.   0-486-60755-0

A HISTORY OF MECHANICS, René Dugas. Monumental study of mechanical principles from antiquity to quantum mechanics. Contributions of ancient Greeks, Galileo, Leonardo, Kepler, Lagrange, many others. 671pp. 5⅜ x 8½.   0-486-65632-2

STABILITY THEORY AND ITS APPLICATIONS TO STRUCTURAL MECHANICS, Clive L. Dym. Self-contained text focuses on Koiter postbuckling analyses, with mathematical notions of stability of motion. Basing minimum energy principles for static stability upon dynamic concepts of stability of motion, it develops asymptotic buckling and postbuckling analyses from potential energy considerations, with applications to columns, plates, and arches. 1974 ed. 208pp. 5⅜ x 8½.   0-486-42541-X

METAL FATIGUE, N. E. Frost, K. J. Marsh, and L. P. Pook. Definitive, clearly written, and well-illustrated volume addresses all aspects of the subject, from the historical development of understanding metal fatigue to vital concepts of the cyclic stress that causes a crack to grow. Includes 7 appendixes. 544pp. 5⅜ x 8½.   0-486-40927-9

CATALOG OF DOVER BOOKS

ROCKETS, Robert Goddard. Two of the most significant publications in the history of rocketry and jet propulsion: "A Method of Reaching Extreme Altitudes" (1919) and "Liquid Propellant Rocket Development" (1936). 128pp. 5⅜ x 8½. 0-486-42537-1

STATISTICAL MECHANICS: PRINCIPLES AND APPLICATIONS, Terrell L. Hill. Standard text covers fundamentals of statistical mechanics, applications to fluctuation theory, imperfect gases, distribution functions, more. 448pp. 5⅜ x 8½.
0-486-65390-0

ENGINEERING AND TECHNOLOGY 1650–1750: ILLUSTRATIONS AND TEXTS FROM ORIGINAL SOURCES, Martin Jensen. Highly readable text with more than 200 contemporary drawings and detailed engravings of engineering projects dealing with surveying, leveling, materials, hand tools, lifting equipment, transport and erection, piling, bailing, water supply, hydraulic engineering, and more. Among the specific projects outlined-transporting a 50-ton stone to the Louvre, erecting an obelisk, building timber locks, and dredging canals. 207pp. 8⅜ x 11¼.
0-486-42232-1

THE VARIATIONAL PRINCIPLES OF MECHANICS, Cornelius Lanczos. Graduate level coverage of calculus of variations, equations of motion, relativistic mechanics, more. First inexpensive paperbound edition of classic treatise. Index. Bibliography. 418pp. 5⅜ x 8½. 0-486-65067-7

PROTECTION OF ELECTRONIC CIRCUITS FROM OVERVOLTAGES, Ronald B. Standler. Five-part treatment presents practical rules and strategies for circuits designed to protect electronic systems from damage by transient overvoltages. 1989 ed. xxiv+434pp. 6⅛ x 9¼. 0-486-42552-5

ROTARY WING AERODYNAMICS, W. Z. Stepniewski. Clear, concise text covers aerodynamic phenomena of the rotor and offers guidelines for helicopter performance evaluation. Originally prepared for NASA. 537 figures. 640pp. 6⅛ x 9¼.
0-486-64647-5

INTRODUCTION TO SPACE DYNAMICS, William Tyrrell Thomson. Comprehensive, classic introduction to space-flight engineering for advanced undergraduate and graduate students. Includes vector algebra, kinematics, transformation of coordinates. Bibliography. Index. 352pp. 5⅜ x 8½. 0-486-65113-4

HISTORY OF STRENGTH OF MATERIALS, Stephen P. Timoshenko. Excellent historical survey of the strength of materials with many references to the theories of elasticity and structure. 245 figures. 452pp. 5⅜ x 8½. 0-486-61187-6

ANALYTICAL FRACTURE MECHANICS, David J. Unger. Self-contained text supplements standard fracture mechanics texts by focusing on analytical methods for determining crack-tip stress and strain fields. 336pp. 6⅛ x 9¼. 0-486-41737-9

STATISTICAL MECHANICS OF ELASTICITY, J. H. Weiner. Advanced, self-contained treatment illustrates general principles and elastic behavior of solids. Part 1, based on classical mechanics, studies thermoelastic behavior of crystalline and polymeric solids. Part 2, based on quantum mechanics, focuses on interatomic force laws, behavior of solids, and thermally activated processes. For students of physics and chemistry and for polymer physicists. 1983 ed. 96 figures. 496pp. 5⅜ x 8½.
0-486-42260-7

Mathematics

FUNCTIONAL ANALYSIS (Second Corrected Edition), George Bachman and Lawrence Narici. Excellent treatment of subject geared toward students with background in linear algebra, advanced calculus, physics and engineering. Text covers introduction to inner-product spaces, normed, metric spaces, and topological spaces; complete orthonormal sets, the Hahn-Banach Theorem and its consequences, and many other related subjects. 1966 ed. 544pp. 6⅛ x 9¼.   0-486-40251-7

ASYMPTOTIC EXPANSIONS OF INTEGRALS, Norman Bleistein & Richard A. Handelsman. Best introduction to important field with applications in a variety of scientific disciplines. New preface. Problems. Diagrams. Tables. Bibliography. Index. 448pp. 5⅜ x 8½.   0-486-65082-0

VECTOR AND TENSOR ANALYSIS WITH APPLICATIONS, A. I. Borisenko and I. E. Tarapov. Concise introduction. Worked-out problems, solutions, exercises. 257pp. 5⅜ x 8¼.   0-486-63833-2

AN INTRODUCTION TO ORDINARY DIFFERENTIAL EQUATIONS, Earl A. Coddington. A thorough and systematic first course in elementary differential equations for undergraduates in mathematics and science, with many exercises and problems (with answers). Index. 304pp. 5⅜ x 8½.   0-486-65942-9

FOURIER SERIES AND ORTHOGONAL FUNCTIONS, Harry F. Davis. An incisive text combining theory and practical example to introduce Fourier series, orthogonal functions and applications of the Fourier method to boundary-value problems. 570 exercises. Answers and notes. 416pp. 5⅜ x 8½.   0-486-65973-9

COMPUTABILITY AND UNSOLVABILITY, Martin Davis. Classic graduate-level introduction to theory of computability, usually referred to as theory of recurrent functions. New preface and appendix. 288pp. 5⅜ x 8½.   0-486-61471-9

ASYMPTOTIC METHODS IN ANALYSIS, N. G. de Bruijn. An inexpensive, comprehensive guide to asymptotic methods—the pioneering work that teaches by explaining worked examples in detail. Index. 224pp. 5⅜ x 8½   0-486-64221-6

APPLIED COMPLEX VARIABLES, John W. Dettman. Step-by-step coverage of fundamentals of analytic function theory—plus lucid exposition of five important applications: Potential Theory; Ordinary Differential Equations; Fourier Transforms; Laplace Transforms; Asymptotic Expansions. 66 figures. Exercises at chapter ends. 512pp. 5⅜ x 8½.   0-486-64670-X

INTRODUCTION TO LINEAR ALGEBRA AND DIFFERENTIAL EQUATIONS, John W. Dettman. Excellent text covers complex numbers, determinants, orthonormal bases, Laplace transforms, much more. Exercises with solutions. Undergraduate level. 416pp. 5⅜ x 8½.   0-486-65191-6

RIEMANN'S ZETA FUNCTION, H. M. Edwards. Superb, high-level study of landmark 1859 publication entitled "On the Number of Primes Less Than a Given Magnitude" traces developments in mathematical theory that it inspired. xiv+315pp. 5⅜ x 8½.   0-486-41740-9

CALCULUS OF VARIATIONS WITH APPLICATIONS, George M. Ewing. Applications-oriented introduction to variational theory develops insight and promotes understanding of specialized books, research papers. Suitable for advanced undergraduate/graduate students as primary, supplementary text. 352pp. 5⅜ x 8½.
0-486-64856-7

COMPLEX VARIABLES, Francis J. Flanigan. Unusual approach, delaying complex algebra till harmonic functions have been analyzed from real variable viewpoint. Includes problems with answers. 364pp. 5⅜ x 8½.
0-486-61388-7

AN INTRODUCTION TO THE CALCULUS OF VARIATIONS, Charles Fox. Graduate-level text covers variations of an integral, isoperimetrical problems, least action, special relativity, approximations, more. References. 279pp. 5⅜ x 8½.
0-486-65499-0

COUNTEREXAMPLES IN ANALYSIS, Bernard R. Gelbaum and John M. H. Olmsted. These counterexamples deal mostly with the part of analysis known as "real variables." The first half covers the real number system, and the second half encompasses higher dimensions. 1962 edition. xxiv+198pp. 5⅜ x 8½. 0-486-42875-3

CATASTROPHE THEORY FOR SCIENTISTS AND ENGINEERS, Robert Gilmore. Advanced-level treatment describes mathematics of theory grounded in the work of Poincaré, R. Thom, other mathematicians. Also important applications to problems in mathematics, physics, chemistry and engineering. 1981 edition. References. 28 tables. 397 black-and-white illustrations. xvii + 666pp. 6⅛ x 9¼.
0-486-67539-4

INTRODUCTION TO DIFFERENCE EQUATIONS, Samuel Goldberg. Exceptionally clear exposition of important discipline with applications to sociology, psychology, economics. Many illustrative examples; over 250 problems. 260pp. 5⅜ x 8½.
0-486-65084-7

NUMERICAL METHODS FOR SCIENTISTS AND ENGINEERS, Richard Hamming. Classic text stresses frequency approach in coverage of algorithms, polynomial approximation, Fourier approximation, exponential approximation, other topics. Revised and enlarged 2nd edition. 721pp. 5⅜ x 8½. 0-486-65241-6

INTRODUCTION TO NUMERICAL ANALYSIS (2nd Edition), F. B. Hildebrand. Classic, fundamental treatment covers computation, approximation, interpolation, numerical differentiation and integration, other topics. 150 new problems. 669pp. 5⅜ x 8½.
0-486-65363-3

THREE PEARLS OF NUMBER THEORY, A. Y. Khinchin. Three compelling puzzles require proof of a basic law governing the world of numbers. Challenges concern van der Waerden's theorem, the Landau-Schnirelmann hypothesis and Mann's theorem, and a solution to Waring's problem. Solutions included. 64pp. 5¾ x 8½.
0-486-40026-3

THE PHILOSOPHY OF MATHEMATICS: AN INTRODUCTORY ESSAY, Stephan Körner. Surveys the views of Plato, Aristotle, Leibniz & Kant concerning propositions and theories of applied and pure mathematics. Introduction. Two appendices. Index. 198pp. 5⅜ x 8½.
0-486-25048-2

CATALOG OF DOVER BOOKS

INTRODUCTORY REAL ANALYSIS, A.N. Kolmogorov, S. V. Fomin. Translated by Richard A. Silverman. Self-contained, evenly paced introduction to real and functional analysis. Some 350 problems. 403pp. 5⅜ x 8½. 0-486-61226-0

APPLIED ANALYSIS, Cornelius Lanczos. Classic work on analysis and design of finite processes for approximating solution of analytical problems. Algebraic equations, matrices, harmonic analysis, quadrature methods, much more. 559pp. 5⅜ x 8½. 0-486-65656-X

AN INTRODUCTION TO ALGEBRAIC STRUCTURES, Joseph Landin. Superb self-contained text covers "abstract algebra": sets and numbers, theory of groups, theory of rings, much more. Numerous well-chosen examples, exercises. 247pp. 5⅜ x 8½. 0-486-65940-2

QUALITATIVE THEORY OF DIFFERENTIAL EQUATIONS, V. V. Nemytskii and V.V. Stepanov. Classic graduate-level text by two prominent Soviet mathematicians covers classical differential equations as well as topological dynamics and ergodic theory. Bibliographies. 523pp. 5⅜ x 8½. 0-486-65954-2

THEORY OF MATRICES, Sam Perlis. Outstanding text covering rank, nonsingularity and inverses in connection with the development of canonical matrices under the relation of equivalence, and without the intervention of determinants. Includes exercises. 237pp. 5⅜ x 8½. 0-486-66810-X

INTRODUCTION TO ANALYSIS, Maxwell Rosenlicht. Unusually clear, accessible coverage of set theory, real number system, metric spaces, continuous functions, Riemann integration, multiple integrals, more. Wide range of problems. Undergraduate level. Bibliography. 254pp. 5⅜ x 8½. 0-486-65038-3

MODERN NONLINEAR EQUATIONS, Thomas L. Saaty. Emphasizes practical solution of problems; covers seven types of equations. ". . . a welcome contribution to the existing literature...."–Math Reviews. 490pp. 5⅜ x 8½. 0-486-64232-1

MATRICES AND LINEAR ALGEBRA, Hans Schneider and George Phillip Barker. Basic textbook covers theory of matrices and its applications to systems of linear equations and related topics such as determinants, eigenvalues and differential equations. Numerous exercises. 432pp. 5⅜ x 8½. 0-486-66014-1

LINEAR ALGEBRA, Georgi E. Shilov. Determinants, linear spaces, matrix algebras, similar topics. For advanced undergraduates, graduates. Silverman translation. 387pp. 5⅜ x 8½. 0-486-63518-X

ELEMENTS OF REAL ANALYSIS, David A. Sprecher. Classic text covers fundamental concepts, real number system, point sets, functions of a real variable, Fourier series, much more. Over 500 exercises. 352pp. 5⅜ x 8½. 0-486-65385-4

SET THEORY AND LOGIC, Robert R. Stoll. Lucid introduction to unified theory of mathematical concepts. Set theory and logic seen as tools for conceptual understanding of real number system. 496pp. 5⅜ x 8¼. 0-486-63829-4

TENSOR CALCULUS, J.L. Synge and A. Schild. Widely used introductory text covers spaces and tensors, basic operations in Riemannian space, non-Riemannian spaces, etc. 324pp. 5⅜ x 8¼. 0-486-63612-7

ORDINARY DIFFERENTIAL EQUATIONS, Morris Tenenbaum and Harry Pollard. Exhaustive survey of ordinary differential equations for undergraduates in mathematics, engineering, science. Thorough analysis of theorems. Diagrams. Bibliography. Index. 818pp. 5⅜ x 8½. 0-486-64940-7

INTEGRAL EQUATIONS, F. G. Tricomi. Authoritative, well-written treatment of extremely useful mathematical tool with wide applications. Volterra Equations, Fredholm Equations, much more. Advanced undergraduate to graduate level. Exercises. Bibliography. 238pp. 5⅜ x 8½. 0-486-64828-1

FOURIER SERIES, Georgi P. Tolstov. Translated by Richard A. Silverman. A valuable addition to the literature on the subject, moving clearly from subject to subject and theorem to theorem. 107 problems, answers. 336pp. 5⅜ x 8½. 0-486-63317-9

INTRODUCTION TO MATHEMATICAL THINKING, Friedrich Waismann. Examinations of arithmetic, geometry, and theory of integers; rational and natural numbers; complete induction; limit and point of accumulation; remarkable curves; complex and hypercomplex numbers, more. 1959 ed. 27 figures. xii+260pp. 5⅜ x 8½. 0-486-63317-9

POPULAR LECTURES ON MATHEMATICAL LOGIC, Hao Wang. Noted logician's lucid treatment of historical developments, set theory, model theory, recursion theory and constructivism, proof theory, more. 3 appendixes. Bibliography. 1981 edition. ix + 283pp. 5⅜ x 8½. 0-486-67632-3

CALCULUS OF VARIATIONS, Robert Weinstock. Basic introduction covering isoperimetric problems, theory of elasticity, quantum mechanics, electrostatics, etc. Exercises throughout. 326pp. 5⅜ x 8½. 0-486-63069-2

THE CONTINUUM: A CRITICAL EXAMINATION OF THE FOUNDATION OF ANALYSIS, Hermann Weyl. Classic of 20th-century foundational research deals with the conceptual problem posed by the continuum. 156pp. 5⅜ x 8½. 0-486-67982-9

CHALLENGING MATHEMATICAL PROBLEMS WITH ELEMENTARY SOLUTIONS, A. M. Yaglom and I. M. Yaglom. Over 170 challenging problems on probability theory, combinatorial analysis, points and lines, topology, convex polygons, many other topics. Solutions. Total of 445pp. 5⅜ x 8½. Two-vol. set.
Vol. I: 0-486-65536-9   Vol. II: 0-486-65537-7

INTRODUCTION TO PARTIAL DIFFERENTIAL EQUATIONS WITH APPLICATIONS, E. C. Zachmanoglou and Dale W. Thoe. Essentials of partial differential equations applied to common problems in engineering and the physical sciences. Problems and answers. 416pp. 5⅜ x 8½. 0-486-65251-3

THE THEORY OF GROUPS, Hans J. Zassenhaus. Well-written graduate-level text acquaints reader with group-theoretic methods and demonstrates their usefulness in mathematics. Axioms, the calculus of complexes, homomorphic mapping, p-group theory, more. 276pp. 5⅜ x 8½. 0-486-40922-8

Math–Decision Theory, Statistics, Probability

ELEMENTARY DECISION THEORY, Herman Chernoff and Lincoln E. Moses. Clear introduction to statistics and statistical theory covers data processing, probability and random variables, testing hypotheses, much more. Exercises. 364pp. 5⅜ x 8½.   0-486-65218-1

STATISTICS MANUAL, Edwin L. Crow et al. Comprehensive, practical collection of classical and modern methods prepared by U.S. Naval Ordnance Test Station. Stress on use. Basics of statistics assumed. 288pp. 5⅜ x 8½.   0-486-60599-X

SOME THEORY OF SAMPLING, William Edwards Deming. Analysis of the problems, theory and design of sampling techniques for social scientists, industrial managers and others who find statistics important at work. 61 tables. 90 figures. xvii +602pp. 5⅜ x 8½.   0-486-64684-X

LINEAR PROGRAMMING AND ECONOMIC ANALYSIS, Robert Dorfman, Paul A. Samuelson and Robert M. Solow. First comprehensive treatment of linear programming in standard economic analysis. Game theory, modern welfare economics, Leontief input-output, more. 525pp. 5⅜ x 8½.   0-486-65491-5

PROBABILITY: AN INTRODUCTION, Samuel Goldberg. Excellent basic text covers set theory, probability theory for finite sample spaces, binomial theorem, much more. 360 problems. Bibliographies. 322pp. 5⅜ x 8½.   0-486-65252-1

GAMES AND DECISIONS: INTRODUCTION AND CRITICAL SURVEY, R. Duncan Luce and Howard Raiffa. Superb nontechnical introduction to game theory, primarily applied to social sciences. Utility theory, zero-sum games, n-person games, decision-making, much more. Bibliography. 509pp. 5⅜ x 8½. 0-486-65943-7

INTRODUCTION TO THE THEORY OF GAMES, J. C. C. McKinsey. This comprehensive overview of the mathematical theory of games illustrates applications to situations involving conflicts of interest, including economic, social, political, and military contexts. Appropriate for advanced undergraduate and graduate courses; advanced calculus a prerequisite. 1952 ed. x+372pp. 5⅜ x 8½.   0-486-42811-7

FIFTY CHALLENGING PROBLEMS IN PROBABILITY WITH SOLUTIONS, Frederick Mosteller. Remarkable puzzlers, graded in difficulty, illustrate elementary and advanced aspects of probability. Detailed solutions. 88pp. 5⅜ x 8½.   65355-2

PROBABILITY THEORY: A CONCISE COURSE, Y. A. Rozanov. Highly readable, self-contained introduction covers combination of events, dependent events, Bernoulli trials, etc. 148pp. 5⅜ x 8¼.   0-486-63544-9

STATISTICAL METHOD FROM THE VIEWPOINT OF QUALITY CONTROL, Walter A. Shewhart. Important text explains regulation of variables, uses of statistical control to achieve quality control in industry, agriculture, other areas. 192pp. 5⅜ x 8½.   0-486-65232-7

Math–Geometry and Topology

ELEMENTARY CONCEPTS OF TOPOLOGY, Paul Alexandroff. Elegant, intuitive approach to topology from set-theoretic topology to Betti groups; how concepts of topology are useful in math and physics. 25 figures. 57pp. $5\frac{3}{8} \times 8\frac{1}{2}$. 0-486-60747-X

COMBINATORIAL TOPOLOGY, P. S. Alexandrov. Clearly written, well-organized, three-part text begins by dealing with certain classic problems without using the formal techniques of homology theory and advances to the central concept, the Betti groups. Numerous detailed examples. 654pp. $5\frac{3}{8} \times 8\frac{1}{2}$. 0-486-40179-0

EXPERIMENTS IN TOPOLOGY, Stephen Barr. Classic, lively explanation of one of the byways of mathematics. Klein bottles, Moebius strips, projective planes, map coloring, problem of the Koenigsberg bridges, much more, described with clarity and wit. 43 figures. 210pp. $5\frac{3}{8} \times 8\frac{1}{2}$. 0-486-25933-1

THE GEOMETRY OF RENÉ DESCARTES, René Descartes. The great work founded analytical geometry. Original French text, Descartes's own diagrams, together with definitive Smith-Latham translation. 244pp. $5\frac{3}{8} \times 8\frac{1}{2}$. 0-486-60068-8

EUCLIDEAN GEOMETRY AND TRANSFORMATIONS, Clayton W. Dodge. This introduction to Euclidean geometry emphasizes transformations, particularly isometries and similarities. Suitable for undergraduate courses, it includes numerous examples, many with detailed answers. 1972 ed. viii+296pp. $6\frac{1}{8} \times 9\frac{1}{4}$. 0-486-43476-1

PRACTICAL CONIC SECTIONS: THE GEOMETRIC PROPERTIES OF ELLIPSES, PARABOLAS AND HYPERBOLAS, J. W. Downs. This text shows how to create ellipses, parabolas, and hyperbolas. It also presents historical background on their ancient origins and describes the reflective properties and roles of curves in design applications. 1993 ed. 98 figures. xii+100pp. $6\frac{1}{2} \times 9\frac{1}{4}$. 0-486-42876-1

THE THIRTEEN BOOKS OF EUCLID'S ELEMENTS, translated with introduction and commentary by Sir Thomas L. Heath. Definitive edition. Textual and linguistic notes, mathematical analysis. 2,500 years of critical commentary. Unabridged. 1,414pp. $5\frac{3}{8} \times 8\frac{1}{2}$. Three-vol. set.
Vol. I: 0-486-60088-2 Vol. II: 0-486-60089-0 Vol. III: 0-486-60090-4

SPACE AND GEOMETRY: IN THE LIGHT OF PHYSIOLOGICAL, PSYCHOLOGICAL AND PHYSICAL INQUIRY, Ernst Mach. Three essays by an eminent philosopher and scientist explore the nature, origin, and development of our concepts of space, with a distinctness and precision suitable for undergraduate students and other readers. 1906 ed. vi+148pp. $5\frac{3}{8} \times 8\frac{1}{2}$. 0-486-43909-7

GEOMETRY OF COMPLEX NUMBERS, Hans Schwerdtfeger. Illuminating, widely praised book on analytic geometry of circles, the Moebius transformation, and two-dimensional non-Euclidean geometries. 200pp. $5\frac{5}{8} \times 8\frac{1}{4}$. 0-486-63830-8

DIFFERENTIAL GEOMETRY, Heinrich W. Guggenheimer. Local differential geometry as an application of advanced calculus and linear algebra. Curvature, transformation groups, surfaces, more. Exercises. 62 figures. 378pp. $5\frac{3}{8} \times 8\frac{1}{2}$. 0-486-63433-7

History of Math

THE WORKS OF ARCHIMEDES, Archimedes (T. L. Heath, ed.). Topics include the famous problems of the ratio of the areas of a cylinder and an inscribed sphere; the measurement of a circle; the properties of conoids, spheroids, and spirals; and the quadrature of the parabola. Informative introduction. clxxxvi+326pp. 5⅜ x 8½.
0-486-42084-1

A SHORT ACCOUNT OF THE HISTORY OF MATHEMATICS, W. W. Rouse Ball. One of clearest, most authoritative surveys from the Egyptians and Phoenicians through 19th-century figures such as Grassman, Galois, Riemann. Fourth edition. 522pp. 5⅜ x 8½.
0-486-20630-0

THE HISTORY OF THE CALCULUS AND ITS CONCEPTUAL DEVELOPMENT, Carl B. Boyer. Origins in antiquity, medieval contributions, work of Newton, Leibniz, rigorous formulation. Treatment is verbal. 346pp. 5⅜ x 8½. 0-486-60509-4

THE HISTORICAL ROOTS OF ELEMENTARY MATHEMATICS, Lucas N. H. Bunt, Phillip S. Jones, and Jack D. Bedient. Fundamental underpinnings of modern arithmetic, algebra, geometry and number systems derived from ancient civilizations. 320pp. 5⅜ x 8½.
0-486-25563-8

A HISTORY OF MATHEMATICAL NOTATIONS, Florian Cajori. This classic study notes the first appearance of a mathematical symbol and its origin, the competition it encountered, its spread among writers in different countries, its rise to popularity, its eventual decline or ultimate survival. Original 1929 two-volume edition presented here in one volume. xxviii+820pp. 5⅜ x 8½.
0-486-67766-4

GAMES, GODS & GAMBLING: A HISTORY OF PROBABILITY AND STATISTICAL IDEAS, F. N. David. Episodes from the lives of Galileo, Fermat, Pascal, and others illustrate this fascinating account of the roots of mathematics. Features thought-provoking references to classics, archaeology, biography, poetry. 1962 edition. 304pp. 5⅜ x 8½. (Available in U.S. only.)
0-486-40023-9

OF MEN AND NUMBERS: THE STORY OF THE GREAT MATHEMATICIANS, Jane Muir. Fascinating accounts of the lives and accomplishments of history's greatest mathematical minds–Pythagoras, Descartes, Euler, Pascal, Cantor, many more. Anecdotal, illuminating. 30 diagrams. Bibliography. 256pp. 5⅜ x 8½.
0-486-28973-7

HISTORY OF MATHEMATICS, David E. Smith. Nontechnical survey from ancient Greece and Orient to late 19th century; evolution of arithmetic, geometry, trigonometry, calculating devices, algebra, the calculus. 362 illustrations. 1,355pp. 5⅜ x 8½. Two-vol. set.   Vol. I: 0-486-20429-4   Vol. II: 0-486-20430-8

A CONCISE HISTORY OF MATHEMATICS, Dirk J. Struik. The best brief history of mathematics. Stresses origins and covers every major figure from ancient Near East to 19th century. 41 illustrations. 195pp. 5⅜ x 8½.   0-486-60255-9

Physics

OPTICAL RESONANCE AND TWO-LEVEL ATOMS, L. Allen and J. H. Eberly. Clear, comprehensive introduction to basic principles behind all quantum optical resonance phenomena. 53 illustrations. Preface. Index. 256pp. 5⅜ x 8½. 0-486-65533-4

QUANTUM THEORY, David Bohm. This advanced undergraduate-level text presents the quantum theory in terms of qualitative and imaginative concepts, followed by specific applications worked out in mathematical detail. Preface. Index. 655pp. 5⅜ x 8½. 0-486-65969-0

ATOMIC PHYSICS (8th EDITION), Max Born. Nobel laureate's lucid treatment of kinetic theory of gases, elementary particles, nuclear atom, wave-corpuscles, atomic structure and spectral lines, much more. Over 40 appendices, bibliography. 495pp. 5⅜ x 8½. 0-486-65984-4

A SOPHISTICATE'S PRIMER OF RELATIVITY, P. W. Bridgman. Geared toward readers already acquainted with special relativity, this book transcends the view of theory as a working tool to answer natural questions: What is a frame of reference? What is a "law of nature"? What is the role of the "observer"? Extensive treatment, written in terms accessible to those without a scientific background. 1983 ed. xlviii+172pp. 5⅜ x 8½. 0-486-42549-5

AN INTRODUCTION TO HAMILTONIAN OPTICS, H. A. Buchdahl. Detailed account of the Hamiltonian treatment of aberration theory in geometrical optics. Many classes of optical systems defined in terms of the symmetries they possess. Problems with detailed solutions. 1970 edition. xv + 360pp. 5⅜ x 8½. 0-486-67597-1

PRIMER OF QUANTUM MECHANICS, Marvin Chester. Introductory text examines the classical quantum bead on a track: its state and representations; operator eigenvalues; harmonic oscillator and bound bead in a symmetric force field; and bead in a spherical shell. Other topics include spin, matrices, and the structure of quantum mechanics; the simplest atom; indistinguishable particles; and stationary-state perturbation theory. 1992 ed. xiv+314pp. 6⅛ x 9¼. 0-486-42878-8

LECTURES ON QUANTUM MECHANICS, Paul A. M. Dirac. Four concise, brilliant lectures on mathematical methods in quantum mechanics from Nobel Prize-winning quantum pioneer build on idea of visualizing quantum theory through the use of classical mechanics. 96pp. 5⅜ x 8½. 0-486-41713-1

THIRTY YEARS THAT SHOOK PHYSICS: THE STORY OF QUANTUM THEORY, George Gamow. Lucid, accessible introduction to influential theory of energy and matter. Careful explanations of Dirac's anti-particles, Bohr's model of the atom, much more. 12 plates. Numerous drawings. 240pp. 5⅜ x 8½. 0-486-24895-X

ELECTRONIC STRUCTURE AND THE PROPERTIES OF SOLIDS: THE PHYSICS OF THE CHEMICAL BOND, Walter A. Harrison. Innovative text offers basic understanding of the electronic structure of covalent and ionic solids, simple metals, transition metals and their compounds. Problems. 1980 edition. 582pp. 6⅛ x 9¼. 0-486-66021-4

HYDRODYNAMIC AND HYDROMAGNETIC STABILITY, S. Chandrasekhar. Lucid examination of the Rayleigh-Benard problem; clear coverage of the theory of instabilities causing convection. 704pp. 5⅜ x 8¼. 0-486-64071-X

INVESTIGATIONS ON THE THEORY OF THE BROWNIAN MOVEMENT, Albert Einstein. Five papers (1905–8) investigating dynamics of Brownian motion and evolving elementary theory. Notes by R. Fürth. 122pp. 5⅜ x 8½. 0-486-60304-0

THE PHYSICS OF WAVES, William C. Elmore and Mark A. Heald. Unique overview of classical wave theory. Acoustics, optics, electromagnetic radiation, more. Ideal as classroom text or for self-study. Problems. 477pp. 5⅜ x 8½. 0-486-64926-1

GRAVITY, George Gamow. Distinguished physicist and teacher takes reader-friendly look at three scientists whose work unlocked many of the mysteries behind the laws of physics: Galileo, Newton, and Einstein. Most of the book focuses on Newton's ideas, with a concluding chapter on post-Einsteinian speculations concerning the relationship between gravity and other physical phenomena. 160pp. 5⅜ x 8½. 0-486-42563-0

PHYSICAL PRINCIPLES OF THE QUANTUM THEORY, Werner Heisenberg. Nobel Laureate discusses quantum theory, uncertainty, wave mechanics, work of Dirac, Schroedinger, Compton, Wilson, Einstein, etc. 184pp. 5⅜ x 8½. 0-486-60113-7

ATOMIC SPECTRA AND ATOMIC STRUCTURE, Gerhard Herzberg. One of best introductions; especially for specialist in other fields. Treatment is physical rather than mathematical. 80 illustrations. 257pp. 5⅜ x 8½. 0-486-60115-3

AN INTRODUCTION TO STATISTICAL THERMODYNAMICS, Terrell L. Hill. Excellent basic text offers wide-ranging coverage of quantum statistical mechanics, systems of interacting molecules, quantum statistics, more. 523pp. 5⅜ x 8½. 0-486-65242-4

THEORETICAL PHYSICS, Georg Joos, with Ira M. Freeman. Classic overview covers essential math, mechanics, electromagnetic theory, thermodynamics, quantum mechanics, nuclear physics, other topics. First paperback edition. xxiii + 885pp. 5⅜ x 8½. 0-486-65227-0

PROBLEMS AND SOLUTIONS IN QUANTUM CHEMISTRY AND PHYSICS, Charles S. Johnson, Jr. and Lee G. Pedersen. Unusually varied problems, detailed solutions in coverage of quantum mechanics, wave mechanics, angular momentum, molecular spectroscopy, more. 280 problems plus 139 supplementary exercises. 430pp. 6½ x 9¼. 0-486-65236-X

THEORETICAL SOLID STATE PHYSICS, Vol. 1: Perfect Lattices in Equilibrium; Vol. II: Non-Equilibrium and Disorder, William Jones and Norman H. March. Monumental reference work covers fundamental theory of equilibrium properties of perfect crystalline solids, non-equilibrium properties, defects and disordered systems. Appendices. Problems. Preface. Diagrams. Index. Bibliography. Total of 1,301pp. 5⅜ x 8½. Two volumes. Vol. I: 0-486-65015-4 Vol. II: 0-486-65016-2

WHAT IS RELATIVITY? L. D. Landau and G. B. Rumer. Written by a Nobel Prize physicist and his distinguished colleague, this compelling book explains the special theory of relativity to readers with no scientific background, using such familiar objects as trains, rulers, and clocks. 1960 ed. vi+72pp. 5⅜ x 8½. 0-486-42806-0

A TREATISE ON ELECTRICITY AND MAGNETISM, James Clerk Maxwell. Important foundation work of modern physics. Brings to final form Maxwell's theory of electromagnetism and rigorously derives his general equations of field theory. 1,084pp. 5⅜ x 8½. Two-vol. set.   Vol. I: 0-486-60636-8   Vol. II: 0-486-60637-6

QUANTUM MECHANICS: PRINCIPLES AND FORMALISM, Roy McWeeny. Graduate student-oriented volume develops subject as fundamental discipline, opening with review of origins of Schrödinger's equations and vector spaces. Focusing on main principles of quantum mechanics and their immediate consequences, it concludes with final generalizations covering alternative "languages" or representations. 1972 ed. 15 figures. xi+155pp. 5⅜ x 8½.   0-486-42829-X

INTRODUCTION TO QUANTUM MECHANICS With Applications to Chemistry, Linus Pauling & E. Bright Wilson, Jr. Classic undergraduate text by Nobel Prize winner applies quantum mechanics to chemical and physical problems. Numerous tables and figures enhance the text. Chapter bibliographies. Appendices. Index. 468pp. 5⅜ x 8½.   0-486-64871-0

METHODS OF THERMODYNAMICS, Howard Reiss. Outstanding text focuses on physical technique of thermodynamics, typical problem areas of understanding, and significance and use of thermodynamic potential. 1965 edition. 238pp. 5⅜ x 8½.
                    0-486-69445-3

THE ELECTROMAGNETIC FIELD, Albert Shadowitz. Comprehensive undergraduate text covers basics of electric and magnetic fields, builds up to electromagnetic theory. Also related topics, including relativity. Over 900 problems. 768pp. 5⅜ x 8¼.   0-486-65660-8

GREAT EXPERIMENTS IN PHYSICS: FIRSTHAND ACCOUNTS FROM GALILEO TO EINSTEIN, Morris H. Shamos (ed.). 25 crucial discoveries: Newton's laws of motion, Chadwick's study of the neutron, Hertz on electromagnetic waves, more. Original accounts clearly annotated. 370pp. 5⅜ x 8½.   0-486-25346-5

EINSTEIN'S LEGACY, Julian Schwinger. A Nobel Laureate relates fascinating story of Einstein and development of relativity theory in well-illustrated, nontechnical volume. Subjects include meaning of time, paradoxes of space travel, gravity and its effect on light, non-Euclidean geometry and curving of space-time, impact of radio astronomy and space-age discoveries, and more. 189 b/w illustrations. xiv+250pp. 8⅜ x 9¼.   0-486-41974-6

STATISTICAL PHYSICS, Gregory H. Wannier. Classic text combines thermodynamics, statistical mechanics and kinetic theory in one unified presentation of thermal physics. Problems with solutions. Bibliography. 532pp. 5⅜ x 8½.   0-486-65401-X

Paperbound unless otherwise indicated. Available at your book dealer, online at **www.doverpublications.com**, or by writing to Dept. GI, Dover Publications, Inc., 31 East 2nd Street, Mineola, NY 11501. For current price information or for free catalogues (please indicate field of interest), write to Dover Publications or log on to **www.doverpublications.com** and see every Dover book in print. Dover publishes more than 500 books each year on science, elementary and advanced mathematics, biology, music, art, literary history, social sciences, and other areas.